# NUTRIGENOMICS IN VETERINARY SCIENCE

# NUTRIGENOMICS IN VETERINARY SCIENCE

**Mukul Anand and**
**Dr. Shalini Vaswani**

Delve
PUBLISHING

www.delvepublishing.com

**Nutrigenomics in Veterinary Science**

*Mukul Anand and Dr. Shalini Vaswani*

**Delve Publishing**
4164 Lakeshore Road
Burlington, ON L7L 1A4
Canada
www.delvepublishing.com
Email: orders@arclereducation.com



ISBN: 978-1-77956-403-0 (Hardcover)

# ABOUT THE AUTHORS

**Mukul Anand** is an Assistant Professor specializing in Reproductive Physiology. He has been working at DUVASU, Mathura since July 2009, first as an Assistant Professor and currently as an Associate Professor. He holds a Bachelor's degree in Veterinary Sciences and Animal Husbandry, a Master's degree in Veterinary Sciences, and a Doctorate in Philosophy. He has been involved in various national assignments related to animal husbandry and dairying, including serving as a member of the National Advisory Committee for Sheep and Goat and an expert member for Bovine Sex Semen Sorting Technology. He has also participated in international conferences and worked as a visiting scientist at Shinshu University, Japan. He has published numerous research papers, with a focus on reproductive physiology and semen quality in various animal species. He has authored or co authored several books related to veterinary sciences, animal husbandry, and animal behavior. He has conducted training programs and workshops on topics such as goat farming and artificial insemination in goats. He has served as a reviewer for academic journals and has presented papers at national seminars and conferences. He has won several awards and served as course coordinator/ director for training at various institutions.

**Dr Shalini Vaswani** obtained the degree of B.V.Sc & A.H. with Chancellor Gold Medal from NDUAT, Kumarganj. She did her postgraduation in discipline of Animal Nutrition and received B.N Gupta Memorial Award, for Academic Excellence from NDRI, Karnal in year 2007. Thereafter, she started her professional career as Veterinary officer in UP State Government, in 2007. After serving in field for three years, she perused her PhD from DUVASU, Mathura. Then, in January, 2014 she joined as Assistant Nutritionist in Department of Animal Nutrition, GADVASU, Ludhiana. Later, she joined as Assistant Professor, DUVASU, Mathura. She received scholarship from VLIR UOS, by the Belgian federal government, for International Training Programme in Dairy Nutrition at Ghent University, Belgium in 2015. She had successfully established silage unit, feed analytical laboratory equipped with recent and advance instruments for analysis, Mineral mixture and fodder block unit and pellet feed plant at university under the RKVY funded projects as Principal Investigator. She had successfully handled projects of more than 8 crores funded by external agencies as Principal Investigator and Co- Principal Investigator. During her career of eight years as Assistant Professor, she had published 06 book, 40 international and national research articles, 2 lead papers, 3 review articles, and assisted 16 students as guide, co-guide and research committee member. Also, she had successfully conducted ICAR sponsored winter school as Course co-ordinator, Organizing Secretory and Course coordinator of more than 13 training programmes for Veterinary Officers, rural youth and farmers across the country. She is recipient of several awards and recognitions from International and national organizations.

# TABLE OF CONTENTS

# LIST OF FIGURES

# LIST OF TABLES

# PREFACE

The field of veterinary science has long recognized the importance of nutrition in maintaining animal health, optimizing productivity, and preventing disease. However, in recent years, the integration of genomic knowledge into the study of nutrition—known as nutrigenomics—has significantly expanded our understanding of how dietary components interact with an animal's genetic makeup. This interaction plays a critical role in shaping not only the immediate health outcomes of animals but also their long-term well-being and productivity.

The purpose of this book, *Nutrigenomics in Veterinary Science*, is to explore this dynamic and evolving field and present a comprehensive overview of its applications in various animal species. The book comprises nine chapters, each of which deals with specific aspects of nutrigenomics, providing a thorough examination of how nutritional interventions can modulate gene expression in animals, from sheep and goats to dogs, horses, and pigs. Throughout, the book aims to show how nutrigenomics can enhance our ability to manage animal health in the face of modern challenges, such as environmental stressors, diseases, and the demand for more efficient and sustainable livestock production.

Chapter 1 provides a foundational understanding of the field, introducing readers to the basic principles of how nutrients can influence gene expression and, conversely, how genetic variation can affect an animal's response to nutrients. This chapter serves as the entry point to a more nuanced discussion about the interaction between diet and genetics, setting the stage for the more specific applications of nutrigenomics explored in later chapters.

As global temperatures rise, animals are increasingly exposed to heat stress, which can severely impact their productivity and health. Therefore, Chapter 2 explores how nutrigenomics can be used to mitigate the effects of heat stress by modulating gene expression and epigenetic responses, providing a glimpse into how future breeding and nutritional strategies can be optimized for a changing environment.

Chapter 3 provides an in-depth look at the relationship between nutrition, gene expression, and productivity in these ruminants. It explores how advancements in nutrigenomics are shaping our understanding of nutrient metabolism and requirements, particularly in the context of improving growth rates, wool production, and disease resistance.

In Chapter 4, we take a closer look at the role of nutrigenomics during the early developmental stages of sheep. Lambs are particularly sensitive to nutritional interventions during their growth phases, and this chapter discusses how specific nutrients can influence gene expression to improve their health outcomes, growth performance, and overall viability.

Chapter 5 broadens the discussion to include other livestock species. The chapter explores how nutrigenomics is transforming traditional livestock management by identifying nutritional strategies that optimize genetic potential, improve disease resistance, and enhance productivity. This chapter is crucial for understanding how nutrigenomics can be applied across various species, from cattle to poultry, and its implications for the future of animal farming.

In Chapters 6 and 7, the focus is on companion animals, specifically dogs. Chapter 6 introduces how nutrigenomics is being used to tailor nutritional recommendations to the genetic profiles of individual dogs, leading to more personalized diets that improve overall health and longevity. Chapter 7 deals with how nutrigenomics can be used to prevent and manage common canine diseases. The chapter explores genetic testing, dietary interventions, and emerging research in disease prevention through nutrigenomic strategies.

Chapter 8 provides an exploration of how nutrigenomics is influencing the nutritional management of horses, particularly in the prevention and treatment of diseases such as, laminitis and metabolic syndrome. This chapter offers insights into how dietary strategies tailored to an individual horse's genetic profile can significantly improve health outcomes and athletic performance.

Chapter 9, the last chapter of the book, explores how nutrigenomics can help optimize feed formulations for pigs, improving growth rates, feed efficiency, and disease resistance. The chapter discusses the specific genetic markers that are being studied in pigs and how these discoveries are shaping modern feeding practices in swine production.

This book provides veterinary professionals, animal nutritionists, and researchers with valuable insights into the application of nutrigenomics across various animal species. Integrating the latest research in genomics and nutrition, *Nutrigenomics in Veterinary Science* seeks to contribute to the ongoing transformation of veterinary practice and animal care in the 21st century.

*Chapter* 1

# INTRODUCTION TO NUTRIGENOMICS

## Contents

# INTRODUCTION

Large-scale studies of genome function have been made possible by the tremendous advancement of molecular genetics techniques in recent years. The development of new fields that would be included under the umbrella term "Nutritional Genomics" has been aided by these advancements as well as a drop in cost and an increase in accessibility. This emerging field investigates the molecular, cellular, and systemic interactions between food and its constituents and the genome. Currently, nutritional genomics research is separated into two areas: "Nutrigenomics" is a new, multidisciplinary field in nutritional science that aims to explain how genes affecting transcription factors, RNA and protein expression, cellular homeostasis, and metabolite production can interact with dietary nutrients (genome, transcriptome, proteome, metabolome). In this field, numerous studies on humans and mice have been conducted with the primary goal of understanding the molecular basis of diseases like obesity, cancer, and cardiovascular diseases and how they are influenced by interactions between gene and diet types. Nutrigenomic studies on domestic animals are much rarer and typically concentrate on a small number of genes, particularly those associated with a specific treatment. These studies primarily address the energy content of the diet or the content of certain of its constituents, such as polyunsaturated fatty acids (PUFA), protein, or L-carnitine. "Nutrigenetics" investigates the relationship between phenotypic effects of diet and structural genetic variation. Thousands of gene polymorphisms may cause slight variations in nutritional biochemistry that affect how a diet affects metabolism.

When it comes to incorporating genomic methods into nutrition research, nutrigenomics is crucial. As a distinct methodological and conceptual approach, nutrigenomics broke away from mainstream nutritional science more than ten years ago. The status as a distinct scientific discipline was solidified, occasionally at odds with "nutrition science." Until now, genetics and nutrition have mostly been applied as separate fields in animal production, with little consideration given to the effects of the interactions between the genome and nutrition on the different physiological and metabolic processes that transcend phenotype and production. The nutrigenomic studies are of interest because they enable us to gain a deeper understanding of the genes and metabolic pathways impacted by diet, as well as the potential regulatory effects of these nutritional interventions.

## 1.1. OVERVIEW OF NUTRIGENOMICS

Nutritional genomics, or "nutrigenomics," is the study of how genes and their products interact with dietary chemicals to change phenotype and, in turn, how genes and their products metabolize nutrients (Kaput et al., 2005).

Nutrigenomics is the study of the molecular interactions between the genome and various dietary bioactive compounds as well as nutrients.

There are two distinct fields within nutritional genomics, namely. The terms "nutrigenomics" and "genetics." Concerning how dietary nutrients can interact with genes affecting transcription factors, RNA and protein expression, cellular homeostasis, and metabolite production (genome, transcriptome, proteome, metabolome), a novel and multidisciplinary research field in nutritional science has emerged called "nutrigenomics." The study of how structural genetic variation influences the phenotypic effects of diet is known as nutrigenetics. Thousands of different gene polymorphisms may cause variations in nutritional biochemistry that affect how a diet affects metabolism.

### 1.1.1. Nutrient Gene Interaction

It has long been understood that the diet is a complex combination of organic materials that provides the body with the protein and energy it needs to grow and thrive. Nutrients will, however, have a range of biological activities. Certain nutrients, referred to as antioxidants, must function as radical scavengers and are involved in disease prevention. It has also been demonstrated that other nutrients function as nutritional hormones and as strong signaling molecules (Muller and Kersten, 2003). Phytochemicals are among the secondary metabolites of plants that regulate animal health and production. Lower nutrition in terms of essential nutrients, macronutrient imbalances, or even toxic concentrations of specific food compounds are linked to several diseases and disorders. Numerous etiological causes of diseases result from the interaction of several genes and different nutrients (Mariman, 2006). Variations in food digestion, nutrient absorption, metabolism, and excretion have been noted, as well as genetic diseases in these processes, due to the remarkable diversity found in all living things. The functional integrity of the gene will mainly depend on metabolic signals that the nucleus receives from various internal factors, e.g., hormones and external factors, e.g., among the most powerful environmental stimuli are nutrients. Nutrition is one of the many environmental factors that cause genomes to change. Thus, nutrients, micronutrients, and phytochemicals present in food can all have a significant impact on how genetic information is expressed (Van Ommen, 2002). The medium-to long-term effects of epigenetic factors and transcriptional regulatory factors (TF) on the interactions between genes and nutrients. TF can be activated or repressed either directly or indirectly by bioactive nutrigenomic compounds. Consequently, rather than nutrients directly binding to the genome, the interaction between genes and nutrients is believed to be TF-mediated. Therefore, knowledge of the transcriptional regulatory factors is crucial to comprehend the interaction between genes and nutrients.

## 1.1.2. Transcriptional Factors with Nutrigenomics in Livestock

Nuclear receptors that are ligand dependent or non-ligand dependent are examples of transcriptional regulatory factors. Ligand dependent nuclear receptors (LdNR) are made up of hepatic nuclear factor 4 (HNF4), liver X receptors (LXR), and peroxisome proliferator activated receptors (PPAR). All of these receptors can bind to and be activated by major nutrients, such as fatty acids for all receptors (Khan and Vanden Heuvel, 2003) and cholesterol metabolites (oxyster-ols) for LXR receptors (Zhao and Dahlman-Wright, 2010). Among the micronutrient responders are the LdNR that are vitamin-specific. Retinoids, which are metabolites of vitamin A, activate retinoic acid receptors (RAR) and retinoid X receptors (RXR) (Minucci et al., 1997), vitamin D receptor (VDR), and vitamin E-activated pregnane X receptor (Landes et al., 2003).

**Table 1.1**. Ligand-Dependent Nuclear Receptors: A Timeline

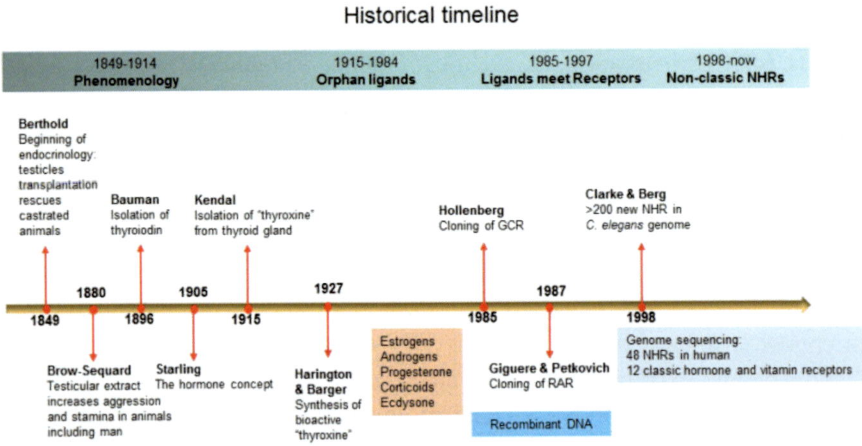


Source: Tao LJ, Seo DE, Jackson B, Ivanova NB, Santori FR. (2020). Nuclear Hormone Receptors and Their Ligands: Metabolites in Control of Transcription. Cells, 9(12), 2606. doi: 10.3390/cells9122606. PMID: 33291787; PMCID: PMC7762034.

When nutrients activate a gene, non-LdNR TF will regulate its expression; however, this activation is not direct; rather, it is mediated by other internal factors. These include the protein known as sterol regulatory element binding protein (SREBP1; gene name SREBF1), which is affected by glucose (Uttarwar et al.) and long-chain fatty acids (LCFA); polyunsaturated fatty acids PUFA); (Cunningham et al., 1998) have an impact on Spot 14 or thyroid hormone responsive protein (THRSP); the protein known as carbohydrate responsive element binding protein (ChREBP; gene name MLXIPL), which is xylulose-5-phosphate and glucose 6-phosphate-dependent (Li et al., 2006); and nuclear factor kappa-light-chain-enhancer of activated B cells, CCAAT/enhancer binding protein, and activating transcription factor 4, which react to amino acid deprivation (Table 1.1).

### 1.1.3. Techniques Used in Nutrigenomics

Small-scale methods (PCR and quantitative real-time PCR) that measure the expression of fewer genes have historically been used in nutrigenomics research. Indeed, more potent solutions to a range of biological issues will be possible thanks to new genomic analysis technologies. Thanks to the development of microarrays and the massive sequencing of the transcriptome (RNAseq), gene expression can now be studied on a large scale. The idea of accumulating large amounts of data through automation and parallelization of DNA/RNA and protein chemistry was brought about by these new technologies. The omics technologies designed to collect transcriptomic, proteomic, and metabolomic data have been the most extensively used. Recently developed bioinformatics methods combined with biological data from transcriptomics and genomics research have already made it possible for biologists to use contemporary systems approaches to examine interactions that take place within living systems.

### 1.1.4. Nutritional Genomics in Ruminants

The two main things that affect ruminant productivity are dietary modifications and nutritional tactics. Reproductive efficiency and fertility in dairy cattle are dependent on genetic variables and dietary approaches. As the animal is especially vulnerable to nutritional imbalances during the transitional period and the early stages of lactation, this is crucial. Byrne (2005) also noted that diet had a definite impact on the expression of particular genes linked to protein turnover, cytoskeletal remodeling, and metabolic homeostasis, as well as nutritional restriction brought on by the consumption of subpar feeds. Numerous of these expression changes were expected based on observations that led to modifications in animal physiology and growth during typical nutrient restriction.

Using rat microarray technology, Jones et al. (2004) reported on the effects of entophyte-infected tall fescue on gene expression in the luteal tissue of heifers. He observed certain alterations in the cell, including distinct variations in the expression of genes linked to neuronal functions, transport functions, control of the cell cycle, and programmed cell death. This demonstrated unequivocally that nutrients affect heifers' transcriptional levels.

### 1.1.5. Application of Nutrigenomics in Ruminant Health and Nutrition

Crucial instruments for affecting ruminant productivity are nutritional approaches and dietary modifications. The reproductive success of milch animals is greatly influenced by their genetic makeup and nutrition. In the transition period and early lactation, when the animal is susceptible to nutritional imbalances, this is crucial. There is very little information about the diet and gene expression associated with livestock productivity and reproductive performance. The significance of the connection between specific nutrients and the control of gene expression may start to become clear. The adverse effect of selenium deficiency is an increase in stress through upregulating the expression of particular genes and signaling pathways. This occurs when animals are fed diets that alter protein synthesis at the transcriptional level. Genes that regulate the detoxification process and defense against oxidative damage lead to phenotypic changes. The identification of gene markers associated with economically significant traits such as the production of milk, meat, wool, etc. Dietary regimens will improve these traits (Kore et al., 2008).

### 1.1.6. Application of Nutrigenomics in Ruminant Reproduction and Fertility

The ability of cattle and other livestock species to reproduce is dependent on dietary modifications that alter the way certain genes function. It is possible to comprehend how nutrients and gene expression regulation interact, for example, by looking at the effects of dietary selenium on mouse gene expression (Rao et al., 2001). He discovered that selenium affects how mice synthesis proteins by controlling transcriptional levels of certain genes. It might be possible to start comprehending the significance of the relationship between specific nutrients and the regulation of gene expression by directly demonstrating the effects of nutritional strategies or diets on the expression of genes related to fertility in either male or female animals. By controlling the transcriptional level of gene expression, deficiency in selenium can affect the patterns of protein synthesis in mice.

## 1.2. THE ROLE OF MIRNAS IN RUMINANTS

The ruminant family comprises one of the most affluent subspecies of terrestrial mammals. They have a significant impact on a variety of ecosystems, and industries, including agriculture, leisure pursuits, and cultures. They are found in a wide range of diverse environments worldwide. A key element in this group of animals' success is their capacity to endure and multiply on high-fiber, low-protein, low-quality plant resources. The majority of stakeholders agree that maintaining animal health is essential to the production of livestock, but they disagree on what level of health is appropriate for consumers, farmers, and veterinarians.

There is no precise scientific definition of "animal health," which can range from the absence of disease to a broad definition of health as a state of unrestricted physical, physiological, and psychological well-being. There are also no established benchmarks by which the health of the animals and the caliber of their feed can be accurately assessed. Ensuring the health and performance of their livestock through the proper provision of feed is one of farmers' primary objectives. Organic farming should try to increase the energy content of cattle diets to improve the efficiency of protein utilization and, consequently, reduce nitrogen loss to the environment. This can be achieved by improving the quality of the forage and making rations with more balanced ratios of forage and grain. miRNAs have a major impact on the physiology, biological development, and cell differentiation of farm animals. These mechanisms regulate early embryonic development, adipose tissue expansion, oocyte maturation, and muscle growth and hypertrophy. Recent studies have shown how important miRNAs are for maintaining intestinal homeostasis and for the development of the rumen in sheep, goats, and cows. As a result, the miRNA expression profiles of sheep and cattle have been determined in the rumen, small intestine (duodenum and jejunum), and large intestine (cecum and colon).

Additionally, the study showed that some miRNAs are only expressed in particular intestinal segments, suggesting that the local microenvironment may influence the roles that these miRNAs play. Furthermore, during the expression and processing phases, taxonomic variations in the way miRNAs control gene expression usually take place. The colon is an essential part of the digestive system in the hind gut of ruminants. In ruminants, the colon ferments and uses cellulose, hemicellulose, and cell wall polysaccharides primarily. In the study by Yan and colleagues, 39 differentially expressed miRNAs and 88 highly expressed miRNAs were screened from a total of 1572 miRNAs found in the colon tissues during the analysis of colon miRNA transcriptomes in preweaning and postweaning goats. This allowed for the discovery of various functions of dynamic miRNAs in the regulatory system governing colon growth in goats. The function of the rumen tissue miRNAome and transcriptome in relation to dietary modifications or the addition of a phytogenic feed additive was reported in a study conducted on cows from our group. We looked into the effects of a phytogenic feed additive on the role of miRNAs in the epithelial transcriptome during the diet switch from forage (FD) to high-grain (HG) feeding. The study identified potential miRNA control mechanisms of gene expression during the transition from FD to HG feeding and phytogenic supplements, demonstrating that miRNAs have a direct role in the host's responses to nutrition. The existence of miRNAs in rumen fluid and the possibility of miRNA-mediated cross-talk within the ruminal ecosystem were investigated in a separate study conducted by our research group. According to the study, the ruminal microbiota may play a role in the interaction between the host and the microbiota. This communication is thought to be two-way, with the host producing particular miRNAs that may help shape the gut bacterial profile and the microbiota influencing the host's pattern of miRNA expression.

### 1.2.1. Potential Regulatory Functions of miRNAs in Ruminant's Milk

An essential organ for producing milk for human infant and adult nutrition is the ruminant mammary gland (MG). Microbial productivity is influenced by a variety of environmental factors, including post-transcriptional regulation of gene expression, breed, genetics, disease pathogens, and nutrition. One of the remarkable results of evolution is lactation, a highly dynamic and intricate process. Parts of lactation include the MG's growth as well as the production and release of milk. Virtually all of the milk in the world comes from camels, goats, sheep, cattle, and buffaloes. Donkeys, yaks, horses, and reindeer are some other uncommon milk-producing animals.

The frequency and importance of each species differ significantly among various countries and geographic regions. The three primary determinants of the dairy species that are preserved are climate, water, and feed. Additional factors that might affect the survival of a dairy species include consumer demand, dietary preferences, and the socioeconomic composition of each household (e.g., less wealthy households typically depend more on small ruminants). Much work has been done to boost milk production and enhance its nutritional value since it has been demonstrated that milk from dairy ruminants, such as cows, goats, and sheep, benefits humans. Both cow and goat milk are frequently used as dairy products, and the majority of baby formulas are made with proteins from cow milk. The findings indicate that substantial amounts of milk-derived miRNAs are absorbed by adults from commercially pasteurized milk. Furthermore, studies have demonstrated that milk exosomes can be incorporated into peripheral, kidney, intestinal, and intestinal cancer cells in addition to macrophages and colon cancer cells. Golan-Gerstl et al. report that bovine milk pasteurization does not seem to remove miRNAs and that 95% of the miRNAs expressed in human milk are also expressed in goat and bovine milk. Furthermore, conserved miRNAs shared 24 verified sites involved in immunomodulatory activities. The addition of feed rich in unsaturated fatty acids to the diet can enhance the health-promoting properties of milk, though the effects on the genes and pathways involved in these processes are still not completely understood. Because the synthesis of milk fat is a complex process regulated by multiple factors, there are several strategies to improve the beneficial components in milk produced by ruminants. These strategies include nutrition, seasonal feed changes, and genomic variation.

The genetic variations of miRNAs expressed in the mammary gland or detected in milk and localized in dairy quantitative trait loci (QTLs) in bovine, caprine, and ovine species were analyzed to identify variations that could potentially explain dairy features. The study identifies miRNA genetic variations of interest in the context of dairy production by using whole genome data to identify miRNA genetic variants expressed in the mammary gland and localized in dairy QTLs.

### *1.2.1.1. Cattle*

Cow milk provides an abundance of essential nutrients and is a good source of proteins, lipids, amino acids, and other bioactive substances like hormones and cytokines. Because cow milk is so nutritious, it has been commercialized and is frequently consumed by people for growth and health benefits. Li et al. present the miRNA expression spectra of three different milk fractions (fat, whey, and cells), compare the miRNome profiles of the milk fractions with the miRNome profiles of mammary gland tissue, and identify the milk fraction whose miRNome profile is most likely to be a good predictor of the mammary gland tissue miRNome profile. Their results showed that miRNAs from milk fat accurately represented the miRNAome of mammary gland tissue. Functional annotation of the most highly expressed miRNAs in milk demonstrated their critical regulatory roles in the functions of the mammary gland and possibly in milk recipients. Udder illnesses are serious problems for the dairy industry worldwide, especially mastitis caused by bacterial infections. Mastitis is still a major problem for the dairy industry worldwide, costing millions of dollars annually in lost milk production (quantity and quality), necessitating costly treatments, and leading to the early culling of animals. Because dry cow antibiotics are so widely used, mastitis also contributes significantly to the development of antibiotic resistance in cattle. With the importance of early disease detection for reducing the significant financial and animal welfare consequences of mastitis worldwide, improved instruments that can precisely identify early mammary inflammation in cattle are desperately needed.

Different approaches are needed to track the health of the mammary glands and identify early inflammation. Somatic cell counting (SCC) is the most commonly used method. It can be done on individual milk samples or in large quantities of milk, directly or indirectly through colorimetric quantification, often using the California Mastitis Test (CMT). A budding mechanism that encloses a crescent of the cytoplasm of the mammary epithelial cells in the plasma membrane causes the cells to release milk fat globules. Determining the role of miRNAs in lactation primarily involves figuring out whether the miRNAs detected in milk are unique to the mammary gland or originate from the blood. Tzelos et al. examined the association between four inflammation-associated miRNAs (bta-miR-26a, bta-miR-142-5p, bta-miR-146a, and bta-miR-223) and CMT scores (0 to 3), which were obtained from numerous individual quarter milk samples (n = 236) obtained from dairy cows at different lactations. For bta-miR-26a, bta-miR-142-5p, and bta-miR-146a, they verified that the significant lactation number effect (P 0.01) was caused by higher miRNA levels during lactation 1 compared to later lactations. Additionally, they demonstrated that early inflammation could be precisely identified (100% sensitivity, >81% specificity) by measuring bta-miR-223 and bta-miR-142-5p levels. They claimed that their results provide more proof of the value of miRNAs as early diagnostic markers for cow mastitis. Izumi et al. found differences between mature milk (53 known miRNAs) and colostrum (100 miRNAs) using microarray and quantitative PCR analysis. They verified that cow's milk contained some mRNA, but that the mRNA and naturally occurring miRNA in raw milk were resistant to both RNase treatment and acidic environments. Degradation of synthesized miRNA spiked in raw milk whey was observed. Wang's research group examined the molecular mechanisms controlling milk secretion and quality in dairy cows using transcriptome analyses of mammary gland tissues from the cows during the high-protein/high-fat, low-protein/low-fat, or dry periods. In comparison to non-lactation, they found that 25 core differentially expressed miRNAs (DE miRNAs) were critical for the development of the mammary gland and the differentiation of the epithelial cell terminal during lactation. Their research revealed that DE miRNAs may have a role in controlling milk secretion and quality during the development of the mammary gland. Xia et alscreened and identified differentially expressed miRNAs, candidate genes, and co-regulatory pathways related to the metabolism of milk fat using miRNA and transcriptome data from the mammary epithelial cells of dairy cattle with high (H, 4.85%) and low (L, 3.41%) milk fat percentages during mid-lactation. They found regulatory candidate genes involved in lipid metabolism as well as functional miRNAs in the co-expression networks of the dairy cattle mammary system (Table 1.2).

Our knowledge of potential genetic element regulatory mechanisms and gene signaling networks involved in milk fat metabolism is advanced by this information. Werry et alpostulated that some variations in bull fertility may be mirrored in the levels of distinct miRNAs in sperm since small non-coding

RNAs have been connected to a variety of phenotypes in bovine sperm. Nevertheless, a strong profile of expressed miRNAs linked to fertility has not yet been found despite extensive searches for sperm-borne molecular biomarkers of male fertility. The miRNAs that were most prevalent in both populations were determined to be miRs -34b-3p, -100-5p, -191-5p, -30d-4p, and -21-5p. Additionally, differences in the overall amounts and specific patterns of isomiR expression were evaluated. The results provide a comprehensive overview of the bovine sperm miRNAome and indicate a wide range of possible functions in reproduction. The abundance of milk fat globules from five cows before and after a lipopolysaccharide (LPS) challenge was compared to support the use of these globules as a source of small non-coding RNAs to diagnose mastitis. We also looked at six miRNAs that are known to be controlled in the mammary gland during inflammation.

The findings suggested that milk fat globules could be a readily available source of miRNAs, which could be used as biomarkers to identify mastitis early on and facilitate the administration of prompt and effective treatment. Depending on the level of inflammation, the abundance of four miRNAs (miR-494-3p, -148a-3p, -99a-5p, and -125b-5p) varied. Using mRNA sequencing, the Cui study team discovered 497 known miRNAs and 49 new ones in the mammary glands of dairy cows that were milking. Among them, miRNA-71 showed variable expression in the milk of cows with high and low protein and fat contents. Based on their previous RNA sequencing data, a total of 21 differentially expressed genes can be referred to as targets for some of the 71 DE miRNAs, indicating that they may be important regulators of the milk protein and fat characteristics in dairy calves. To gain a better understanding of the molecular regulatory mechanisms of milk protein production in dairy cows, the miRNAomes of five essential metabolic tissues (rumen, duodenum, jejunum, liver, and mammary gland tissues) involved in protein synthesis and metabolism from 18 dairy cows fed high- and low-quality diets were studied. Specifically, the tissues of the rumen, duodenum, jejunum, liver, and mammary gland expressed 340, 338, 337, 330, and 328 miRNAs. According to the research, miRNAs expressed in these tissues might be involved in regulating the transfer of amino acids needed for the production of milk downstream. This is a crucial mechanism that could be connected to low milk protein when forage feed is inadequate. Upon comparing the mammary gland miRNomes of two dairy cow breeds, namely Holstein and Montbéliarde, which have distinct mammogenic potentials linked to variations in dairy performance, 754 unique miRNAs were discovered in the mammary glands of mid lactating Holstein (n = 5) and Montbéliarde (n = 6) cows. The results show differences in the mammary gland miRNomes of two breeds of dairy cattle, and they imply that miRNAs might play a role in the synthesis of milk components and the flexibility of the mammary gland, both of which are related to the amount and makeup of milk. MiRNAs were found to play a regulatory role in the early development of the gastrointestinal tract (GIT) using Illumina RNA-sequencing. The results further support the theory that host–microbial interactions regulate gut development through an alternative mechanism. These findings included temporal and regional variations in miRNA expression as well as a correlation between miRNA expression and the microbial population in the GIT during early life. An integrated study of miRNA and mRNA expression profiling between bulls and steers was conducted based on differences in meat quality attributes and 90 differently expressed mRNAs; 18 DE miRNAs were found. The results provided strong evidence of cattle gene interactions and molecular genetic regulation.

### *1.2.1.2. Goats*

In Latin America and the Mediterranean region, goat milk is usually made into cheese; in Africa and South Asia, it is usually consumed raw or acidified. A prominent member of the livestock is the goat (Caprahircus). Many studies on goats have examined how different feeding strategies affect the amount of fat in their milk; however, the physiological basis for this response remains unclear.

Levels of miR-183, which targets MST1 (Macrophage Stimulating 1), were found to be positively correlated with the fatty acid content of goat milk in a study that looked at the relationship between differentially expressed miRNAs in goat mammary tissue and the fatty acid composition of goat milk. Milk fat metabolism is inhibited when miR-183 targets the MST1 gene in the cytoplasm of goat mammary epithelial cells. The lipid composition of goat milk is one of the most important factors determining its nutritional quality. Ollier and colleagues, for example, found that while whole intact rapeseeds or sunflower oil in high-forage or high-concentrate diets affected milk yield and composition, these effects did not extend to mammary mRNA expression of key genes involved in lactose (e.g., α-lactalbumin), protein (e.g., β-casein), and lipid metabolism (e.g., lipase, lipoprotein).

On lactating goats fed a supplement of sunflower seed oil, the response to dietary interventions did not appear to be mediated by changes in the mRNA abundance of genes encoding essential lipogenic enzymes and the associated transcription factors. Through high-throughput sequencing, thirty highly expressed miRNAs were identified in the mammary glands of lactating goats. Among these, miR-103 is one whose expression correlates with lactation. The findings of this study shed new light on the functions of miR-103 and the molecular mechanisms regulating the synthesis of milk fat. Breastfeeding goats fed diets containing extruded linseed either alone or in combination with fish oil showed significant changes in the fat content of their milk as well as effects on mRNA related to protein metabolism and transport pathways as opposed to lipid metabolism. Goat mammary epithelial cells treated with MiR-25 mimics reduced the expression of genes related to lipid metabolism, which was negatively correlated with milk production during different stages of lactation. The results of the study indicated that miR-25 may be involved in lipid metabolism and demonstrated the function of the miR-25/PGC-1beta regulatory axis during lactation (Table 1.2). Between early and late lactation, a total of 1487 unique miRNAs were discovered, comprising 45 novel miRNA candidates, 1442 known and conserved miRNAs, 378 differentially expressed miRNAs, and 758 co-expressed miRNAs. Researchers will gain a better understanding of how genes control mammary gland development and lactation thanks to the study's findings, which suggested that miRNAs may be essential to both early and late lactation throughout the development of the dairy goat mammary gland. Following the sequencing of the goat genome, 487 miRNAs were found. On chromosome 21, the largest miRNA clusters were found to exist. Using the Illumina-Solexa high-throughput sequencing method, goat MG tissues during early lactation were analyzed, yielding 131 novel and 300 conserved miRNAs. Furthermore, in goat MG tissues from dry off, 346 conserved and 95 novel miRNAs were found, and peak lactation does employ the same method (Illumina-Solexa sequencing).

### *1.2.1.3. Sheep*

The majority of sheep worldwide reside in developing countries. They are even more prevalent than goats in colder climates. Sheep farming can yield a wide range of products, such as milk, meat, fiber, skin, and dung, though the majority of small-scale farmers in developing countries raise sheep primarily for meat or to sell as livestock in nearby markets. Genetic selection in dairy sheep has not resulted in a significant increase in milk production or lactation duration. Among the sheep breeds used to produce dairy products are Awassi, East Friesian, and Lacaune.

Using miRNA-Seq, 155 DE miRNAs, comprising 78 up-regulated and 77 down-regulated miRNAs, were discovered between the tail fat tissue of Tibetan sheep (short-thin-tailed) and Hu sheep (short-fat-tailed) in a study involving several breeds of fat-tailed sheep. The results may provide a helpful theoretical framework for future investigations into the molecular mechanisms underlying sheep tail adipogenesis. When working with a complex process like NGS, RNA sequencing and cell-level validation—an error-based approach to design and optimization—are essential. By closely examining the possibility of potential failure spots, one can determine the level of validation and quality control necessary for particular process steps. It also aids in validating modifications made to different test system components and troubleshooting any errors. When the function of miRNA in the deposition of intramuscular fat (IMF) was studied, 59 DE-miRNAs were found in Aohan fine-wool sheep (AFWS) ranging in age from 2 months (Mth-2) to 12 months (Mth-12). The study identified lists of miRNAs associated with intramuscular lipid deposition in sheep and their potential targets in an attempt to improve the quality of sheep meat. In a study, small-tailed Han and Gansu Alpine Merino sheep with different milk production characteristics had their mammary gland (MG) tissues' microstructures and miRNA expression profiles compared at the height of lactation. The two breeds differed in the expression of eighteen of the twenty-four mature miRNAs that were produced. The results showed that it is possible to enhance the roles of miRNAs in MG growth and lactation in sheep. Table 1.2 shows that the majority of the targeted genes of differentially expressed miRNAs were implicated in metabolic and signaling pathways linked to the development of MG, milk protein, and fat synthesis.

**Table 1.2**. MiRNAs Involved in the Regulation Processes of Mammary Lipid Metabolism of Ruminants

| miRNAs | Targeted Genes and Pathways | Regulating Functions |
|---|---|---|
| miR-103 | PKAN3, AMPKα pathway | Accelerates de novo synthesis of fatty acids/Unsaturated or saturated fatty acids ratio |
| miR-224 | ALOX15, PTGS1, ACADM | Milk fat metabolism/Increased apoptosis rate |
| miR-221 | FASN, NR1H3, ACSL1 | Droplet of lipid formation |
| miR-24-3p | MEN1 | Regulates synthesis of milk proteins |
| miR-24 | ACACA, TIP47, GPAM | Droplet of fat formulation, concentration of fatty acid, synthesis fatty acid |
| miR-486 | PTEN | Regulates phosphoacyl alcohol signal transduction |
| miR-124a | PECR | Metabolizes fatty acid |
| miR-135a | PPLR | Regulates prolactin secretion |

| | | |
|---|---|---|
| miR-106b | ABCA1 | Accumulates triglycerides and cholesterol in epithelial cells of the mammary gland |
| miR-145 | INSIG1 | Stimulates the production of milk fat |

Source: Ojo OE, Kreuzer-Redmer S. (2023). MicroRNAs in ruminants and their potential role in nutrition and physiology. Vet Sci., 10(1):57. doi: 10.3390/vetsci10010057. PMID: 36669058; PMCID: PMC9867202.

### 1.2.2. miRNAs Involved in Disease and Health in Ruminants

Because ruminant diseases increase mortality and lower productivity in dairy herds, they result in significant financial losses for farmers around the world. Mastitis, other production-related illnesses, and both internal and external parasites typically do not kill the animal but always reduce the efficiency of the system. Diseases can affect the productivity of dairy farms by decreasing feed conversion, lowering milk quality, lowering milk production, lowering fertility, and delaying puberty. Humans may be exposed to health risks related to dairy animal diseases (e.g., brucellosis, tuberculosis). In developing countries, small-scale dairy farming is susceptible to several health risks. This is caused by a number of factors, such as inadequate knowledge of disease prevention, treatment, and control; a high incidence of infections; and the cost, accessibility, or suitability of services related to animal health. Even one animal dying from illness can have a major financial impact on a small-scale dairy farmer with limited resources. Generally speaking, small-scale dairy farmers don't make significant investments in animal health, especially when it comes to illness prevention. The different dairy species and breeds have different health needs based on their anatomical and physiological characteristics. Concerns about animal health and welfare can be significantly reduced by selecting dairy animals that are compatible with the surrounding environment. Relevant traits for a dairy animal to possess include climate adaptation, resource grazing, resistance to endemic illnesses, and immunity to local parasites. Dairy animals housed in intensive systems are more susceptible to agents that cause transmissible diseases, but animals kept in extensive systems are more susceptible to parasite diseases. Due to their immunization against endemic diseases, animals may be more vulnerable when brought to a new environment.

#### *1.2.2.1. Mastitis*

Mastitis, an inflammation of the mammary glands, is the most costly disease in the dairy industry. It is primarily caused by a broad range of bacteria that fall into the categories of environmental and infectious bacteria. In cattle, sheep, and goats that are extensively managed to produce meat or milk, mastitis, especially the sub-clinical variety, is the most common production-related disease (Figure 1.1).

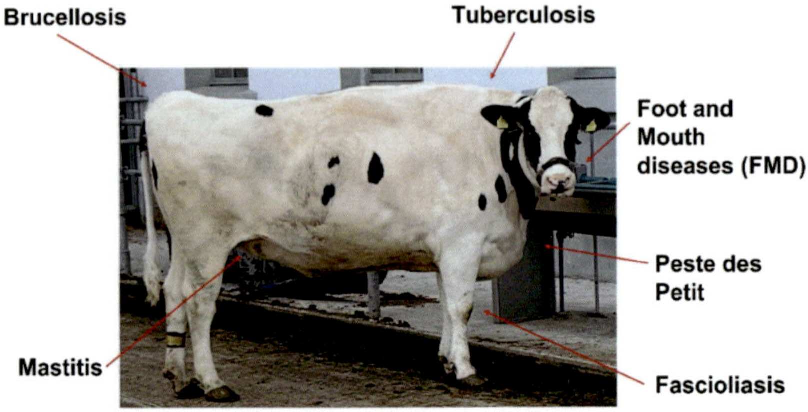


**Figure 1.1.** *Important infectious diseases in cows that are known to be at least partially regulated through miRNA action.*

Source: Ojo OE, Kreuzer-Redmer S. MicroRNAs in Ruminants and Their Potential Role in Nutrition and Physiology. Vet Sci. 2023 Jan 14;10(1):57. doi: 10.3390/vetsci10010057. PMID: 36669058; PMCID: PMC9867202.

Mastitis has detrimental effects on the economy in addition to significant negative effects on the health and welfare of the animals. The quality of dairy products, like cheese, and food safety (diseases transmitted through food) are pertinent viewpoints as well, since milk from impacted animals may contain harmful bacteria and have changed compositions that the dairy industry finds undesirable. *Streptococcus agalactiae*, *Staphylococcus aureus*, and Strept spp. coliform agents (*E. coli* and *Klebsiella pneumonia*), *Mycoplasma* spp., and *Corynebacterium bovis* (summer mastitis of heifers, dry cows, and beef breeds) are the most common pathogens causing bovine mastitis, particularly *M.bovis* (California mastitis). Numerous infections have the potential to cause severe systemic disorders, such as fever, anorexia, and both acute and chronic mastitis. Similar symptoms and treatment options apply to cattle mastitis as they do to small ruminants.

Certain miRNAs might play significant roles in the course of mastitis. Different levels of expression of several miRNA target genes, such as granulocyte-macrophage colony-stimulating factor (GM-CSF) and interleukin-8 (IL-8) were found in the mammary tissue of cows that had mastitis. Furthermore, Li et al. observed that in S-infected mammary epithelial tissue, miR-31 and miR-205 were down-regulated. miR-223 was upregulated in contrast to aureus in healthy controls (Table 1.3).

*S. aureus* is the main cause of mastitis in dairy cows; however, some studies have looked into the dynamic changes in miRNA expression in peripheral blood during the onset and progression of mastitis in these animals. The type, dosage, and duration of infections of pathogenic bacteria all have a significant impact on the expression level of miRNAs associated with mastitis in dairy cows. Understanding the changes in miRNA during the pathogenesis is essential for the molecular diagnosis and biological therapy of mastitis. In a recent study, a mastitis cow model was created by infecting the mammary gland tissues of dairy cows with 5 mL of 105 CFU/mL of *S. aureus*. Therefore, it is critical to identify biomarkers for early detection of *S. aureus*-infected mastitis in peripheral blood from dairy cows. The results offered a fresh foundation for research on molecular diagnostics and biological treatment for dairy cow mastitis caused by *S. aureus*. Given the unknown genetic basis of mastitis, it is imperative to understand the regulation of immunological mechanisms during the disease, with a particular focus on the function of miRNAs.

**Table 1.3**. MiRNAs Involved in Important Diseases of Ruminant Animals

| Potential miRNA Biomarkers | Diseases | Pathogens | Sample Tissue |
|---|---|---|---|
| miR-2339, miR-21-3p, miR-423-5p, miR-499, miR-92a, miR-193a-3p, miR-23a, miR-99b, miR-21-3p, miR-193a-3p, miR-365-3p, miR-30c, and miR-30b-5 pmiR-31, miR-205, miR223 | Mastitis | *Staphylococcus aureus* | BMEC |
| miR-144, miR-451 and miR-7863 | | *Escherichia coli and Staphylococcus aureus* | BMEC |
| miR-21, miR-146a, miR-155, miR-222, and miR-383 | | *Streptococcus agalactiae* | Milk |
| let-7i, miR-21, miR-27, miR-99b, miR-146, miR-147, miR-155 and miR-223 | | California mastitis test (CMT) | Milk |
| miR-17-5p, miR-31 and miR-1281 | Foot and Mouth disease | Foot and Mouth disease virus | Serum |
| miR-21-5p, miR-101, miR-126-3p, miR-145, miR-197, bta-miR-223 | | | Serum |
| bta-miR-142-5p, bta-miR-146a and bta-miR-423-3p | Tuberculosis | *Mycobacterium bovis* | Lung |
| mir-19b, mir-19b-2, mir-1271, mir-100, mir-301a, mir-32, mir-6517 and mir-7857 | | | Blood |

| | | | |
|---|---|---|---|
| miR-21-3p, miR-1246, miR-27a-5p, miR-760-3p, miR-320a and miR-363 | Peste des petits ruminant's virus infection | Peste des petits ruminant's virus | Spleen and lung |
| miR-204-3p, miR-338-3p, miR-30b-3p, miR-199a-5p, miR-199a-3p and miR-1 | | | Peripheral blood mononuclear cells |
| miR-let-7f, miR-151, miR-30e, miR-191, miR-150 and miR-339b | Brucellosis | *Brucella abortus* | Serum |
| miR-87, miR-71 | Fascioliasis | *Fasciola gigantica* | Serum |

Source: Ojo OE, Kreuzer-Redmer S. (2023). MicroRNAs in ruminants and their potential role in nutrition and physiology. Vet Sci., 10(1):57. doi: 10.3390/vetsci10010057. PMID: 36669058; PMCID: PMC9867202.

Through the use of next-generation sequencing, it was discovered that the type of pathogen causing mastitis alters the parenchyma's miRNA profile. It was discovered to vary based on whether coagulase-negative (CoNS) or coagulase-positive (CoPS) staphylococci were infecting the glands. They concluded that their findings about the role of the identified miRNAs in the etiology of mastitis constitute a novel step in our understanding of the molecular mechanisms underlying the disease and may pave the way for future research on the role of miRNAs in the regulation of bacterial infection-related molecular pathways, as well as more effective preventative and therapeutic measures.

### *1.2.2.2. Tuberculosis*

Tuberculosis (TB) is a zoonotic disease that can infect humans from animals. The usual causes of tuberculosis in cattle and sheep are Mycobacteriu *M. bovis* (M bovis) or *Mycobacterium avium*. Pleomorphic, aerobic, nonmotile, nonspore-forming bacilli are known as mycobacteria. Cases of goats have been connected to *M. avium*, *M bovis*, or tuberculosis. Cattle that are infected with tuberculosis by *M. bovis* have a wide variety of hosts. Because of its significance, animal TB, particularly in cattle, has a disastrous effect on both public health and animal productivity. Cattle tuberculosis has led to severe restrictions on the trade of live animals and their products, resulting in substantial economic losses. Early identification of animals afflicted with *M. avium* subspecies paratubercolosis (MAP) would be highly beneficial in preventing the virus's spread and reducing its transmission. Circulating miRNA profiles and gene expression have been proposed as putative disease markers in both human and veterinary medicine. Therefore, to identify potential biomarkers to aid in the diagnosis of MAP infection, gene expression and related miRNA levels were investigated in cows that had tested positive for MAP using ELISA and culture (Table 1.3).

The results showed that miRNA expression is impacted by MAP infection and that miRNAs play a critical role in regulating the host's response to infection. In comparison to the currently available ELISA-based diagnostic techniques, the gene expression and miRNA profiles may function as biomarkers of infection and an earlier means of diagnosing MAP infection. In response to Mycobacteriu *M. bovis* (M bovis) infection, a study details the first next-generation sequencing method for temporally profiling miRNA expression in primary bovine alveolar macrophages (AMs). The cause of bovine tuberculosis, *Mycobacterium bovis*, travels through the air and is ingested by lung alveolar macrophages. The findings indicate that miRNAs play a critical role in controlling the complex interplay between *M. bovis* survival strategies and the host immune response.

### 1.2.2.3. Brucellosis

Brucellosis is a serious and economically important zoonotic disease that affects domesticated animals, people, and even wildlife. The illness is brought on by gram-negative bacteria belonging to the genus Brucella. It is believed that four of the six known classical Brucella species are dangerous to people. The most common source of human brucellosis is Brucella abortus, which mainly infects cattle, buffalo, elk, yaks, and camels. The second most common cause of infection in humans is *Brucella melitensis*, which mainly infects goats and sheep. Animal brucellosis primarily affects the genitalia of females and results in abortions, though there have also occasionally been reports of orchitis or joint symptoms.

The potential use of the miRNAs as biomarkers in vaginal secretions was investigated in a recent study that characterized the serum miRNA signature associated with brucellosis in water buffaloes. RT-qPCR was used to confirm dysregulated miRNAs in blood serum and vaginal fluids, and next-generation sequencing was used to evaluate the miRNA expression profiles in blood sera from Brucella-positive and Brucella-negative individuals. The results provided an overview of miRNA expression levels and suggested potential markers for Brucella infection in water buffaloes (Table 1.3). If additional functional and mechanistic studies of these miRNAs are taken into consideration, a better understanding of the molecular processes underlying Brucella infection and host immune response may be possible.

### 1.2.2.4. Foot and Mouth

Foot and mouth disease (FMD) is one of the most serious and urgent livestock diseases in the world (Figure 1.1). Its annual economic impact on observable production losses and vaccine costs is estimated to be between US$6.25 billion and US$21 billion in endemic areas of the world. Animals with cloven hooves are susceptible to foot-and-mouth disease (FMD), which is caused by the FMD virus (FMDV), a type of aphthovirus that belongs to the family Pirogenaviridae. Among the domesticated species vulnerable to FMD are water buffalo (Bubalus bubalis), pigs, sheep, goats, and cattle. The control of FMD is particularly difficult due to the existence of subclinical disease types. In a proof-of-concept study using miRNA PCR array plates, the relative abundance of 169 miRNAs in bovine serum collected at three distinct FMDV infection stages was ascertained. Subclinical FMDV persistence caused an alteration in the circulating miRNA profile in animals that had recovered from infection. Table 1.3 shows that bta-miR-17-5p was more abundant during acute infection while bta-miR-31 was more abundant during FMDV persistence.

These findings suggest that FMDV infection in cattle involves non-coding regulatory RNAs. Subsequent investigations will delineate the distinct roles played by the detected miRNAs in FMDV replication. Additionally, it will explore whether this miRNA profile is conserved across all FMDV serotypes and whether it holds promise for inventive diagnostic applications. A study of the host miRNA response following FMDV infection was conducted on cattle, which are a natural host for FMDV. SerummiRNA expression was found to be significantly altered early in the infection process. The findings indicate that early pathogenesis is characterized by changes in miRNA expression. Identifying putative miRNA target genes may help to disentangle the molecular mechanisms underlying the virus-host interaction, which could facilitate the creation of new therapeutic strategies.

### *1.2.2.5. Fascioliasis*

The ancient food-borne zoonotic disease known as fascioliasis is caused by certain species of leaf-like digenetic trematodes in the genus Fasciola (Figure 1.1). This disease is often ignored. Fasciolaemiasis has an asymptomatic incubation phase after parasite ingestion, which is succeeded by an acute and a chronic clinical phase. The acute phase of the Fasciola infection is brought on by the juvenile worms' penetration of the intestinal wall and peritoneum. They then proceed to the bile ducts and the surface of the liver. Fascioloasis affects a significant number of countries worldwide, with the Middle East and Latin America reporting the highest rates of infection. Together with *F. gigantica*, it is estimated that *F. hepatica* infects 250 million sheep and 300 million cattle globally, and that this causes an annual economic loss of USD 3 billion. Osinowo et al. carried out a 90-day study to establish a baseline for the prevalence of *Fasciola gigantica* infestation in cattle butchered at the Minna metropolitan abattoir in the Chanchaga Local Government area of Niger State, Nigeria. This study established the presence of *Fasciola gigantica* in animals killed in Minna Metropolis. Only areas with lower rates of snail infestations—particularly those close to rivers and streams—should be suitable for cattle to graze.

In this study, two groups of twenty-four (8–10-month-old) buffaloes showed differential expression of 121 host circulating miRNAs, of which 44 were significantly down-regulated and 77 were up-regulated. The sera of buffaloes infected with *F gigantica* contained four miRNAs derived from the parasite, and during the infection, the buffalo sera displayed dysregulation of the host's circulating miRNAs (Table 1.3). As a result, circulating parasite-derived miRNAs will be more effective as diagnostic targets for parasite infection, and the molecular mechanisms underlying host-parasite interactions will be better understood.

### *1.2.2.6. Peste des Petit*

The genus Morbillivirus, belonging to the family Paramyxoviridae, is the cause of peste des petits ruminants. The official name of this virus was Peste des petits ruminants virus (PPRV) until 2016, at which point it was renamed as Small Ruminant Morbillivirus (SRM). Nonetheless, experts in the domain continue to call it PPRV. The virus is a pleomorphic particle with an RNA genome encased in a lipoprotein membrane and a ribonucleoprotein core. The genome is a single-stranded, roughly 16 kilobases (kb) long negative polarity negative sense RNA. The endemic small ruminant plague affects Africa, the Middle East, Central Asia, and East Asia. In recent times, PPRV lineage IV has expanded greatly throughout Asia and Africa, including China, Nepal, India, and Pakistan (from the north to Tanzania). Epidemics are common in Turkey; the most recent one occurred between 2011 and 2016.

In June 2018, an incident was reported for the first time in Europe, in Bulgaria. Many researchers have used PPRV-specific antibodies in a range of tests and assays to find virus antigens in tissue, swabs, conjunctival smears, and formalin-fixed tissues. Currently, virus isolation is the "gold standard" for PPR diagnosis. The PPRV can be isolated and cultured *in vitro* using primary sheep and cattle cells, as well as established cell lines like Vero (African green monkey kidney) cells and Marmoset B-lymphoblastoid-B95a cells. Sequencing of miRNAs and proteomics data were used to ascertain the role of differentially expressed miRNA (DEmiRNA) in PPR virus (PPRV)-infected lung and spleen tissues of sheep and goats.

As per the findings, the combination of PPRV-induced miR-21-3p, miR-320a, and miR-363 may enhance viral pathogenesis in sheep by down-regulating numerous immune response genes in the lung and spleen (Table 1.3).

Important information about the molecular pathogenesis of PPR is provided by the fact that the PPRV–Izatnagar/94 isolate induces a greater host response in goats than in sheep. The deep sequencing method was utilized to determine the cellular miRNA expression profile in goat peripheral blood mononuclear cells (PBMC) infected with the widely used vaccine strain Nigeria 75/1 for mass immunization campaigns against Peste des petits ruminants. The results of this study provide a crucial starting point for future investigations into the roles of miRNA in PPRV pathogenesis and replication.

## 1.3. DEVELOPMENT OF MIRNAS AS BIOMARKERS OF DIET, NUTRITIONAL STATUS, AND THEIR POTENTIAL FOR THERAPEUTIC USE IN RUMINANTS

Biological molecules known as biomarkers are employed to understand physiological processes, recognize aberrant processes or pinpoint diseases. miRNA can be utilized as biomarkers for therapeutic, prognostic, or diagnostic purposes in the management of livestock diseases. They are also referred to as signature molecules, biochemical markers, or molecular markers. A miRNA biomarker is defined as a specific type of miRNA that is generated or enriched in a given tissue, and whose levels in circulation may reveal physiological or pathological alterations in that tissue. In nutrigenomics, changes in homeostasis that indicate the connection between a nutrient-dense diet and health or illness must be taken into account when selecting biomarkers.

MiRNAs from plasma, leucocytes, serum, and feces, however, may be useful biomarkers to measure the physiological effects of dietary or intervention lifestyle studies, according to recent research that this review cited. Current studies described in this section suggest that miRNAs from plasma or serum, PBMC, and feces may be useful biomarkers to measure the physiological effects of dietary or lifestyle intervention studies. The majority of small-scale RNA biomarker discovery studies use high-throughput profiling techniques, like PCR arrays or sequencing, to identify potential sequences associated with specific pathological or physiological conditions. Cell-free biofluid samples can be analyzed in two ways: as isolated fractions containing short RNAs, or as whole samples (like plasma) lipoprotein or exosomal fractions). Studies have shown that different fractions in plasma contain different sequences of miRNA, suggesting that unfractionated biofluid samples may be more effective for achieving full screening of miRNA populations. Since individual RT-qPCR assays provide some control over enzymatic inhibition and sample contamination, they are typically utilized for validation. After validation, the biomarkers can be tested on larger subject cohorts and/or their function can be investigated *in vitro* or *in vivo* using methods like gain-of-function or loss-of-function assays and luciferase reporter assays. A recent study found that the miRNA expression of bovine serum could be influenced by diet and that this could have an impact on immunity. The results indicated that manipulating an animal's physiological state could be accomplished through effective dietary strategies.

Research on domestic animals, especially ruminants, is scarce. Circulating miRNAs have been used as biomarkers for a range of diseases and physiological conditions in mice and humans. The identification of deregulated miRNAs and their targets presents opportunities for the development of prognostic, therapeutic, and diagnostic strategies for diseases in animals. MiRNAs are easily quantifiable and are naturally produced by cells into bodily fluids where they persist in relatively stable protein or lipid complexes (Figure 1.2). This provides the opportunity to use miRNAs as non-invasive markers of tissue function associated with various physiological conditions and states (like pregnancy). In addition, certain miRNAs are specific to particular tissues or developmental stages (e.g., neoplasia, osteoarthritis, sepsis, and cardiovascular disease). Eight non-pregnant heifers were given plasma samples on pregnancy days using Illumina small-RNA sequencing to explore the possibility of circulating miRNAs serving as markers of early pregnancy in cattle. A genome-wide analysis revealed that miR-26a may be a circulating biomarker of early pregnancy, and it was possible to characterize the plasma miRNA populations associated with early pregnancy in cattle. Age and genetic background had an impact on the expression patterns of 306 plasma miRNAs in 18 animals, and it was found that these characteristics were associated with dairy cow productivity and well-being.

The results of the study suggest that circulating miRNAs may be useful markers for dairy cows to aid with improved health, welfare, and production outcomes. The miRNA profiles in plasma and cell fractions of blood were analyzed in six non-pregnant Holstein–Friesian cross cows in order to identify tissue-derived miRNAs that could be helpful as markers of tissue function in this important food animal species. An important finding in the context of postpartum negative energy balance in dairy cows was the identification of miR802, a circulating miRNA that has not been found in cattle before and may control lipid metabolism and insulin sensitivity. It might therefore offer a particular biomarker of liver function. Circulating miRNAs are suitable as biomarkers for tracking different physical situations in animals, as evidenced by the fact that their profiles change in response to diseases, bacterial or viral infections, and physiological states (e.g., pregnancy). It is increasingly evident that dietary feed components are essential for both the regulation of miRNAs and nutrition and health. Given that a range of bioactive feed ingredients influences the production of miRNAs, it makes sense that some of these components may also influence the susceptibility of diseases to onset, exacerbation, and progression.

The foot and mouth disease virus (FMDV) naturally infects cattle, thus these animals were used to investigate the host miRNA response after FMDV infection. There was a noticeable shift in serum miRNA expression during the early phases of infection. The research discovered that early pathogenesis is associated with modifications in miRNA expression and that identifying putative miRNA target genes could help decipher the molecular mechanisms behind virus–host interaction and, in turn, aid in the creation of therapeutic strategies.

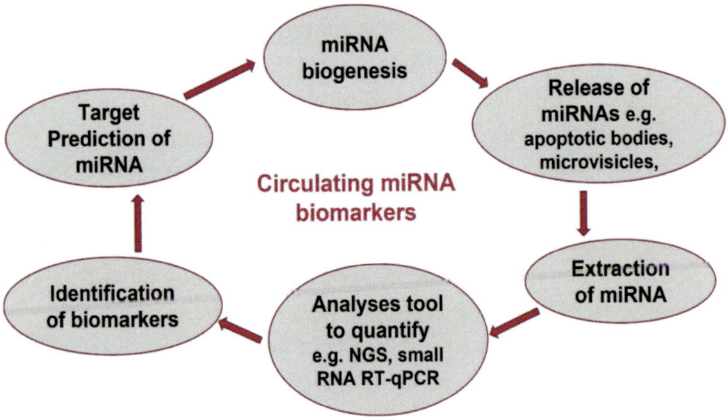


**Figure 1.2**. *Circulating miRNAs as biomarkers.*

***Source***: *Ojo OE, Kreuzer-Redmer S. (2023). MicroRNAs in ruminants and their potential role in nutrition and physiology. Vet Sci., 10(1):57. doi: 10.3390/vetsci10010057. PMID: 36669058; PMCID: PMC9867202.*

## 1.4. COMPOUNDS WITH NUTRIGENOMIC POTENTIAL IN DAIRY COWS

### 1.4.1. Short-Chain Fatty Acids

Butyrate is the primary short-chain fatty acid (SCFA) obtained from rumen fermentation that has the most nutrigenomic information when it comes to dairy cows. In MDBK cells, butyrate modulates the expression of numerous genes linked to immune response, signaling, and cell cycle arrest. It modifies the amount of alternative splicing in MDBK cells and histone deacetylase activity, both of which are epigenetic functions. Butyrate supplementation stimulated the expression of cyclin D1, a protein involved in cell proliferation, in ruminal papillae cultures of dairy goats. In dairy cow ruminal papillae, butyrate also had an impact on the expression of genes involved in lipogenesis and glycolysis.

Remarkably, data from a recent study showed that prolactin (PRL) and GH were downregulated in dairy cow anterior pituitary cells by SCFA, specifically propionate. These results appear to be at odds with the well-established benefits of energy in the diet, which are typically linked to propionate production and an increase in protein synthesis in the mammary glands of cows that is partially induced by growth hormone. More recently, expression of SCD1 and ACACA was significantly upregulated in MacT cells cultured with 5 mM acetate, 5 mM β-hydroxybutyrate, or a combination of the 2 (5 mM each) (Jacobs et al., 2013). The possibility of an effect via PPAR of the two saturated fats from acetate produced through de novo synthesis was ruled out by the observations that PPARG expression was not detected in MacT cells and that treatment with either C16:0 or C18:0 did not affect the expression of any PPAR-related gene.

Acetate's previously noted nutrigenomic effect is most likely transduced through receptors on the plasma membrane. It is established that certain proteins on the plasma membrane known as free fatty acid receptors (FFAR) are bound by SCFA. Among their many important functions, the FFAR members of the G protein-coupled receptor (GPR) mediate the impact of fatty acids on insulin secretion, adipocyte differentiation, and metabolism. Four receptors play this role: GPR40 (FFAR1), GPR120 (FFAR4), and GPR41 (FFAR3), which bind SCFA, GPR43 (FFAR2), and GPR43 (FFAR2).

There aren't many studies on FFAR in ruminants. Yonezawa et al. (2009) have identified the expression of FFAR2 (GPR43) and FFAR3 (GPR41) in mammary epithelial cells and bovine mammary tissue during lactation. According to the study's findings, those receptors mediate the rise in intracellular $Ca^{2+}$, fall in cAMP, and raise mitogen-activated protein kinase (MAPK) phosphorylation. Hosseini et al. (2012) found that FFAR3 and FFAR2 increased during *in vitro* bovine adipogenesis, but propionate, insulin, or β-hydroxybutyrate had no effect on expression. However, propionate was found to increase the expression of FFAR3 in goat white adipose tissue (Mielenz et al., 2008) and in cow adipose tissue from pregnancy to lactation, even though it is well-known that the rate of adipogenesis decreases during the early stages of lactation (McNamara, 1991). In mice, GPR43 has a significant role in the microbiota-host communication that influences insulin sensitivity and obesity (Kimura et al., 2013); given the elevated level of SCFA in the blood, this discovery may have significant implications for ruminants.

Less research has been done on the GPR40 in ruminants. Hidalgo et al.

(2011) found that oleic acid via GPR40 increased intracellular $Ca^{2+}$, MAPK phosphorylation, superoxide production, and granule release in bovine neutrophils. Yonezawa et a. (2008) found FFAR1 mRNA in primary bovine mammary epithelial cells and mammary tissue. The authors hypothesized that GPR40 was probably involved in the oleate-induced stimulation of mammary cell proliferation. As far as we are aware, there are no GPR120 data on dairy cows or other ruminants.

### 1.4.2. In Vitro Nutrigenomic Effects of Saturated Versus Unsaturated Long-Chain Fatty Acids

We suggested that using saturated LCFA can be a practical and affordable method of delivering nutrigenomic interventions because of their stronger nutrigenomic effect and strong PPAR agonistic activity seen *in vitro* (Bionaz et al., 2013). However, it's important to remember that applying *in vitro* methods to *in vivo* conditions can have drawbacks when studying nutrigenomics. The *in vitro* conditions do not replicate the *in vivo* scenario because cells are removed from their original niche and lose the surrounding cross-talk (both chemical and physical) with other cells. The state of cultured cells is actually quite static when contrasted with the dynamic fluctuations in hormones and metabolites that occur in their native environment. Using immortalized cells highlights the limitations of extrapolating from *in vitro* to *in vivo* conditions. Cultured cells frequently alter who they are and probably how they react to outside stimuli. For instance, the transcriptome profile of MacT cells differs significantly from that of mammary tissue (Hosseini et al., 2013). The study specifically allowed for the conclusion that, while MacT cells are probably less responsive than mammary tissue to the regulation of milk fat synthesis, they can still be regarded as a reasonably adequate model to study the regulation of protein synthesis. The latter conclusion was partially corroborated by the significantly lower expression of PPARγ and several of its candidate target genes (e.g., when comparing lipoprotein lipase) in MacT cells to mammary tissue (Hosseini et al., 2013). Consequently, *in vitro* studies are helpful in the nutrigenomic framework for highly accurate testing of a particular TF agonist's dose-responses. This can produce preliminary data needed to conduct *in vivo* experiments. Before any useful nutrigenomic applications are suggested, more *in vivo* research needs to be completed.

There is a long history of feeding fat supplements to lactating dairy cows to boost their calorie density or energy content. Saturated fats are better for you to use than oil because they are easier for ruminants to digest than unsaturated fats, and oil tends to cause milk fat depression among other things. In addition to giving dairy cows energy from food, LCFAs have a number of biological effects on them. Dietary LCFA, especially PUFA, generally improves fertility and has a beneficial effect on many reproductive organ components. Saturated LCFA supplements to dairy cows are generally beneficial, according to a recent analysis of data from multiple studies that supplemented the animals with C16:0 and/or C18:0. Specifically, the authors noted that the majority of the studies improved the transition of dairy cows from pregnancy to lactation while also increasing milk fat percentage and yield. The rise in the percentage of supplemented LCFA in milk fat supports the theory that the increased milk fat is caused by greater availability of LCFA for the mammary tissue.

Recently, milk fat production was found to increase with an increase in the yield of most fatty acids, including SCFA, when midlactation dairy cows were fed a fat supplement containing 98% C18:0. This suggests that increased de novo fatty acid synthesis contributed to the rise in milk fat. However, it's not always the case that adding saturated LCFA supplements results in an increase in milk fat. The nutrigenomic effects of those saturated LCFA are probably responsible for the improvement in overall metabolism and increase in milk fat synthesis that have been reported in the literature upon supplementation. There is growing evidence of the nutrigenomic effects of long-chain fatty acids (LCFA) *in vivo*, albeit with differing outcomes and in different tissues.

### *1.4.2.1. Mammary Tissue*

In a study conducted *in vivo*, midlactation dairy cows supplemented with 40% C16:0 and 40% C18:0 saturated lipid for 21 days did not experience an increase in milk fat synthesis when compared to the control group; however, a combination of soy and fish oil caused a decrease in milk fat. In comparison to the combination of fish and soy oil, saturated lipids increased the expression of lipogenic genes in mammary tissue. However, because the control group's gene expression data were missing from the study, conclusions regarding the transcriptomic impact of saturated fat in breast tissue cannot be drawn. Likely, the mixture of oil well-known ability to lower milk fat was the primary cause of the observed nutrigenomic differences. This isn't always the case, though. Overall, it is evident that some unsaturated LCFA has a strong negative nutrigenomic effect, while saturated LCFA may have a positive nutrigenomic effect in mammary tissue.

### *1.4.2.2. Adipose Tissue*

In the adipose tissue of midlactation dairy cows, there were no nutrigenomic effects from saturated LCFA, but there was a higher expression of lipogenic genes with a mixture of oil compared with a control. Conversely, in a more current investigation, Schmitt et al. (2011) found that in adipose tissue at 11 days postpartum, the expression of multiple lipogenic genes, associated transcription factors, and nuclear receptor coactivators was higher in cows supplemented for 30 days with a commercial lipid high in saturated LCFA as opposed to a control or unsaturated fat.

### *1.4.2.3. Liver*

In the identical study that Schmitt et al. Hepatic expression was evaluated as well in 2011. The findings showed that prepartum supplementation with saturated fat had a greater nutrigenomic effect than unsaturated fat (i.e, one week after beginning the fat supplementation), but postpartum fish oil supplementation showed a greater nutrigenomic effect (i.e,>4 weeks after the addition). Remarkably, the data showed that postpartum fat supplementation, whether it be saturated or unsaturated, caused an overall reduction in the expression of genes, including ACOX1, CPT1A, and FGF21, which one would expect to be upregulated if PPARα was one of the target TFs. These data suggest that supplementing with saturated fat may benefit the liver before childbirth, but they do not appear to support a beneficial role for such supplements postpartum. Carriquiry et al., 2009 found that the liver of transition dairy cows did not exhibit any nutrigenomic effects from ω3 PUFA is the year. The liver tissue of dairy calves also showed evidence of the absence of the nutrigenomic effect of fish oil via PPARα.

### *1.4.2.4. Differentially Nutrigenomic Response to LCFA by Tissues*

The type of fat supplementation had different nutrigenomic effects on the liver and adipose tissue of nursing dairy cows, according to a study (Hiller et al., 2013). Like the postpartum phase described in Akbar et al.'s work. (2013b), in contrast to saturated fat, unsaturated fat supplementation by Hiller et al. (2013) revealed a general decline in the expression of genes linked to the metabolism of fatty acids (mainly those involved in triglyceride synthesis) in the liver, but nearly the opposite was seen in adipose tissue. In a different study, milk yield was reduced and the expression of ACACA in both the liver and adipose tissue, as well as the mRNA expression of leptin and its receptor, were unaffected by supplementing dairy cows for 12 weeks before giving birth with fat that mostly contained C16:0 and C18:1.

The aforementioned information provides evidence for a tissue-specific nutrigenomic response to long-chain fatty acids (LCFAs), whereby unsaturated LCFAs have positive nutrigenomic effects on adipose tissue (though they also have some positive effects on it) and negative nutrigenomic effects on mammary tissue. Saturated LCFAs typically have some positive nutrigenomic effects on the liver. Given the potential practical applications, this tissue-specific nutrigenomic response to LCFA is highly significant and warrants further investigation.

Although there aren't many published studies that examine the nutrigenomic effects of fat supplementation in dairy cows *in vivo*, more research is necessary before firm conclusions can be made. Saturated LCFA appears to have a weak nutrigenomic effect *in vivo*, despite the extremely strong nutrigenomic effects seen *in vitro*. The absence of exact standards to establish the right dose and duration is one of the main problems with *in vivo* fat supplementation. An important factor in the activation of TF is the agonist dose, particularly when taking into account the fine-tuning of TF activity.

### 1.4.3. Considerations in Using an In Vitro Approach to Study Nutrigenomic Effect of LCFA in Dairy Cows

It is still unclear why LCFA has such strong nutrigenomic effects *in vitro* but relatively little *in vivo*, but one possible explanation is that the amount of LCFA (especially saturated) that cells can absorb *in vivo* is already almost at the maximum binding capacity of the LdNR. This is somewhat corroborated by the fact that fetal bovine serum (FBS) is not used in the media during *in vitro* nutrigenomic studies utilizing LCFA; nonetheless, PPAR is activated with a comparable or greater magnitude when 10% FBS is added to culture media in comparison to 100 μM C16:0 or the synthetic PPARγ agonist rosiglitazone. Very few attempts have been made to ascertain the fatty acid composition of FBS. As an illustration, Lagarde et al. (1984) examined the fatty acid profile of media containing 20% FBS and found that C16:0 and C18:0 had estimated total fatty acid profiles of 67.5 and 32.7 μM (with 40.2 and 13.4 μM in the FFA fraction), respectively. From this perspective, it's intriguing that CPT1A expression in MDBK cells required a dose of 100 μM of C16:0 combined 4:1 with albumin. This final observation suggests that the high activation of PPAR seen in our study may not be adequately explained by the presence of C16:0 and C18:0 in FBS.

We used data from three independent studies to estimate the amount of each LCFA available *in vivo* at a cellular level for mammary cells. In the computation, we made the assumption that the tissues could only absorb free fatty acids (LCFA) from NEFA and triacylglycerol utilizing information from Enjalbert et al. (1998) found that the mammary gland extracts varying amounts of LCFA from plasma, ranging from 30% in control cows to 56% in cows that receive fat supplements. Based on these parameters, we calculated that in comparison to cows that have received fat supplementation (whose approximate range is from 145 to 650 μM), the LCFA availability for mammary cells from cows that have not received fat supplementation is significantly lower (ranging from 60 to 130 μM total LCFA). In comparison to Holstein cows, the data show that Jersey cows have a lower availability of LCFA.

In Holstein cows supplemented with soybean oil, the concentration of C16:0 and C18:0 ranges from 13 to 54 μM, reaching a maximum of >300 μM for C18:0, and approximately 200 μM for C16:0 in cows duodenally infused with 500 g/d of C16:0-enriched fat mixture. Thus, taking into account saturated LCFA, the 100 μM concentration of LCFA used in several *in vitro* nutrigenomic studies falls within a similar range, although it is marginally higher than that which is available for cells *in vivo* in cows that are not supplemented with fat, and it is lower than that in cows that are supplemented with fat. In light of Table 1.2's data, doses of unsaturated LCFA—particularly PUFA—between 0 and 10 μM appear to be more physiologically relevant. The study conducted by Enjalbert et al. estimated the available C16:0 concentration to be 195 μM based on a 500 g/d duodenal infusion of C16:0. (1998) seems overly elevated and might be harmful.

These observations have significant practical implications. It might be worthwhile to focus the nutrigenomic effort on increasing the abundance of LdNR or their coactivators if the concentration of LCFA is already close to or sufficient to maximize the activation of LdNR. Many LCFA, additional nutritional substances, and management techniques can influence TF expression; this seems to be the case with PPAR. A recent study using saturated fat supplementation in transition cows increased the expression of several PPAR coactivators in adipose tissue, lending support to the idea that PPAR activity can be increased *in vivo*. We don't yet know if this has an impact on PPAR's nutrigenomic activity.

### 1.4.3.1. Amino Acids

Bionaz et al. (2012a) offered a thorough analysis of the research on AA's effects on milk protein synthesis up until 2012. The review concluded that, in bovine mammary tissue, the activity of the primary pathway regulating protein synthesis, with mTOR as its central hub, is essentially inhibited and induced by a cooperative relationship between insulin, IGF-1, GH, and AA (e.g., leucine), and glucose resulting in increased mammary protein translation. Furthermore, the information that was available suggested that while AA, insulin, and glucose appear to fine-tune protein synthesis, changes in the mRNA expression of genes linked to glucose transport and AA uptake have a major impact on milk protein synthesis.

At the time of the Bionaz et al review (2012a), the hypothesis regarding the potential inhibition of the mTOR pathway by adenosine monophosphate-activated protein kinase (AMPK) was derived from nonruminant literature. The inhibition of glucose metabolism by activating AMPK via 2-deoxyglucose has been shown in recent work with MacT cells to cause a rapid and significant reduction in the global rates of protein synthesis (Burgos et al., 2013). In a later investigation, Appuhamy et al. (2014) examined the effects of different essential AA by themselves or in conjunction with glucose and acetate on the phosphorylation of mTOR and AMPK in MacT cells.

The research supported the beneficial effect of energy on the synthesis of milk proteins, as noted by Burgos et al. (2013) and verified the beneficial effects on mTOR pathway activation through phosphorylation of essential AA, which was accompanied by an increase in casein synthesis. Thus, evidence currently suggests a physiological relationship between glucose and the mammary availability of essential amino acids (AA), which are factors that influence the phosphorylation status and functionality of mTOR and AMPK. These findings need to be confirmed by more *in vivo* research.

The nutrigenomic impact of individual AA on milk protein synthesis *in vitro* using primary bovine mammary cells has been the main focus of recent molecular studies. Lys and Met, which are believed to be the most limiting AA for milk synthesis, have received some attention. Nan et al(2014) demonstrated that increased mTOR phosphorylation, as well as upregulation of MTOR's mRNA expression, the transcription regulator E74-like factor 5 (ELF5), and the genes encoding casein and lactalbumin, were the primary drivers of peak casein synthesis at a Lys to Met ratio of 3:1. Another study showed that when supplemented at a level equivalent to twice the concentration found in casein, Arg, a conditionally essential amino acid, can also increase the expression of casein genes along with MTOR, RPS6KB1, and STAT5, and decrease the expression of the translation inhibitor 4EBP1.

## 1.5. NUTRIGENOMIC LEVEL OF FEED INTAKE

### *1.5.1. Liver*

In the liver of dairy cows, Velez and Donkin (2005) found that pyruvate carboxylase gene expression is upregulated during brief feed restriction. Prepartum feed intake (restricted and high feed intake) has a strong temporal nutrigenomic effect on dairy cow livers. Shahzad et al. (2014) conducted a more recent analysis of the same data using systems biology revealing a significant "priming" effect of feed intake level on dairy cows' livers. Specifically, it was found that the livers of cows that were fed less before giving birth were better prepared to handle the metabolic demands of the postpartum period. When early postpartum dairy cows were fed restricted grazing (60% equated to the optimal level of forage), extensive transcriptomic effects on the liver were also observed in comparison to animals that were fed adequately. A general decrease in liver metabolism was found by the functional analysis, which is probably because the liver needs to save energy for other tissues. Furthermore, PPAR signaling was activated but cholesterol synthesis was strongly inhibited. Similar overall nutrigenomic effects were found in the liver of feed-restricted midlactation dairy cows, albeit to a lesser extent. Sigl et al. (2013) reported that short-term feed restriction was enough to induce change in the expression of lipid-related genes in the liver of early postpartum dairy cows.

### 1.5.2. Adipose Tissue

Prepartum dietary feed intake level had a transcriptomics effect on subcutaneous adipose tissue that was comparable to that of the liver, but there did not appear to be a priming effect on the adipose tissue. Additionally, the high intake nutrigenomic effect (i.e., high dietary energy level) prepartum was more severe in fat tissue than in the liver and significantly induced the gene networks responsible for triglyceride accumulation in fat tissuc as compared to a control group. The higher BCS was also consistent with increased triglyceride synthesis in cows fed a high prepartum diet.

### 1.5.3. Mammary Tissue and Immune Cells

Boutinaud et al. (2008) observed a reduction in the expression of two glucose transporters' genes (SLC2A1 and, to a lesser extent, SLC5A1) in mammary epithelial cells taken from the milk of dairy cows fed a restricted diet. Furthermore, Sigl et al. (2014) found that a brief feed restriction in dairy cows during the early and middle stages of lactation had a significant impact on the expression of genes related to the synthesis of milk proteins. It was shown that feed restriction reduced the expression of several genes related to milk fat in mammary tissue in two studies involving sheep and goats. Neutrophils from cows with high prepartum feed intake expressed more genes related to inflammation and metabolism at two weeks postpartum, and they also showed reduced phagocytosis.

Overall, the information covered above points to a nutrigenomic role for feed quantity or dietary energy intake in dairy cows; however, the exact mechanism underlying this effect is still unknown and requires further investigation. Changes in total feed intake also cause changes in other dietary components, in addition to energy content. As a result, the nutrigenomic impact of feed intake level is intricate and probably involves numerous TF. This and the interactions between tissues should be taken into consideration using a systems biology approach. The priming effect found in the liver of feed-restricted cows is extremely interesting and could be used to improve the overall "capacity" of tissues, especially the liver, to face metabolic and inflammatory challenges. However, the extensive transcriptomics effects as a consequence of changes in the level of feed intake, as observed in liver and adipose tissue, do not allow for a practical approach to fine-tune metabolism, as is the case for the use of specific nutritional compounds (Ojo, 2023).

## REFERENCES

1. Abete, I., Navas-Carretero, S., Marti, A., & Martinez, J. A. (2012). Nutrigenetics and nutrigenomics of caloric restriction. *Progress in Molecular Biology and Translational Science*, *108*, 323–346.
2. Adan, R. A., Walther, N., Cox, J. J., Ivell, R., & Burbach, J. P. (1991). Comparison of the estrogen responsiveness of the rat and bovine oxytocin gene promoters. *Biochemical and Biophysical Research Communications*, *175*, 117–122.
3. Akhtar, M. M., Micolucci, L., Islam, M. S., Olivieri, F., & Procopio, A. D. (2016). Bioinformatic tools for microRNA dissection. *Nucleic Acids Research*, *44*, 24–44. https://doi.org/10.1093/nar/gkv1221.
4. Bionaz, M., & Loor, J. J. (2008). Gene networks driving bovine milk fat synthesis during the lactation cycle. *BMC Genomics*, *9*, 366.
5. Bionaz, M., Chen, S., Khan, M. J., & Loor, J. J. (2013). Functional role of PPARs in ruminants: Potential targets for fine-tuning metabolism during growth and lactation. *PPAR Research*, 2013, Article 684159.
6. Cameron, A., & McAllister, T. A. (2016). Antimicrobial usage and resistance in beef production. *Journal of Animal Science and Biotechnology*, *7*, 68. https://doi.org/10.1186/s40104-016-0127-3.
7. Do, D. N., Dudemaine, P.-L., Fomenky, B. E., & Ibeagha-Awemu, E. M. (2019). Integration of miRNA weighted gene co-expression network and miRNA-mRNA co-expression analyses reveals potential regulatory functions of miRNAs in calf rumen development. *Genomics*, *111*, 849–859. https://doi.org/10.1016/j.ygeno.2018.05.009.
8. Dysin, A. P., Barkova, O. Y., & Pozovnikova, M. V. (2021). The role of microRNAs in the mammary gland development, health, and function of cattle, goats, and sheep. *Non-Coding RNA*, *7*, 78. https://doi.org/10.3390/ncrna7040078.
9. Filipowicz, W., Bhattacharyya, S. N., & Sonenberg, N. (2008). Mechanisms of post-transcriptional regulation by microRNAs: Are the answers in sight? *Nature Reviews Genetics*, *9*, 102–114. https://doi.org/10.1038/nrg2290.
10. Gardner, P. P., & Vinther, J. (2008). Mutation of miRNA target sequences during human evolution. *Trends in Genetics*, *24*, 262–265. https://doi.org/10.1016/j.tig.2008.03.009.
11. Hiller, B., Angulo, J., Olivera, M., Nuernberg, G., & Nuernberg, K. (2013). How selected tissues of lactating Holstein cows respond to dietary polyunsaturated fatty acid supplementation. *Lipids*, *48*, 357–367.
12. Hou, L., Ji, Z., Wang, G., Wang, J., Chao, T., & Wang, J. (2018).

Identification and characterization of microRNAs in the intestinal tissues of sheep (*Ovis aries*). *PLoS ONE*, *13*, e0193371. https://doi.org/10.1371/journal.pone.0193371.

13. Hu, Y., Lan, W., & Miller, D. (2017). Next-generation sequencing for microRNA expression profile. *Methods in Molecular Biology*, *1617*, 169–177. https://doi.org/10.1007/978-1-4939-7046-9_12.
14. Hussein, M., Harvatine, K. H., Weerasinghe, W. M. P. B., Sinclair, L. A., & Bauman, D. E. (2013). Conjugated linoleic acid-induced milk fat depression in lactating ewes is accompanied by reduced expression of mammary genes involved in lipid synthesis. *Journal of Dairy Science*, *96*, 3825–3834.
15. Jacobs, A. A., Dijkstra, J., Liesman, J. S., Vandehaar, M. J., Lock, A. L., van Vuuren, A. M., Hendriks, W. H., & van Baal, J. (2013). Effects of short- and long-chain fatty acids on the expression of stearoyl-CoA desaturase and other lipogenic genes in bovine mammary epithelial cells. *Animal*, *7*, 1508–1516.
16. Janovick, N. A., & Drackley, J. K. (2010). Prepartum dietary management of energy intake affects postpartum intake and lactation performance by primiparous and multiparous Holstein cows. *Journal of Dairy Science*, *93*, 3086–3102.
17. Kaur, S., & Kumar, S. (2020). Nutriepigenomics: Need of the day to integrate genetics, epigenetics and environment towards nutritious food for healthy life. *Food Science and Nutrition Technology*, *5*, 000239.
18. Kumar, S., Chinnusamy, V., & Mohapatra, T. (2018). Epigenetics of modified DNA bases: 5-Methylcytosine and beyond. *Frontiers in Genetics*, *9*, 640. https://doi.org/10.3389/fgene.2018.00640.
19. Lee, R. C., Feinbaum, R. L., & Ambros, V. (1993). The *C. elegans* heterochronic gene lin-4 encodes small RNAs with antisense complementarity to lin-14. *Cell*, *75*, 843–854. https://doi.org/10.1016/0092-8674(93)90529-Y.
20. Leroy, J. L., Sturmey, R. G., Van Hoeck, V., De Bie, J., McKeegan, P. J., & Bols, P. E. (2014). Dietary fat supplementation and the consequences for oocyte and embryo quality: Hype or significant benefit for dairy cow reproduction? *Reproduction in Domestic Animals*, *49*, 353–361.
21. Liang, G., Malmuthuge, N., McFadden, T. B., Bao, H., Griebel, P. J., Stothard, P., & Guan, L. L. (2014). Potential regulatory role of microRNAs in the development of bovine gastrointestinal tract during early life. *PLoS ONE*, *9*, e92592. https://doi.org/10.1371/journal.pone.0092592.
22. Matias-Garcia, P. R., Wilson, R., Mussack, V., Reischl, E., Waldenberger,

M., Gieger, C., Anton, G., Peters, A., & Kuehn-Steven, A. (2020). Impact of long-term storage and freeze-thawing on eight circulating microRNAs in plasma samples. *PLoS ONE*, *15*, e0227648. https://doi.org/10.1371/journal.pone.0227648.

23. Mendes, N. D., Freitas, A. T., & Sagot, M.-F. (2009). Current tools for the identification of miRNA genes and their targets. *Nucleic Acids Research*, *37*, 2419–2433. https://doi.org/10.1093/nar/gkp145.
24. Mendes, R. E. (2012). *Ruminants: Anatomy, behavior, and diseases*. Hauppauge, NY: Nova Biomedical.
25. Michlewski, G., & Cáceres, J. F. (2019). Post-transcriptional control of miRNA biogenesis. *RNA*, *25*, 1–16. https://doi.org/10.1261/rna.068692.118.
26. Ojo, O. E., & Kreuzer-Redmer, S. (2023). MicroRNAs in ruminants and their potential role in nutrition and physiology. *Veterinary Sciences*, *10*(1), 57. https://doi.org/10.3390/vetsci10010057.
27. Pacífico, C., Ricci, S., Sajovitz, F., Castillo-Lopez, E., Rivera-Chacon, R., Petri, R. M., Zebeli, Q., Reisinger, N., & Kreuzer-Redmer, S. (2022). Bovine rumen epithelial miRNA-mRNA dynamics reveals post-transcriptional regulation of gene expression upon transition to high-grain feeding and phytogenic supplementation. *Genomics*, *114*, Article 110333. https://doi.org/10.1016/j.ygeno.2022.110333.
28. Ricci, S., Petri, R. M., Pacífico, C., Castillo-Lopez, E., Rivera-Chacon, R., Sener-Aydemir, A., Reisinger, N., Zebeli, Q., & Kreuzer-Redmer, S. (2022). Characterization of presence and activity of microRNAs in the rumen of cattle hints at possible host-microbiota cross-talk mechanism. *Scientific Reports*, *12*, 13812. https://doi.org/10.1038/s41598-022-17445-z.
29. Ross, S. A., & Davis, C. D. (2014). The emerging role of microRNAs and nutrition in modulating health and disease. *Annual Review of Nutrition*, *34*, 305–336. https://doi.org/10.1146/annurev-nutr-071813-105729.
30. Stefani, G., & Slack, F. (2006). MicroRNAs in search of a target. *Cold Spring Harbor Symposia on Quantitative Biology*, *71*, 129–134. https://doi.org/10.1101/sqb.2006.71.032.
31. Sun, H.-Z., Chen, Y., & Guan, L. L. (2019). MicroRNA expression profiles across blood and different tissues in cattle. *Scientific Data*, *6*, Article 190013. https://doi.org/10.1038/sdata.2019.13.
32. Thornton, J. E., & Gregory, R. I. (2012). How does Lin28 let-7 control development and disease? *Trends in Cell Biology*, *22*, 474–482. https://doi.org/10.1016/j.tcb.2012.06.001.

33. Vella, M. C., & Slack, F. J. (2005). *C. elegans* microRNAs. *WormBook*, *1*, 1–9. https://doi.org/10.1895/wormbook.1.26.1.
34. Wang, X., Gu, Z., & Jiang, H. (2013). MicroRNAs in farm animals. *Animal*, *7*, 1567–1575. https://doi.org/10.1017/S1751731113001183.
35. Yan, Q., Tian, L., Chen, W., Kang, J., Tang, S., & Tan, Z. (2022). Developmental alterations of colonic microRNA profiles imply potential biological functions in kid goats. *Animals*, *12*, 1533. https://doi.org/10.3390/ani12121533.
36. Zhong, T., Wang, C., Hu, J., Chen, X., Niu, L., Zhan, S., Wang, L., Guo, J., Cao, J., Li, L., et al. (2020). Comparison of microRNA transcriptomes reveals the association between miR-148a-3p expression and rumen development in goats. *Animals*, *10*, Article E1951. https://doi.org/10.3390/ani10111951.

Chapter 2

# GENETIC AND EPIGENETIC MODULATION OF ANIMAL EXPOSURE TO HIGH TEMPERATURE

## Contents

# INTRODUCTION

Pathogen attacks and heat stress are two examples of environmental stressors (abiotic and biotic factors) that affect both terrestrial and aquatic organisms. Climate change is predicted to cause an increase in the severity of thermal stress in the upcoming decades. Animals' core body temperatures rise when temperatures rise above the prescriptive zone; this can lead to hyperthermia and change several biological processes, including the breakdown of biological regulatory mechanisms, a decline in immunity, a decline in health, and an increase in mortality. According to Ritossa (1996), the heat-shock response (HSR) was first identified as a transcriptional reaction to an elevated temperature shock. Heat shock proteins and heat shock factor 1 were identified as a result of its discovery (HSF1). HSF1, the main actor in HSR, is mediated through cell-autonomous and non-autonomous signals following particular cellular requirements, according to mounting evidence (Li et al., 2017). HSF1 protects the structural stability of refolding or breaking down intracellular proteins by increasing transcriptional activity, particularly that of the molecular chaperones heat-shock protein family (HSP). To disperse heat and lessen heat illness, organisms must repeatedly be exposed to heat stress, a process known as heat acclimation. Researchers are examining heat acclimation and thermal death in insects in addition to mammals. The findings indicated that there is a significant correlation between heat acclimation and HSR. Heat acclimation causes HSR to occur, but HSR encourages the process of heat acclimation.

Furthermore, calcium is a crucial second intracellular messenger that plays a key role in heat acclimation. It can translate extracellular stimuli into intracellular signals and influence signaling pathways to control gene expression, development, survival, and differentiation of cells (Berna-Erro et al., 2012). In addition, calcium and HSF1 are essential for biological cell stress response to heat stress, including endoplasmic reticulum (ER) stress and reactive oxygen species (ROS) stress (Ryu et al., 2013), as well as cell death (Manucha et al., 2011). Thus, we primarily focused on HSF1 and calcium when discussing the genetic mechanisms of heat acclimation and HSR in this section. A summary of the key genes, HSF1, calcium, and other fundamental mechanisms involved in HSR is provided.

The process of thermoregulation is frequently not fully explained by genetic regulation or environmental exposure alone. Epigenetics regulates environmental stress without altering the structure of DNA, and it is essential for transcription, imprinting, nuclear organization, and genome stability (Bartels et al., 2018). Various epigenetic modifications that alter the transcription level rather than the DNA sequence have been studied to date, such as histone modifications, non-coding RNA, and DNA methylation (Arányi et al., 2005; Fan and Huang, 2010). DNA methylation is one of the most well-studied epigenetic regulatory mechanisms. It commonly takes place at the CpG dinucleotide in eukaryotes. Recently, methylation profiling across the genome of economically significant animals has been carried out. Specifically, heat stress causes a lot of genes to undergo DNA methylation, which modifies a lot of genes' expression. Differential patterns in tissues or cell types are exhibited by dynamic changes in DNA methylation. The development of heat acclimation and alterations in the mitochondrial respiratory chain (OXPHOS) may be influenced by epigenetic mechanisms (Dai et al., 2018). The interactions between genetic and epigenetic controls during heat stress, however, have not been the subject of many investigations.

## 2.1. GENES MODULATE ANIMAL RESPONSES TO HEAT STRESS

### 2.1.1. Heat and Oxidative Stress

Oxidative stress is caused by ROS production, which is triggered by heat stress. Under heat stress, human umbilical vein endothelial cells have been shown to produce two time-dependent oxygen free radicals: $O^{2-}$ and $H_2O_2$. Under heat stress, $O_2$ increases noticeably right away, followed by an increase in $H_2O_2$, suggesting that excess $O_2$ is a major factor in the generation of primary ROS following heat stress.

It is commonly known that HSP genes play a role in the synthesis of HSF1 during oxidative stress. Research has been done on the protective stress response to heat stress involving small HSPs (sHSPs, molecular weight 8.5–40 kDa) and the HSP60, HSP70, and HSP90 protein families. The HSP70 family eliminates ROS production brought on by heat. In rat histiocytoma, HSP70 inhibits mitochondrial damage by lowering ROS levels. By preventing $H_2O_2$-induced ASK1-p38 activation in human umbilical vein endothelial cells, HSP90 lowers ROS damage. Under heat stress, HSP60, a typical eukaryotic mitochondrial protein, prevents apoptosis through the outer mitochondrial membrane.

A few sHSPs are known to take part in oxidative stress reactions that lower the degree of oxidative damage by preserving the redox state, such as HSPB1 (HSP25/27) and HSPB5 (αB-crystallin). Reduced GSH is formed in murine L929 fibrosarcoma cells as a result of increased antioxidant glutathione (GSH) concentration and increased glucose 6-phosphate dehydrogenase (G-6-P-DH) activity caused by sHSP overexpression.

HSF1 and nuclear factor erythroid 2-related factor 2 (Nrf2) have been demonstrated to work in concert under oxidative stress to bind to promoters and transcriptionally upregulate the expression of many target genes, including heme oxygenase-1 (HMOX1) and sequestosome-1 (SQSTM1, also known as p62), activating transcription factor 3 (ATF3), HSPA1A/B/L (HSP70), and HSPB1 (HSP25/27). Thus, in response to oxidative stress, HSF1 and HSP can trigger a range of cytoprotective mechanisms and shield related genes.

Furthermore, the metabolic rate is elevated by heat stress. Increased oxygen consumption by the organism causes electron leakage in the respiratory chain and raises ROS levels. Calcium stimulates the Krebs cycle enzymes and oxidative phosphorylation in the mitochondria to promote ATP synthesis. Complex I and complex III are the primary sources of ROS; the former increases ROS through glutathione, the latter through ubisemiquinone radical intermediate (QH) at the QO site of complex III. When complexes I and III are inhibited, calcium can increase the production of ROS. In normal circumstances, calcium decreases ROS from these complexes. To raise the level of ROS, calcium may regulate the three-dimensional conformational changes in the respiratory chain complex. The precise mechanisms behind this process remain unclear, although calcium has a significant impact on changes in ROS content.

## 2.1.2. Heat and Endoplasmic Reticulum Stress

The ER is a dynamic organelle involved in calcium buffering, lipid and carbohydrate metabolism, and protein folding. The unfolded protein response (UPR) is triggered by a variety of cellular stresses that arise under heat stress, including disruptions in calcium homeostasis, redox imbalance, and protein folding defects. These stresses lead to an accumulation of misfolded and unfolded proteins in the ER lumen. The genes that code for the ER resident protein needed for protein folding are activated by UPR. Through the conserved intracellular signaling pathway of UPR, this regulatory response sends signals from the ER lumen to the nucleus via the inositol-requiring protein 1α/spliced X-box binding protein 1 (IER1α/XBP1). The splicing of the mRNA encoding the transcription factor XBP1 is regulated by the UPR signals via the sensor IRE1α. HSR shields eukaryotic cells from the harm that protein toxicity causes in conjunction with UPR. Research has indicated that two HSF1-regulated chaperones, HSP72 (HSPA1A) and HSP90 (HSPC1), interact with the cytoplasmic domains of IRE1α to improve the IRE1α/XBP1 complex at the ER. This process, in turn, impacts UPR in cell culture. To prevent ER stress-induced phosphorylation and oligomerization of IRE1α, the ATPase domain of Hsp72 attaches itself to the monomeric, nonphosphorylated cytoplasmic tail of IRE1α. Furthermore, under ER stress conditions, increased XBP1 protein is necessary for enhanced cell survival induced by Hsp72. Moreover, it has been demonstrated that Hsp90 binds to the cytoplasmic domains of IRE1α to extend its half-life. Further research is necessary to understand the molecular mechanisms by which HSP72 and HSP90 inhibit ER stress.

It has been determined that calcium, besides HSF1, is connected to ER stress. When the ER experiences heat stress, the buildup of unfolded proteins in the ER allows calcium to escape via Bak/Bax and enter the cytoplasm, which triggers the ER to trigger apoptosis. IP3R's redox sensitivity is primarily responsible for the release of calcium. Furthermore, cytotoxicity brought on by an excess of cytoplasmic calcium can trigger endogenous or mitochondrial-dependent apoptotic pathways.

### 2.1.3. Heat and Apoptotic Pathway

Transducers and central integrators of pro-apoptotic signals are the mitochondria. The heat-shock proteins HSP90, HSP70, and sHSPs may be able to prevent heat-induced cell death by interfering at different stages of the mitochondrial-dependent apoptotic pathway, according to recent research. In cell culture, sHSPs (HSP27) and HSP70 both block the intrinsic pathway of apoptotic cell death by preventing the release of cytochrome c. Because ROS-mediated caspase 3 activation partially depends on the downregulation of HSP70 and HSP90 in L-02 hepatocytes, additional experiments revealed that HSP70 and HSP90 may inhibit apoptosis through the caspase 3 pathway. Furthermore, to further prevent apoptosis in *Plutella xylostella* sperm and ovarian cells, HSP69s and HSP20 mediate ROS through caspase 7 and caspase 9.

Heat-induced apoptosis in cell culture is inhibited by HSP70-enhanced miR-23a, which also inhibits NOXA (Bcl-2 family proteins), which activates Bax (Bcl-2 family proteins) and releases cytochrome c from the mitochondria. These mechanisms work in concert to prevent apoptosis caused by heat. Human umbilical vein endothelial cells' $H_2O_2$-induced ASK1-p38/JNK signaling and cell apoptosis are reduced by the HSP90–Akt ASK1 complex. On the other hand, HSF1 overexpression can, in a Tdag-51-dependent manner, initiate apoptotic cell death programs (T-cell death-related genes). Heat stress causes HSF1 to directly activate Tdag51-related regulatory regions, which in turn causes HeLa cells to start transcription. Nonetheless, several investigations have revealed that contingent upon the cellular environment, TDAG51 may possess both pro- and anti-apoptotic properties. Regarding its pro-apoptotic role, the data that is currently available is consistent with the idea that apoptosis is enhanced when TDAG51 expression is elevated. In reference to TDAG51's anti-apoptotic activity, TDAG51 may regulate the expression of antioxidant enzymes or non-enzymatic components in order to reduce the generation of excess ROS in response to oxidative stress, which in turn lowers the rate of apoptosis in mouse embryonic fibroblasts (MEFs). The physiological significance of TDAG51 expression in the oxidative stress response necessitates further investigation into its regulatory mechanisms.

The hypothesis that calcium mediates TDAG51 under heat stress is not supported by any concrete data. But in human renal cubular epithelial cells (HRPTECs), calcium imbalance can mediate TDAG51 upregulation and prime the cells for mesenchymal transformation during the mesenchymal transition (EMT) response, a developmental stage in which cells lose epithelial characteristics and gain mesenchymal properties. This suggests that calcium may regulate the expression of Tdag51 in a similar way under heat stress. Researchers looking into how calcium affects the apoptotic pathway found that heat stress can cause apoptosis in rabbit corneal cells by means of the calcium-mediated mitochondrial apoptosis pathway, such as IP3R-regulated cytoplasmic calcium elevation. This pathway then influences activations of Apaf-1, caspase 9, and caspase 3, ultimately mediating apoptosis. Furthermore, ER calcium release and mitochondrial calcium loading are enhanced by the redox sensitivity of IP3R and the mitochondria permeability transition pore (MPTP). Figure 2.1 provides a concise overview of research on complex regulatory mechanisms produced by organisms under thermal stress, including oxidative stress, calcium signaling, UPR, and their combinations. Table 2.1 provides an overview of how HSF1 and calcium are regulated during heat stress.

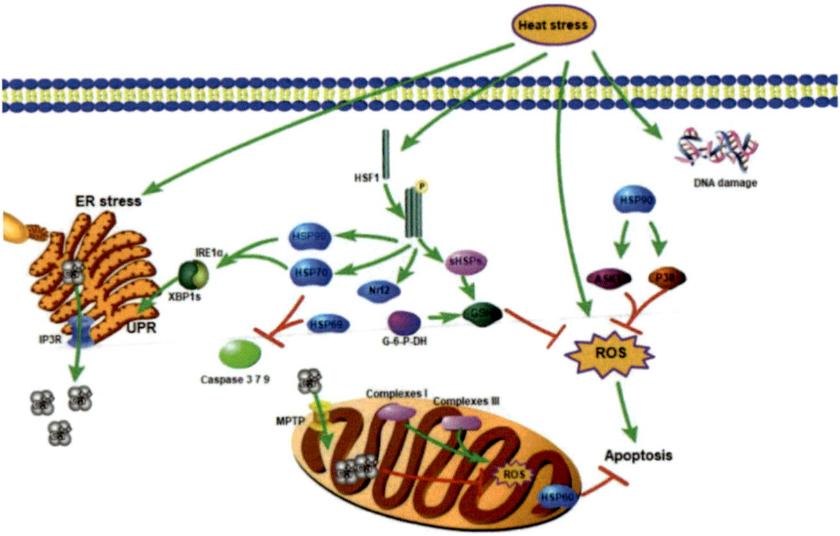


**Figure 2.1**. *Genetic mechanisms of animal responses to heat.*

***Source***: *Wu J, Zhang W and Li C (2020). Recent Advances in genetic and epigenetic modulation of animal exposure to high temperature. Front. Genet. 11:653. doi: 10.3389/fgene.2020.00653.*

**Table 2.1**. Regulatory Effects of Heat Shock Factor 1 (HSF1) and Calcium (Ca) Under Heat Stress

| | | | |
|---|---|---|---|
| HSF1 | Synergized with nuclear factor erythroid 2-related factor 2 (Nrf2) to regulate sequestosome-1 (SQSTM1)/p62, heme oxygenase-1 (HMOX1), activating transcription factor 3 (Atf3), heat-shock protein family (Hsp)70 and Hsp25/27 | | Over-expressed HSF1 triggered apoptosis in a Tdag-51-dependent manner |
| Hsp90 | Hsp90 inhibited hydrogen peroxide ($H_2O_2$)-induced ASK1–p38 activation | Interact with inositol-requiring protein 1α (IRE1α) to regulate unfolded protein response (UPR) | Hsp90–Akt–ASK1 complex decreased cell apoptosis |
| Hsp70 | Abolished heat-induced reactive oxygen species (ROS) production, but the mechanism was unknown | Interact with IRE1α to regulate UPR | HSP70-enhanced miR-23a prevented the release of cytochrome c |
| Small Hsps (sHsps) | Increased glutathione (GSH) and G6PD contributed to reduced state; abolished ROS by TNFα | | Interrupted apoptosis by preventing the release of cytochrome c |
| Calcium | Reduced ROS by regulating complexes I and III | Leaked from the endoplasmic reticulum (ER) into the cytoplasm, which causes ER to induce apoptosis | Cytoplasmic calcium elevation activated Apaf-1, caspase 9 and caspase 3 |

Source: Wu J, Zhang W and Li C (2020) Recent advances in genetic and epigenetic modulation of animal exposure to high temperature. Front. Genet., 11:653. doi: 10.3389/fgene.2020.00653.

## 2.1.4. Heat Stress and Heat Acclimation

It's critical to comprehend how organisms respond to heat stress and how to mitigate it. When stress is applied, heat acclimation changes the rate and magnitude of HSP expression. Heat-acclimated organisms have a higher HSP70 level than the non-acclimated rat heart under normal circumstances.

Larger HSP reserves in acclimation rats in a normal environment compared to the normal group may be a factor in delayed thermal injury (Maloyan et al., 1999). A different pathway than the one involved in heat stress may be involved in an elevated HSP level. HSP70, HSP90, and HSP110 are necessary for fish to survive at high temperatures right away, while HSP60, HSP70, and HSP78 are necessary for fish to survive at high temperatures over the long term, according to studies on the effects of the HSP family on heat acclimation. In rice leaf folder larvae, HSP90 is only expressed in the first two generations of heat selection, whereas HSP70's mRNA expression rises in each of the five heat selection generations (Gu et al., 2019). These findings corroborate the notion that HSP70 is the primary factor in heat acclimation and HSR, while also serving as an indirect proof of the point.

## 2.2. EPIGENETIC MECHANISMS OF ANIMAL RESPONSES TO HEAT STRESS AND HEAT ACCLIMATION

Through DNA methylation, histone modifications, and other regulatory pathways, epigenetic regulation alters the expression of genes or the rate at which they express themselves. We use HSP as an illustration (Figure 2.2).

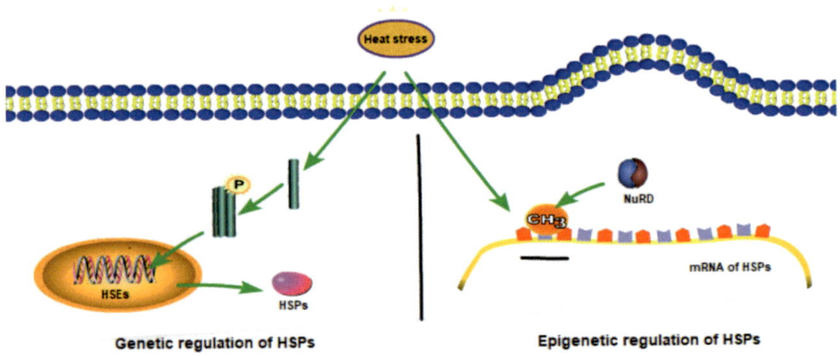


**Figure 2.2**. *Genetic and epigenetic mechanisms of heat-shock protein family (HSP) synthesis.*

***Source***: *Wu J, Zhang W and Li C (2020) Recent advances in genetic and epigenetic modulation of animal exposure to high temperature. Front. Genet. 11:653. doi: 10.3389/fgene.2020.00653.*

According to the genetic mechanism underlying HSP synthesis, heat stress causes HSF1 to become trimerized and phosphorylated. This enables HSF1 to attach to HSR DNA elements in the nucleus and activate HSP transcriptional expression (Kukreti et al., 2013). In contrast, the HSP70 promoter is hypermethylated in the epigenetic regulation of HSP (using HSP70 as an example) following acclimation to high temperatures due to a decrease in both POU class 2 homeobox 1 (POU2F1, also known as octamer-binding transcription factor-1, is a ubiquitous transcription factor that plays a key role in the regulation of genes related to inflammation and cell cycles) binding and recruitment of the nucleosome remodeling deacetylase (NuRD, combining both deacetylase and remodeling enzymatic activities in a single macromolecular complex) chromatin-remodeling complex, consequently affecting HSP70 expression (Matsuda et al., 2006). Furthermore, the regulation of epigenetic modifications under environmental stressors has been studied. Next, we'll talk about how heat stress affects histone and DNA methylation.

### 2.2.1. Epigenetic Mechanisms of Histone Methylation Under Heat Stress

Histone modifications, in addition to DNA methylation, can impact chromosomal domains, which in turn can impact the transcriptional expression of individual genes. Research has primarily examined the histone modifications found in H2B and H3. Methylproline at the N-terminus of H2B has been found to be the amino acid residue responsible for H2B methylation. Heat shock lowers the amount of H3 methylation while simultaneously raising the level of H2B methylation. In rat astrocyte and cortical neuronal cultures, many particular methylated amino acid residues of H3, including H3 at lysine 4 (H3K4) and H3 at lysine 9 (H3K9), have been found. While H3K9 methylation is associated with gene repression, H3K4 methylation is associated with the activation of gene transcription. In addition, H3 methylation is heritable. A study of *Caenorhabditis elegans* expression profile revealed that trimethylation of histone H3 lysine 9 can pass down temperature-induced expression of endogenously inhibited repeats for several generations.

### 2.2.2. Epigenetic Mechanisms of Heat Acclimation

The molecular program is essential for heat adaptation to occur. Growing research suggests that epigenetic mechanisms play a significant role in heat acclimation processes. The accessibility of DNA to transcription factors is influenced by epigenetic mechanisms, which in turn control gene expression and phenotypic. A significant subset of epigenetic markers implicated in epigenetic mechanisms are miRNAs, of which the heat-acclimated Rattus norvegicus has upregulated miR-297. Furthermore, reduced levels of miR-1 and miR-206 stimulate the synthesis of HSP70. Apart from methylation, phosphorylation and acetylation, two other posttranslational modifications in histones, are crucial for cytoprotective acclimatory memory. Following research on the heat shock element (HSE) binding site of the HSP70 and HSP90 promoters, it was discovered that constitutively acetylated histone H4 and a preserved euchromatin state result in acclimatory memory. Early histone H3 phosphorylation also contributes to histone H4 acetylation.

Research on oxidative phosphorylation has demonstrated that metabolic rate is a key measure of heat tolerance. Several markers, including an organism's metabolism, vary greatly depending on the degree of heat stress and heat acclimation. Research on metabolic capacity has shown that heat stress increases the sensitivity and activation of cytochrome c oxidase (COX-respiratory complex IV). These findings supported the theory that, in hot conditions, cells prefer OXPHOS. Conversely, in organisms acclimated to heat, the expression of proliferator-activated receptor γ coactivator 1α-a (PGC-1α), citrate synthase, and cytochrome c oxidase decreases. Researchers propose a theory that HSP70 inhibits OXPHOS in the acclimation organism while also enhancing glycolysis to make up for the ATP imbalance, in conjunction with the larger HSP reserves in the acclimation organism previously discussed and the observation that HSP70 inhibits OXPHOS. Different regulatory mechanisms are displayed by the OXPHOS complex in response to heat stress and heat acclimation. According to a different theory, heat acclimation lessens the sensitivity of freshwater and marine animals' respiration to temperature changes, which lowers mitochondrial energy metabolism. Elevated calcium content in the mitochondria is another indicator of heat acclimation. The levels of ROS, metabolic rate, and calcium content are all closely correlated. Evidence currently available indicates that calcium has varying effects on mitochondrial ROS levels depending on the metabolic state of the mitochondria (Adam-Vizi and Starkov, 2010). Both the RYR and CLIC2 genes have been found to be differentially methylated, and under heat stress, calcium homeostasis is mediated by the chloride intracellular channels (CLICs) via the ryanodine receptor gene (RYR) in the ER membrane (Board et al., 2004). In addition to this route, by inhibiting IP3R in tumors in MDA-MB-231 cells, methylated pyruvate kinase M2 (PKM2) isoforms can also change the calcium flux from the ER to the mitochondria. Here, we propose a theory that suggests DNA methylation may control the movement of calcium through different organs by methylating particular genes, aiding in the process of heat acclimation.

## 2.3. THE USE OF ANIMALS MODELS TO ADVANCE EPIGENETIC SCIENCE

Environmentally-induced epigenome modifications have been observed in a wide range of animal phyla, including humans, rodents, and insects. These include DNA methylation, histone tail modifications, and chromatin remodeling. More recently, noncoding RNA and microRNA gene regulation have been added to the list. As a result, it is becoming more widely acknowledged that exposure to environmental, dietary, behavioral, and physical factors modifies the epigenome in addition to causing mutations, which in turn impacts gene expression and impacts health and disease. Furthermore, through the dysregulation of epigenetic mechanisms, these exposures have been directly associated with the development of subsequent diseases. These epigenetic modifications offer a special opportunity to enhance human health because they may be reversible, in contrast to genetic mutations.

Developmental biologist Conrad Waddington (1940) coined the term "epigenetics" to describe "the interactions of genes with their environment, which bring the phenotype into being." Waddington's theory was first put forth as a molecular explanation in the 1970s by Holliday and Pugh (1975), who suggested covalent chemical DNA modifications, such as methylation of cytosine-guanine (CpG) dinucleotides. A revised definition of epigenetics—which now defines it as heritable changes in gene expression that take place without a change in DNA sequence—was developed several decades after it was discovered that complex and multifactorial mechanisms regulate X inactivation in mammals and genomic imprinting (Wolffe and Matzke 1999). The term "epigenomics" was coined to describe the study of the "effects of chromatin structure including the higher order of chromatin folding and attachment to the nuclear matrix, packaging of DNA around nucleosomes, covalent modifications of histone tails (acetylation, methylation, phosphorylation, ubiquitination), and DNA methylation" (Murrell et al., 2005). The genomics revolution inspired the investigation of genome-wide rather than local gene analyses). Last but not least, research showing some gene loci's resistance to methylation reprogramming during embryogenesis showed that epigenetic changes can be passed down both mitotically and transgenerationally.

The most extensively researched type of epigenetic modification is DNA methylation, which takes place in the one-carbon metabolism pathway. This pathway depends on many enzymes and micronutrient cofactors, such as dietary folate, choline, and betaine. Mammals' cytosines in CpG dinucleotides are the primary site of stable repressive mark DNA methylation; however, its regulation is more dynamic than previously thought (Maunakea et al., 2010). For instance, methylation at non-CpG sites may be crucial for developmental homeostasis, according to recent evidence of methylation of non-CpG cytosines in human embryonic stem cells (Lister et al., 2011). According to Holliday and Grigg's 1993 study, spontaneous deamination of 5-methylcytosine to thymine and subsequent fixation in a population over evolutionary timescales are the reasons behind the significant underrepresentation of CpG dinucleotides in mammalian genomes. As a result, most unmethylated CpG sites are found in CpG islands, which are isolated areas with a high concentration of CpG (Deaton and Bird 2011). It is believed that uniform genomic CpG site deamination and conversion, in conjunction with the regeneration of new CpG islands found in repetitive elements with expansion by retrotransposition, are the causes of the resulting uneven distribution of CpG islands (Xing et al., 2004). CpG islands are typically found in the first exons of housekeeping genes or within or close to gene promoters. On the other hand, the body and regulatory regions of repetitive DNA sequences, like transposable elements, are methylated. As a result, transcriptional replication of the parasitic transposable and repetitive elements is prevented. Crucially, it should be noted that not all animals utilize DNA methylation as a mechanism for gene repression. For instance, the model organisms *Drosophila melanogaster* and *Caenorhabditis elegans*, which are fruit flies, exhibit significant levels of DNA methylation, while other insects and nematodes maintain their DNA methylation machinery.

The modification of chromatin structure through covalent histone modifications and the incorporation of histone variants into the nucleosome is another mechanism that drives epigenetic manipulation of cellular phenotype (Saha et al., 2006). A complex of nucleoproteins called chromatin uses a variety of nucleosomes to encapsulate linear genomic DNA. About 147 base pairs of DNA are coiled around an octamer of histone proteins to form a nucleosome. There are two copies of each of the four core histones—H2A, H2B, H3, and H4—in each octamer. In addition to the numerous posttranslational modifications of histone proteins, such as histone acetylation, methylation, ubiquitination, phosphorylation, and ADP-ribosylation, chromatin can also be further altered by association with linker histones, histone variants, and nonhistone proteins (Caiafa and Zampieri 2005; Cheung and Lau 2005). Since chromatin packing is relaxed and histone proteins' affinity for DNA is decreased, histone acetylation is typically linked to transcriptional activation. Depending on the amount of methylation and the altered lysine residue, different transcriptional outcomes are caused by histone methylation (Kouzarides 2007). The histone code can become extremely complex due to the possibility of mono-, di-, or trimethylation of each lysine residue.

Histone modifications also recruit multi-subunit chromatin-protein complexes by interacting with DNA methylation patterns, which adds even more complexity to the regulation of epigenetic gene expression. Kim and Kim (2012), for instance, look at protein complexes that influence the location of epigenetic marks in this issue. Particularly, the mechanisms by which two histone marks—

H3K9 trimethylation and H3K27 trimethylation—are targeted remain unclear. These marks are well-characterized repressive chromatin marks significant in both genic and nongenic regions of metazoan genomes. Here, Kim and Kim present evidence that H3K27 and H3K9 trimethylation in mammals marks specific genomic regions, while the repressive polycomb repressive complex 2 histone-modifying complex targets histone modifications by cooperating with DNA-binding proteins like JARID2, AEBP2, and YY1. In particular, deep sequencing techniques—such as chromatin immunoprecipitation-seq sequencing—are used to assess the distribution of histone modification marks throughout the mammalian genome.

## 2.3.1. Vulnerable Time Points

Throughout life, alterations in DNA methylation and other epigenetic patterns are common, particularly during reprogramming processes linked to typical development and aging. For instance, during embryogenesis, the epigenome is especially dynamic due to extensive DNA synthesis, and early development is when the complex DNA methylation patterning necessary for healthy tissue development is established. Gradual genome-wide DNA hypomethylation and locus-specific promoter increases in DNA methylation at normally unmethylated CpG islands occur as people age. This can result in either gene-specific suppression or genome instability. Furthermore, hypomethylated DNA and noticeable hypermethylation of tumor suppressor genes are frequently linked to cancer in comparison to normal tissue. DNA methylation patterns unique to each tissue are produced by these reprogramming events that occur throughout the life course. Variations in these epigenetic patterns have a significant impact on tissue homeostasis and cellular differentiation.

According to the developmental origins of health and disease hypothesis, epigenetic changes like DNA methylation and chromatin modifications shape an individual's increased susceptibility to disease following early life experiences. In this issue, Ganu et al. (2012) describe various methods for examining epigenetic marks in rodent models, nonhuman primates, and humans as a mechanism connecting early origins to adult disease. Concentrating on both prenatal shortage (famine) and surplus (i.e., diets heavy in fat and calories), they examine new and intriguing evidence that suggests histone modifications, in particular, may play a role in mediating the relationship between adult metabolic disease and the effects of early experiences. Seelan et al. (2012) offer an alternative strategy that focuses on a particular window of vulnerability connected to epigenetic mechanisms. About one to two out of every 100 live births have an orofacial cleft, which has a complicated etiology involving both genetic and epigenetic mechanisms. In particular, they examine the body of research pertaining to the theory that normal orofacial ontogeny is the consequence of the coordinated action of the early embryonic palatal methylome, transcriptome, and repertoire of microRNAs; dysregulation of this process can result in secondary palate defects.

### 2.3.2. Nutritional and Environmental Epigenetics

An emerging field called Nutri-epigenomics studies how diet affects gene expression. In the end, disease phenotypes may be influenced by dietary habits, particularly micronutrient intake, DNA methylation, and other epigenetic processes. In order to identify critical times throughout the life course where diet may significantly alter epigenetic marks, we have previously emphasized the significance of an interspecies approach to synthesize the existing Nutri-epigenomic literature (Anderson et al., 2012). Now, Niculescu (2012) presents the fascinating idea that researchers can understand the extent and kind of influence that each nutrient has on one-carbon metabolism and, consequently, DNA methylation and other epigenetic events by thoroughly examining different levels of nutrient exposure during critical time points. Ho et al. (2012) describe the timing, dose, duration, chemical composition, and significant factors leading to epigenetic consequences affecting disease risk, with a focus on life-course environmental exposures. Once these epigenetic "memories" are understood, they can be valuable biomarkers for identifying people who are at risk for developing future diseases as well as for assessing chemical risk and past exposure.

### 2.3.3. Behavioral and Social Epigenetics

From insects to mammals, epigenetic changes brought on by stress and behavior are common. Examples of animals that are thought to be linked to epigenetic adaptations are the desert locust (*Schistocerca gregaria*), which produces more offspring of the gregarious swarming phenotype when breeding in crowded conditions, and the pea aphid (*Acyrthosiphon pisum*), which produces more winged offspring when under stress from crowded conditions or predators. In a similar vein, throughout the first week of life, rodents display long-lasting changes in the glucocorticoid receptor's DNA methylation as well as numerous other loci in the hippocampus linked to high or low levels of maternal grooming. Here, Jašarević et al. (2012) use male-male competition and female choice as examples of sexually selected traits as a basic conceptual framework to evaluate behavioral epigenetics. They suggest extending the scope of behavioral modification in relation to mate choice by incorporating additional model organisms beyond those that are commonly employed. This innovative approach may help clarify the origins of steroid-induced epigenetic programming because sexually selected behaviors are programmed during early embryonic and postnatal development through endogenous hormone exposure and because xenobiotic endocrine-disrupting chemicals like bisphenol A have been shown to affect the fetal epigenome. This issue also includes an analysis by Gudsnuk and Champagne (2012) of animal models of early-life stress and social experience over the lifespan, with a particular emphasis on laboratory rodents and the relationships between epigenetic markers and social stress, abusive caregiving, maternal separation, prenatal stress, and maternal care.

## 2.4. EPIGENETIC REGULATORS

### 2.4.1. DNA Methylation

Mammals rely on DNA methylation to control a variety of biological processes, such as X-chromosome inactivation, genomic imprinting, and chromosomal stability. The three conserved enzyme families DNA methyltransferase (DNMT) 1, DNMT3a, and DNMT3b catalyze this process, which adds methyl groups to the cytosine of CpG islands (Figure 2.3). In DNA methylation, an epigenetic modification process that adds a methyl group to the DNA molecule, these three enzymes are essential. DNA methylation is involved in many cellular processes, such as development, differentiation, and genomic stability. It also aids in the regulation of gene expression. The primary enzyme involved in preserving DNA methylation patterns during DNA replication is called DNMT1 (DNA methyltransferase 1). It ensures that DNA methylation patterns are faithfully passed down to daughter cells by identifying hemimethylated DNA and appending a methyl group to the unmethylated strand.

The de novo DNA methyltransferase DNMT3a (DNA methyltransferase 3a) is one example. During cellular differentiation and development, it plays a role in creating new DNA methylation patterns. Hematopoiesis, neuronal differentiation, and embryonic development all depend on DNMT3A, which can methylate previously unmethylated DNA regions. Similar in function to DNMT3a, DNMT3b (DNA methyltransferase 3b) is involved in the establishment of DNA methylation patterns during cellular differentiation and development. When it comes to methylating repetitive sequences like transposable elements in the genome, DNMT3b is especially crucial. Together, these three enzymes support the dynamic regulation of DNA methylation patterns in cells, which is essential for healthy development, gene expression, and disease processes.

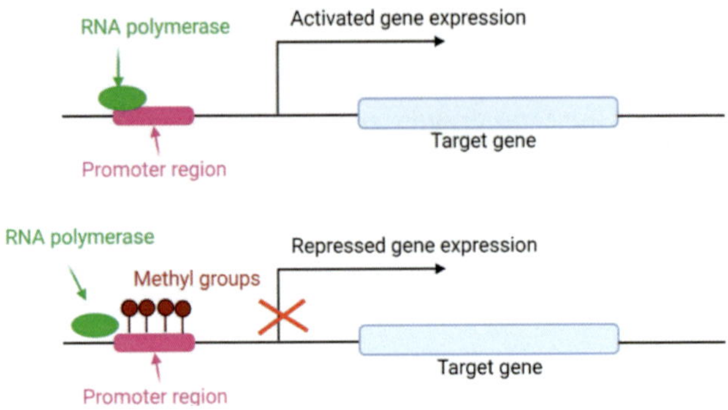


**Figure 2.3.** *Gene expression regulated by DNA methylation. DNMT enzymes add methyl groups in cytosine residues of CpG island.*

***Source****: Lesta, A.; Marín-García, P.J.; Llobat, L. (2023). How Does nutrition affect the epigenetic changes in dairy cows? Animals, 13, 1883. https://doi.org/10.3390/ani13111883.*

These enzymes need a methyl donor to function, and the methyl groups (choline, betaine, methyl-folate, or methionine) that are obtained from the diet can have an impact on this donor. These compounds function as precursors of S-adenosylmethionine (SAM), the universal methyl donor. The control of the insulin-like growth factor II (IGF2) gene is one instance of how methyl groups affect an enzyme's ability to function. IGF2 expression is reliant on the methylation of a particular region found in the H19 gene called Igf2DMR2. Hypermethylation in these areas happens when pregnant rats are given a diet low in choline, which inhibits H19 and increases the expression of IGF2. An excellent illustration of this EGF2 regulation and its connection to nutrition took place during the 1944–1945 Dutch Hunger Winter. The progeny of pregnant women who experienced famine showed a decrease in the IGF2 gene's DNA methylation. Numerous studies show how diet plays a significant role in determining livestock's epigenetic marks. Murdoch et al. (2016) released a thorough analysis that investigates the impact of nutrient quantity and quality on methylation in various livestock species. Methyl-group donors like choline, folate, and betaine are among the nutrients that have a big impact on DNA methylation. Injecting betaine into eggs has been demonstrated to elevate DNA methylation, thereby controlling the metabolism of hepatic cholesterol in chicks. Piglets with altered methylation patterns were also discovered to have been produced by pregnant sows fed diets supplemented with betaine. Hypomethylation in the progeny of sheep fed diets low in vitamin B12, folate, and Met has been associated with birth weight, immunological response, and blood pressure levels.

Quantitative trait loci (QTL) are associated with the most economically significant traits in dairy cows, such as milk yield and milk components. A portion of these QTLs may have an impact on phenotypic variation in livestock production because they are controlled by epigenetic mechanisms, such as methylation regulation. Research on the epigenetic regulation—specifically, DNA methylation—is therefore required because these are characteristics that are relevant to the production of livestock. These investigations ought to concentrate on the role that animal nutrition plays in these methylation shifts.

### 2.4.2. Histone Modifications

The primary building blocks of chromatin, the complex of DNA and proteins that makes up chromosomes, are histone proteins. Histone post-translational modifications are chemical changes that take place on these proteins. These alterations (acetylation, phosphorylation, and methylation) are essential for controlling chromatin structure and gene expression, which in turn affects several biological functions like transcription, DNA replication, and repair (Figure 2.4). These roles can control the expression of genes, contribute to the emergence of genetic diseases, and influence mammalian early development.

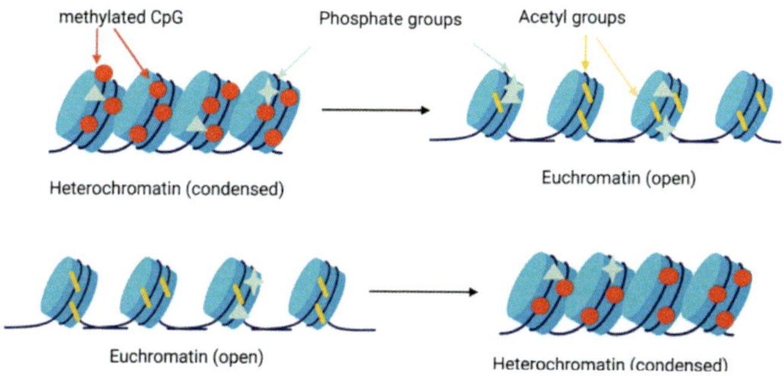


**Figure 2.4**. *Schematic diagram of the gene expression regulation by histone modifications. Methyltransferases add methyl groups in lysine (Lys) and arginine residues, favoring chromatin condensation and preventing gene expression (up). Acetyltransferases and deacetylases add or remove acetyl groups in Lys residues, decreasing or increasing the condensation of chromatin, respectively (up and down). The kinases and phosphatases add phosphate groups in different locations, changing the condensation of chromatin.*

***Source**: Lesta, A.; Marín-García, P.J.; Llobat, L. (2023). How does nutrition affect the epigenetic changes in dairy cows? Animals, 13, 1883. https://doi.org/10.3390/ani13111883.*

Histone acetyltransferases (HATs) and histone deacetylases (HDACs), two families of enzymes that alter the Lys residues of histones, are responsible for controlling the acetylation of histones. HDACs work in the opposite way from HATs in terms of transferring an acetyl group to the Lys side chain. Histones get phosphorylated on serine, threonine, and tyrosine, and kinases and phosphatases control this process. Histones are methylated on the side chains of arginine and Lys by particular methylases and demethylases. Many biological processes are regulated by histone modifications, but one of the most significant is the development of mammals in their early stages. For instance, changes in the 4 and 27 Lys residues of histone 3 (H3K4 and H3K27) control the development of the inner cell mass and trophectoderm in mice. Other histone modifications are linked, among other things, to the control of HOX gene expression, pluripotency in mammalian early development, and genomic imprinting.

### 2.4.3. Small Non-Coding RNA: miRNAs as Epigenetic Regulators

The nematode *Caenorhabditis elegans* is home to the first microRNA (miRNA), lin-4, which was discovered in 1993. However, its significance was not fully understood until let-7, another miRNA, was discovered. MiRNAs are tiny RNA molecules that don't code for proteins; they are typically 20–24 nucleotides long. Numerous organisms, including humans, contain microRNAs, which are involved in a number of biological processes, including differentiation, apoptosis, cell division, and development. By attaching to messenger RNA (mRNA) molecules and either destroying them or preventing their translation into proteins, miRNAs primarily control the expression of genes. Primary miRNA (pri-miRNA) molecules are created by transcription of DNA sequences in the cell nucleus, marking the beginning of the biogenesis of miRNAs. An enzyme known as Drosha then breaks down these pre-miRNAs to create precursor miRNA (pre-miRNA) hairpin structures. The protein Exportin-5 is responsible for exporting the pre-miRNAs from the nucleus to the cytoplasm. Mature miRNAs are created in the cytoplasm when an enzyme known as Dicer further processes the pre-miRNAs. Subsequently, the mature miRNAs are integrated into the RNA-induced silencing complex (RISC), a protein complex. The target mRNA molecules' complementary sequences are guided by the miRNAs within the RISC. Depending on the degree of complementarity between the miRNA and the target mRNA, the binding of miRNAs to mRNA can result in translational repression or mRNA degradation (Figure 2.5). It was shown that additional non-coding RNAs acted as epigenetic modulators. Through a variety of mechanisms, such as the formation of heterochromatin and the inhibition of translation, piwi-interacting RNA (piRNA), small interfering RNA (siRNA), small nucleolar RNA (snoRNA), circular RNAs (circRNAs), and long non-coding RNAs (lncRNAs) can regulate gene expression. It has been demonstrated that these RNA molecules are crucial in determining the epigenetic landscape. It's still unclear, though, how precisely nutrition and the control of these non-coding RNAs are related.

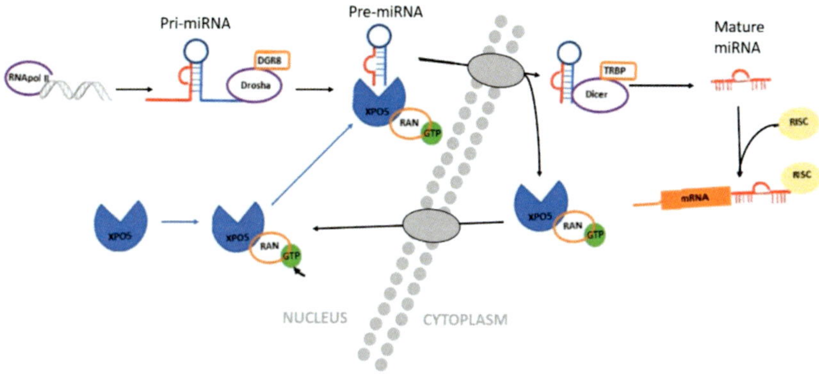


**Figure 2.5.** *Schematic diagram of miRNA biogenesis. In the nucleus, miRNA is transcribed by RNA polymerase II as primary transcripts (pri-miRNA). The Drosha enzyme cuts this pri-miRNA to form a pre-miRNA, which is actively transported to the cytoplasm by the nuclear transport receptor Exportin-5 (XPO5). In the cytoplasm, the pre-miRNA is cut by a second enzyme, Dicer, to form a mature and short double-stranded miRNA molecule. The miRNA duplex is incorporated into the RISC protein complex.*

***Source:*** *Lesta, A.; Marín-García, P.J.; Llobat, L. (2023). How does nutrition affect the epigenetic changes in dairy cows? Animals, 13, 1883. https://doi.org/10.3390/ani13111883.*

## 2.4.4. Epigenetic Regulation and Nutrition

The production of milk in mammals is significantly influenced by epigenetic regulation. Mammary gland cells exhibit significant epigenctic alterations during lactation, which aid in activating the genes needed for milk secretion and production. DNA methylation is one of the main epigenetic processes underlying the production of milk. Expression of genes involved in milk synthesis and secretion may be suppressed as a result of methylation of the promoter regions of those genes. The methylation profile is linked to the amounts of fat and protein in milk and controls its production. Lately, Wang et al. (2021) showed that QTLs for milk fat and protein were co-located with differentially methylated CpG sites. Casein expression is regulated by hypomethylation of the activator of transcription (STAT) 5-binding lactation enhancer in the lactating mammary glands of cows. In specifics, as the post-milking period lengthens, there is an increase in the methylation of the CpG island of STAT5 binding, which may be connected to mammary involution and the drop in milk protein levels. Milk production may be impacted by nutrition's influence on the mammary gland's epigenetic regulation. Research has demonstrated that a mother's diet during pregnancy and lactation can affect the mammary gland's epigenetic modifications, which can alter gene expression and the composition of milk. For instance, it has been demonstrated that maternal protein restriction during lactation and pregnancy modifies the DNA methylation patterns in the offspring's mammary gland, resulting in modifications to the expression of genes related to milk production and secretion in goats and cattle. When specific nutrients, like choline and folic acid, are supplemented, the mammary gland's DNA methylation patterns and gene expression alter. A few AAs, primarily Met, take part in the one-carbon metabolism that controls methylation and purine synthesis. The usage of other metabolites, such as choline, betaine, and folate, is connected to the AA requirements since they supply methyl donors for methyltransferases. Each of them plays a crucial part in controlling S-adenosylmethionine (SAM).

Both *in vivo* and *in vitro* research studies have been done extensively on the effects of Met supplementation on cows. Met is an important AA that is required for many biological functions, including growth and the production of milk. It is also crucial for preserving the general health of the cow because it boosts immunity and functions as an antioxidant to shield cells from harm from free radicals. Increased phosphorylation of the antioxidant transcription regulator (NFE2L2) and decreased protein abundance of its negative regulator (KEAP1) in mammary tissue are the results of increased Met supply during the peripartum period. The antioxidant protein (HMOX1) and several target genes are consequently overexpressed. The information points to the presence of transcriptional and posttranslational processes in breast tissue during the first stages of lactation, which most likely regulate Met's antioxidant effect. Met supplementation has been demonstrated to upregulate the expression of cystathionine β-synthase (CBS) and phosphatidylethanolamine methyltransferase (PEMT), leading to an increase in the synthesis of antioxidants and phosphatidylcholine. AKT serine/threonine kinase 1 phosphorylation status rises in response to an increased Met supply, which sets off a series of intracellular events that upregulate AAs and glucose transporters. Coleman et al. (2019) examined the effects of elevated post-ruminal choline supply during feed restriction-induced negative energy balance (NEB) events, and discovered that these effects were advantageous for the metabolism of liver fatty acids and the production of milk.

Met plays a crucial role in the immune response, and its deficiency can weaken the immune system's capacity to fight infections and diseases, according to research on the impact of Met supplementation on immunity. In addition to improving general health and productivity, a robust immune system in cows can aid in the prevention of illnesses and infections. Lower veterinary expenses, increased milk production, and improved milk quality follow from this. To maximize cow health and productivity, dietary recommendations and management techniques are informed by the valuable insights gained from *in vivo* studies regarding the effects of Met on mammary cell function and milk production. When bovine mammary epithelial cells are supplemented with Met, the amount of milk protein synthesis increases. Gene expression associated with multiple signaling pathways, such as cGMP-PKG, Rap1, calcium, cAMP, PI3K-AKT, MAPK, JAK-STAT, and mTOR, is upregulated, which facilitates this effect. Qi et al. (2018) showed how Met activates the SNAT2-PI3K signaling pathway to support the synthesis of milk proteins and fats as well as the growth of BMECs.

Histone changes have an impact on milk production in addition to DNA methylation. A variety of modifications, including acetylation, methylation, phosphorylation, and ubiquitination, can change the way DNA is packaged around histone proteins, which can have an impact on gene expression. For instance, histone acetylation is typically linked to active gene transcription, whereas histone methylation, depending on the precise modification and location, can result in either the activation or repression of genes. Nutrition can also have an impact on histone modifications. Research has indicated that certain dietary elements, like omega-3 fatty acids and polyphenols, can modify the mammary gland's histone acetylation patterns, which can impact the expression of certain genes and the composition of milk in pigs and goats. It is well known that more carbohydrates can raise the availability of glucose, which the mammary gland can use for the synthesis of milk proteins and the production of milk in cows. However, an imbalance in the cow's metabolism may arise if the animal consumes an excessive amount of energy from carbohydrates. Increased fat synthesis can result from the liver's lipogenesis process, which turns the extra carbohydrates into fatty acids. Research on the relationship between gene expression and energy and fat intake indicates that high-energy diets, whether they come from fats or carbohydrates, may change the methylation patterns of genes involved in the biosynthesis of fatty acids in the mammary tissue of dairy cows. This would lead to a decrease in the amount of milk produced and its fat content.

Non-coding RNAs control gene expression during lactation in relation to histone modifications and DNA methylation. By attaching to complementary sequences in the 3′-untranslated regions (UTR) of mRNAs, miRNAs can inhibit or degrade translation, thereby suppressing the translation of target genes. There is evidence that nutrition affects these non-coding RNAs. For instance, a mother's diet during pregnancy and lactation can affect the patterns of miRNA expression in her offspring's mammary gland, which can change the expression of certain genes and the composition of their milk in rats. Furthermore, miRNAs are linked to the regulation of DNA methylation. Studies conducted *in vitro* show that miRNA-152 regulates DNMT1, which is involved in lactation and mammary gland development. Additionally, inhibition of miRNA-29s results in the hypermethylation of lactation gene promoters. Regarding the composition of milk, prolactin inhibits miRNA-183, which controls the metabolism of milk fat, and miRNA-200 is required for the development of the mammary gland. miRNAs also control the composition of milk. Cui et al. (2020) discovered that in dairy cows with varying milk proteins and fats, 71 miRNAs were expressed differently in the mammary glands. In addition, Billa et al. (2021) discovered eight miRNAs connected to QTLs pertaining to fat composition and lactalbumin content. Lastly, the development and operation of the mammary gland was linked to additional non-coding RNAs. For instance, Shore et al. (2012) discovered pregnancy-induced non-coding RNA (PINC), a non-coding RNA, as a potential milk production regulator. When PINC expression is elevated during pregnancy, alveolar cells are inhibited and milk production and secretion are stopped. Lately, Sun and colleagues published a thorough analysis of circular RNA's functions in milk production and other aspects of livestock production in 2023. Table 2.2 lists the nutrients that were investigated along with the corresponding physiological endpoints and epigenetic effects.

**Table 2.2**. Different Epigenetic Marks and Changes Due to Changes in Diet, and Related Physiological Endpoints

| Nutrient Examined | Epigenetic Mark or Molecule | Physiological Endpoint |
|---|---|---|
| Protein restriction | DNA methylation patterns altered | Milk production in cattle and goats |
| Methionine supplementation | Phosphorylation of NFE2L2 | Antioxidant effect in cattle |
| | Upregulation of gene expression | Protein and fat synthesis in milk in cattle |
| Choline supplementation | DNA methylation patterns altered | Milk production and liver fatty acid metabolism in cattle |
| Polyphenol and omega-3 fatty acid supplementation | Histone acetylation patterns | Milk composition in pigs and goats |

| | | |
|---|---|---|
| Carbohydrates and fat supplementation | DNA methylation patterns altered | Reduction in milk fat content; milk production in cattle |
| High fat supplementation | MiRNA expression patterns altered | Milk production in rats |
| Energy restriction | MiRNA expression patterns altered | Milk production and composition in cattle |

Source: Lesta, A.; Marín-García, P.J.; Llobat, L. (2023). How does nutrition affect the epigenetic changes in dairy cows? Animals, 13, 1883. https://doi.org/10.3390/ani13111883.

## 2.5. EPIGENETIC PROCESSES INFLUENCE LIVESTOCK PRODUCTIVITY AND GROWTH

In livestock animals, epigenetic processes induce changes in phenotypic expression in response to a variety of stimuli, such as nutrition, pathogens, maternal and paternal behavior or stress, environmental stressors, management practices, chemicals, and pollutants.

### 2.5.1. Nutritional Stimuli and Environmental Perturbations

Nutrition is one of the most important factors that affects farm profitability in animal production on farms. The most important environmental factor that influences a person's growth, development, and phenotypic expression after conception is their diet. It has been demonstrated that the epigenome is influenced by the type and amount of nutrients that animals receive. Growing research suggests that permanent changes to an embryo's or germline's epigenome may be passed down to progeny; this is known as intergenerational or transgenerational epigenetic inheritance.

Recent research has demonstrated that regular feeding stimulus induces epigenetic changes in somatic tissues, which may have generational effects on livestock health and disease. Furthermore, it is known that both an excessive lack of nutrition and an excessive amount of nutrients can change the epigenetic state of germ cells and their transmission to the next generations. Numerous studies show that dietary stimuli containing various nutrient categories induced epigenetic modifications that changed the expression of genes and phenotypes. These data suggest that epigenetic changes brought on by nutrition may be heritable, though the underlying mechanisms are unclear. It is evident that changes in DNA methylation in response to nutritional stimuli undoubtedly affect production capacity and may increase the risk of disease.

**Table 2.3**. From: Epigenetics and Inheritance of Phenotype Variation in Livestock

| Organism | Imprinted | Paternally Expressed | Maternally Expressed | Other[a] |
|---|---|---|---|---|
| Cow | 20 | 12 | 8 | – |
| Sheep | 16 | 6 | 8 | 2 (1× isoform dependent, 1× unknown) |
| Dog | 1 | – | 1 | – |
| Pig | 22 | 14 | 6 | 2 (1× tissue dependent, 1× biallelic) |
| Rabbit | 1 | 1 | – | – |
| Chicken | – | – | – | – |
| Opossum | 2 | 1 | 1 | – |
| Lab Opossum | 6 | 2 | 4 | – |
| Human | 97 | 61 | 29 | 7 (4 × isoform dependent, 2 × random, 1 × unknown) |
| Mouse | 124 | 50 | 62 | 12 (5 × isoform dependent, 7 × unknown) |
| Rat | 6 | 3 | 3 | – |
| Wallaby | 5 | 4 | 1 | – |

[a]Other: non-paternal or maternal form of allele expression in the zygote.

Source: Triantaphyllopoulos, K.A., Ikonomopoulos, I. and Bannister, A.J. (2016) Epigenetics and inheritance of phenotype variation in livestock. Epigenetics and Chromatin, 9, 31. https://doi.org/10.1186/s13072-016-0081-5.

Distinctly methylated region (DMR); differentially methylated cytosine (DMC); histone tri-methylation at the 27th lysine residue; histone tri-methylation at the 4th lysine residue; histone tri-methylation at the 36th lysine residue; and histone tri-methylation at the 12th lysine residue of the histone H4 protein (H4K12ac).

For instance, feeding cows methionine supplements during the periparturient phase decreased the liver's overall DNA methylation state, increased the promoter methylation of the nuclear receptor for long-chain fatty acids, peroxisome proliferator-activated receptor alpha (PPARA), and increased the expression of the PPARA gene in supplemented cows relative to non-supplemented cows. Furthermore, there was an increase in the expression of the PPARA gene targets, PCK1 [phosphoenolpyruvate carboxykinase 1], FGF21 [fibroblast growth factor 21], and ANGPTL4 [angiopoietin-like 4], suggesting that extra methionine triggered the PPARA-regulated signaling pathway. The hepatic PPARA pathway has been linked to enhanced immunological and lipid metabolism through activation or upregulation. Additionally, the calves' hepatic metabolism was programmed by the mother's methionine supply during the latter stages of pregnancy. This was achieved by preserving DNA methylation, methionine homeostasis, and energy metabolism, all of which may have improved development performance, growth promotion, and nutrient utilization efficiency. Other research has compared feeding high- vs. diets low in concentration, grain vs. Pigs, cows, or chickens' diets supplemented with grass, minerals/vitamins, or fatty acids changed the epigenetic state of different tissues or cells, leading to different phenotypes (Table 2.3). Livestock productivity is restricted by environmental disturbances such as heat stress, weaning stress, transportation stress, pathogens, dietary modifications, and so forth. These impacts have an immediate effect on health, welfare, and productivity by influencing gene expression, metabolism, and epigenetic factors. For instance, it has been discovered that epigenetic mechanisms cause alterations in blood DNA methylation in Nellore cattle during heat-stressed and recovery periods, or in a variety of tissues (kidney, liver, muscle, lungs, heart, and brain) when hypoxic stress is applied to Tibetan and Yorkshire pigs, Tibetan goats, Tibetan sheep, Chuanzhong goats, and small-tailed Han sheep. Furthermore, both the blood of the cows and their progeny had epigenetic changes brought on by the stress of transporting Brahman cows. Weaning, the process of separating young from their mothers, is a period of stress (e.g., dietary and social stress) linked to changes in the immune response and susceptibility to diseases in livestock production. Recently, weaning stress was found to induce DNA methylation and gene expression changes in weaned piglet peripheral blood mononuclear cells.

## 2.5.2. Epigenetic Alterations are Associated with Differences in Livestock Product Yields

High-value livestock products, such as milk, beef, eggs, and wool, are regulated by epigenetic factors in addition to other factors (e.g., environmental factors, farm management, health, diet, genetics, etc.).

### *2.5.2.1. Milk and Its Products*

Recent studies have demonstrated the significant regulatory roles that epigenetic modifications play in livestock dairy animals, with consequential effects on milk and dairy products (Wang and Ibeagha-Awemu, 2021). After the groundbreaking discovery that aberrant DNA methylation at the STAT5-binding enhancer of the CSN1S1 (αS1-casein) promoter unfavorably controlled the synthesis of αS1-casein in milk during lactation, other research has discovered the impact of changed DNA methylation, chromatin modifications, and ncRNA regulation on cattle milk production.

Dechow and Liu (2018) discovered that lactating dairy cows with high and low milk yields had different blood DNA methylation levels. Additionally, several genes related to milk were found to have altered DNA methylation levels., neuropeptide Y [NPY], retinoid X receptor alpha (RXRα), and peroxisome proliferator-activated receptor alpha (PPARα) in the mammary glands of nursing goats during the dry and lactation phases. Reports analyzing the expression of microRNAs (miRNAs) have revealed that miRNAs control the lactation curve in cows or particular lactation stages. Furthermore, it has been reported that goats with decreased miR-145 expression exhibit fatty acid synthesis due to elevated methylation levels of several lipid-related genes, including FASN (fatty acid synthase), SCD1 (stearoyl-CoA desaturase), PPARG (peroxisome proliferator-activated receptor gamma), and SREBF1 (sterol regulatory element-binding transcription factor 1).

#### *2.5.2.2. Meat, Wool, and Egg*

Meat and wool production development are regulated by DNA methylation and miRNA regulation, two significant epigenetic processes.

Research on sheep longissimus dorsi muscle DNA methylation profiles discovered differentially methylated genes (e.g., ADIPOQ [adiponectin, C1Q, and collagen domain containing], ITGA2 [integrin subunit alpha 2], MYOG [myogenin], CCN2 [cyclin A2], MAP [microtubule-associated protein tau], NR4A1 [nuclear receptor subfamily 4 group A member 1], DLK1 [delta-like non-canonical Notch ligand 1], CO1A2 [collagen type I alpha 2 chain], and DIAPH1 [diaphanous-related formin 1]) with critical roles in muscle function (Fan et al., 2020). ACAD11 [acyl-CoA dehydrogenase family member 11], FASN, FADS6 [fatty acid desaturase 6], CAPA15 [calpain 15], NAT9 [N-acetyltransferase 9], and SPDEF [SAM pointed domain-containing ETS transcription factor] are among the genes whose DNA methylation variations have been linked to variations in muscle development and meat quality in beef cattle breeds (Chen et al., 2019). In chickens, the DNA methylation profiles of juvenile hens and later laying-period hens with varying intramuscular fat deposition and water-holding capacities differ, suggesting that DNA methylation regulates the expression of important genes (e.g., implicated in this process are ATP-binding cassette subfamily A member 1 (ABCA1), glutathione S-transferase theta 1-like (GSTT1L), and collagen type VI alpha 1 chain (COL6A1).

Recent research suggests that epigenetic processes control the production of cashmere wool, which is important for the goat industry because of the growing economic value of wool production. For instance, the degree of DNA methylation affects the genetic stability of cashmere goat wool traits across generations. Moreover, DNA methylation controls the growth of wool fibers and the transformation of fur into unique characters. A recent study also linked the development of secondary hair follicles in cashmere goats to modified miRNA expression profiles, looking at the expression of miRNA in seven stages of fetal development. Furthermore, during the anagen period, there is a difference in the expression of circular RNA (circRNA) in the skins of cashmere goats from Liaoning, which produces high-quality fiber, and Inner Mongolia, which produces high-fiber. Table 2.4 provides more detailed examples of epigenetic influences on various traits, including egg production, cashmere fiber growth, muscle development, beef tenderness, meat quality, fat deposition, fatty acid composition, muscle metabolism, and boar taint. Taken together, these findings point to significant regulatory roles for changes in DNA methylation in lactation, milk and component yields, and traits related to meat, eggs, and wool.

**Table 2.4**. Differing Cattle Phenotypes Have Altered Epigenetic Profiles

| Epigenetic category | Gene function | Number of transcripts tagged | Number of SNP tagging transcripts* | | |
|---|---|---|---|---|---|
| | | | CAT1 | CAT2 | CAT3 |
| a | DNA methyltransferases (DNMT) | 3 | 2 | 33 | 6 |
| b | Methyl-binding domain proteins (MBD) | 6 | 2 | 43 | 8 |
| c | Histone acetyltransferases (HAT) | 11 | 8 | 164 | 13 |
| d | Histone deacetylases (HDAC) | 8 | 2 | 318 | 9 |
| e | Histone methyltransferases (HMT) | 32 | 26 | 637 | 37 |
| f | Histone demethylase | 1 | 0 | 12 | 2 |
| g | Histones | 30 | 3 | 18 | 43 |
| h | Polycomb-group proteins | 3 | 0 | 14 | 6 |
| i | Chromatin-remodeling factors | 99 | 70 | 1499 | 116 |
| | TOTAL | 193 | 113 | 2738 | 240 |

*CAT1, SNP at transcribed regions (exon, splice region, 3'- and 5'-UTR), CAT2, SNP at transcripts intron; CAT3, SNP down- or up-stream of a transcript (up to 3 Kb).

Source: Porto-Neto LR, Fortes MRS, McWilliam SM, Lehnert SA and Reverter A (2014). Variation in genes involved in epigenetic processes offers insights into tropically adapted cattle diversity. Front. Genet. 5:89. doi: 10.3389/fgene.2014.00089.

Histone H3 protein acetylation at residue 27 (H3K27ac) and tri-methylation at residue 36 (H3K36me3) are examples of differentially methylated histone proteins. DMR stands for differentially methylated region.

## 2.5.3. Disease Resistance is Influenced by Epigenetic Factors in Farm Animals

There are two categories of disease resistance: broad sense and narrow sense. In a broad sense, the terms "disease resistance" and "disease resilience" refer to the ability to tolerate and adapt to harsh environments. Disease resistance, when used narrowly, primarily refers to an animal's capacity to slow down the rate at which pathogens proliferate by blocking their mechanisms of infection. This ability is closely linked to host-pathogen interactions as well as the host's resistance to a specific pathogen.

Low heritability characterizes the majority of disease-resistant traits, such as specific or general antibody levels. Apart from genetic variables, environmental factors such as pathogens, management techniques, nutrients, etc.) and epigenetic changes are all factors that affect the health and resistance to disease of animals (Figure 2.6). The regulatory mechanisms and potential role of host epigenetic modifications on animal health and disease resistance traits have received increasing attention in recent years.

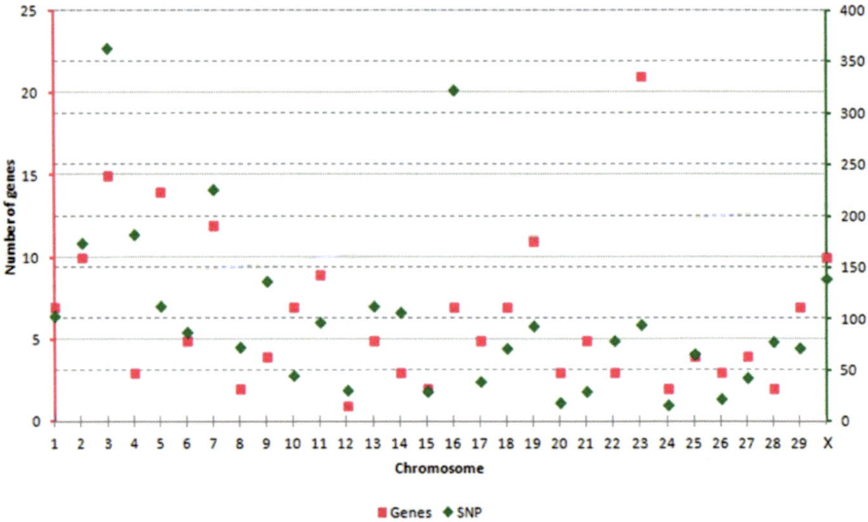


**Figure 2.6**. *Epigenetic factors respond to external and internal environment factors (the exposome) and interact with the underlying genotype to influence phenotypic outcomes.*

***Source**: Porto-Neto LR, Fortes MRS, McWilliam SM, Lehnert SA and Reverter A (2014) Variation in genes involved in epigenetic processes offers insights into tropically adapted cattle diversity. Front. Genet. 5:89. doi: 10.3389/fgene.2014.00089.*

### *2.5.3.1. Tackling Mastitis Resistance of Dairy Livestock as an Example*

When it comes to animal husbandry and the food industry, dairy is one of the sectors with the highest levels of activity and fastest growth. Given that milk and dairy products are now considered to be essential sources of nutrition worldwide, there is a close relationship between the quality of milk and public health. Mastitis, an inflammation of the mammary gland, is one of the many factors that can easily reduce milk production and quality.

One of the most significant illnesses affecting the dairy industry's ability to grow sustainably is mastitis. Pathogenic microorganisms that enter the mammary tissue through the nipple duct or skin trauma are the primary cause of mastitis. Mastitis can be classified as either clinical or subclinical based on its clinical manifestations. The modern dairy industry suffers significant financial losses as a result of mastitis in dairy cows, which also has an impact on abortion and culling rates in addition to milk yield and quality.

Bovine mastitis resistance is influenced by a wide range of factors, such as the environment, pathogens, host genetic background, and epigenetic factors, as is the case with most complex diseases. Since the use of antibiotics in the prevention and treatment of cow mastitis can lead to drug resistance, efforts are being made to eradicate the disease through disease resistance breeding. Breeding for mastitis resistance is challenging due to its low heritability ($<0.1$). However, by combining genetic and epigenetic information, mastitis resistance may be further improved (Figure 2.7).

Recent research has produced a wealth of evidence suggesting that epigenetic modification is a major factor in the onset and progression of inflammation. New concepts and approaches to studying regulatory mechanisms in mastitis resistance have been made possible by studies of epigenetic regulation on inflammatory diseases. The host reacts to inflammation by producing immune-related genes or cytokines that control the spread of inflammation. One important factor in controlling the expression of genes is DNA methylation. Pilot studies in dairy cattle have shown a connection between the development and occurrence of acute and clinical mastitis and changes in DNA methylation. In the mammary gland tissue of dairy cows with acute mastitis caused by *E. coli*, there is evidence of promoter hypermethylation of the αs1-casein gene and sudden suppression of gene expression. Furthermore, in clinically mastitis Holstein cows, hypermethylation of the CD4 gene's promoter region in peripheral blood cells may prevent the gene's expression (Figure 2.7).

To identify useful biomarkers for mastitis resistance selection and molecular diagnosis, more research has recently been conducted on the transcriptome and epigenetic regulatory mechanisms of *S. aureus* (*Staphylococcus aureus*) mastitis in dairy cattle. One of the primary causes of subclinical mastitis is *S. aureus*. When compared to other bacterial infections, *S. aureus* mastitis symptoms are relatively mild, sometimes even nonexistent, last a long time, and exhibit problematic drug resistance. Furthermore, *S. aureus* can proliferate in the cow barn during milking, which complicates the management and prevention of *S. aureus* mastitis. Song et al. (2016) found variations in genome-wide DNA methylation in peripheral blood lymphocytes between healthy and *S. aureus*-mastitis-affected Holstein cows.

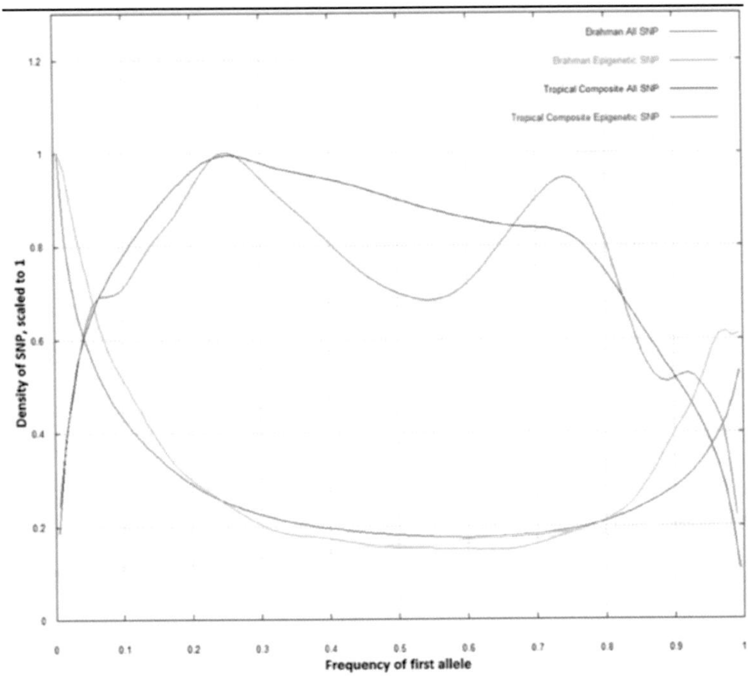


**Figure 2.7**. *Density plot of allelic frequencies for all SNP and epigenetic SNP in brahman and tropical composite.*

***Source****: Porto-Neto LR, Fortes MRS, McWilliam SM, Lehnert SA and Reverter A (2014) Variation in genes involved in epigenetic processes offers insights into tropically adapted cattle diversity. Front. Genet. 5:89. doi: 10.3389/fgene.2014.00089.*

The authors also reported that there was a strong correlation between the development of *S. aureus* subclinical mastitis and the DNA methylation levels of three genes: NRG1 (neuregulin 1), MST1 (macrophage stimulating 1), and NAT9 (N-acetyltransferase 9). This suggests that these genes may be useful epigenetic markers for enhancing cows' resistance to *S. aureus* mastitis. Maps and analyses of genome-wide DNA methylation patterns in the mammary gland quarters of Holstein cows that were experimentally infected with *S. aureus* bacteria revealed that the patterns were CCGG (CpG)/CCWGG (where W = A or T) (non CpG). The outcomes additionally demonstrated that one of the processes controlling gene expression in the processes of the bovine mammary gland during *S. aureus* mastitis is DNA methylation. Furthermore, a number of genes in the "Th17 cell differentiation" and "B-cell receptor signaling" pathways that are immune-related and differentially methylated may be potential candidates for *S. aureus* mastitis. Numerous pathogens, including *M. avium Mycobacteriu M. avium* subsp., that cause different livestock diseases cause alterations in DNA methylation. Table 2.5 provides an overview of epigenetic alterations in response to nutritional stimulus in livestock.

**Table 2.5**. Epigenetic Alterations in Response to Nutritional Stimulus in Livestock

| Parameter | Breed | Organ | Epigenetic alteration | *Potential epigenetic marker genes | References |
|---|---|---|---|---|---|
| **Cattle** | | | | | |
| Methionine supplementation throughout the peripartal period | Holstein cows | Liver | Global DNA methylation was lower, promoter methylation of *PPARA* was higher | *PPARA* | Osorio et al., 2016 |
| Grass-fed and grain-fed | Angus cattle | Rumen tissue | 217 DMRs | *ADAMTS3, ENPP3* | Li Y. et al., 2019 |
| High- vs low-concentrate corn straw (HCS/LCS) and low-concentrate mixed forage diet | Chinese Holstein cows | Mammary tissue | H3 acetylation reduced in HCS. Methylation of *STAT5A* reduced in HCS. Methylation of *SCD* increased in HCS | *STAT5A* | Dong et al., 2014 |
| High- vs low-concentrate diet | Holstein cows | Liver | Chromatin loosening at the promoter region | *TLR4, LBP, HP, SAA3* | Chang et al., 2015 |
| Methyl donor supplementation of pregnant animals | Holstein cows and calves | Blood | More than 2,000 DMG between offspring | *CEND1, VSIG2, B3GNT8*, etc. | Bach et al., 2017 |
| 85 vs 140% feed restriction | Angus-Simmental crossbred cows | *Longissimus dorsi* and *semitendinosus* muscles | One DMR (contains 23 to 24 CpGs) in *IGF2* | *IGF2* | Paradis et al., 2017 |
| Butyrate treatment | Holstein cows | Rumen primary epithelial cells from 2-week-old bull calves | Increased genome coverage (%) of CTCF, H3K27me3, and H3K4me3 and decreased coverage of H3K27ac, H3K4me1, and ATAC | Weak enhancers and flanking active transcriptional start sites | Fang et al., 2019a |
| **Swine** | | | | | |
| Restricted diet | Pig | Endometrium and embryos | Altered DNA methylation of selected genes | *ACP5, RGS12, EDNRB, TLR3, ADIPOR2, DNMT1* | Zglejc-Waszak et al., 2019 |
| Vitamin C supplementation | Pig | Oocyte | Methylation erasure on 5mC, m$^6$A, and H3K27me3, but establishment of H3K4me3 and H3K36me3 | *TET2, DNMT3B, HIF-1α, HIF-2α, TET3, METTL14, KDM5b*, and *EED* | Yu et al., 2018 |
| Omega-3 fatty acid supplementation | Piglet | Leukocyte | One hypomethylated DMR in chromosome 4, two hypermethylated DMRs in chromosomes 4 and 12 | *RUNX1T1* | Boddicker et al., 2016 |
| Maternal bisphenol A and methyl donor supplementation | Piglets from Landrace × Yorkshire sows | Piglet intestinal samples | Altered DNA methylation level of *PEPT1* in jejunum of offspring | *PEPT1, DNMT3a, LCT, DNMT1, MTHFR* | Liu H. et al., 2017 |
| Maternal methyl donor dietary supplementation | Piglets from White × Landrace sows | Liver samples from piglets | Increased methylation at the promoter of *IGF-1* in the methyl donor group | *IGF–1* | Jin C. et al., 2018 |
| Maternal betaine supplementation | Piglets from Landrace × Yorkshire sows | Liver samples from piglets | DNA hypermethylation and enriched H3K27me3 on the promoter region | *GALK1* | Cai et al., 2017 |
| **Poultry** | | | | | |
| Maternal genistein supplementation | Laying broiler breeder hens | Liver from pullets | Induced H3K36me3 and H4K12ac at the *PPARD* promoter | *PPARD* | Lv et al., 2019 |

| Parameter | Breed | Organ | Epigenetic alteration | *Potential epigenetic marker genes | References |
|---|---|---|---|---|---|
| Prenatal betaine supplementation | Juvenile chicken | Liver from juvenile chicken | Increased CpG methylation at promoter regions | *LXR* and *CYP27A1* | Hu et al., 2017b |
| Maternal betaine supplementation | Rugao Yellow chicken breeder hens | Hypothalamus of F1 cockerels | Modification of promoter CpG methylation | *DNMT1*, *BHMT* *SREBP-1*, *SREBP-2*, and *APO-A1* | Idriss et al., 2018 |
| Maternal betaine supplementation | Rugao yellow-feathered laying hens | Blood from pullets | Hypomethylated promoter regions of steroidogenic genes | *AHCYL*, *GNMT1*, and *BHMT* | Abobaker et al., 2019 |
| Methionine and betaine supplementation | Geese | Liver | Altered DNA methylation of *LOC106032502* | *LOC106032502* | Yang et al., 2018 |

**Potential epigenetic marker genes: These are genes harboring differentially methylated CpG sites or regions; vs, versus; DMR, differential methylated region; DMG, differential methylated gene; DMC, differential methylated CpG; HCS, high-concentrate corn straw; LCS, low-concentrate corn straw; CTCF, CCCTC-binding factor; ATAC, Ada-Two-A-Containing complex; H3K27me3, the trimethylation at the 27th lysine residue of the histone H3 protein; H3K4me3, the trimethylation at the fourth lysine residue of the histone H3 protein; H3K27ac, the acetylation at the 27th lysine residue of the histone H3 protein; H3K4me1, the monomethylation at the fourth lysine residue of the histone H3 protein; H3K36me3, the trimethylation at the 36th lysine residue of the histone H3 protein; H4K12ac, the acetylation at the 12th lysine residue of the histone H4 protein.*

Source: Wang M and Ibeagha-Awemu EM (2021) Impacts of epigenetic processes on the health and productivity of livestock. Front. Genet. 11:613636. doi: 10.3389/fgene.2020.613636.

DMR stands for differentially methylated region; DMC stands for differentially methylated cytosine; H3K4me3, or tri-methylation at histone H3 protein's fourth lysine residue; H3K27me3, or tri-methylation at histone H3 protein's 27th lysine residue; *E. coli*, *Staphylococcus aureus*, LPS, Newcastle disease virus, acetylation at lysine residue 27 of the histone H3 protein, and mono-methylation at lysine residuc 4 of the histone H3 protein are all terms used to describe various biological entities. Another significant epigenetic modification that post-transcriptionally controls gene expression, in addition to DNA methylation, is miRNA.

By interrogating genome-wide expression of mRNAs (messenger RNA) and miRNAs in bovine mammary gland cells at 24 h after intra-mammary infection with high or low concentrations of *S. aureus*, Fang et al. (2016) found that, compared to the uninfected quarters, significant upregulation of bta-mir-223 and bta-mir-21-3p were observed in the quarters of cows infected with a high concentration of *S. aureus*. According to network analysis, these miRNAs likely play a major role in defending the host against *S. aureus* infection by inhibiting the genes CXCL14 [C-X-C motif chemokine ligand 14] and KIT [KIT proto-oncogene, receptor tyrosine kinase] (Fang et al., 2016). Furthermore, in *S. aureus*-challenged bovine mammary gland epithelial cells (MAC-T cells), transcriptome sequencing also discovered four differentially expressed miRNAs (bta-miR-2339, miR-499, miR-23a, and miR-99b; Jin et al., 2014). In-depth presentations of significant miRNAs linked to mastitis and other livestock illnesses can be found in Do et al. (2021). These results confirm that a complex network of epigenetic marker-gene-pathway interplay regulates the response of the mammary gland and other tissues to pathogenic infection. They also show how multiple regulation of some key DNA methylation markers, hub miRNAs, and vital genes may influence multiple important pathways leading to phenotypic changes.

## 2.6. EPIGENETIC PROCESSES BE EXPLOITED TO IMPROVE LIVESTOCK HEALTH AND PRODUCTIVITY

While knowing how a factor affects a process is important, knowing how to use that factor to positively influence a process is even more crucial. The effects of epigenetic processes on the health and productivity of livestock vary. How to best utilize these diverse effects for enhanced livestock productivity and health management will be crucial for the productivity of livestock in the future.

Epigenetic factors reflect the evolution of individual phenotypic variations at particular time points or growth stages because they react to various environmental signals (Figure 2.6) and interact with the underlying genetic composition throughout an individual's lifetime. If the relationship between traits and epigenetic markers in livestock is carefully studied, it can lead to the development of epigenetic biomarkers for better management of livestock and further increase genetic gain in the breeder's equation, which will improve livestock. For instance, the effects of environmental factors captured by epigenetic association studies—such as feed quantity and quality, farm environment, and so forth—can be developed and incorporated into livestock management strategies.

### 2.6.1. Potential Application of Epigenetic Data in Livestock Management

Some potential applications of epigenetic biomarkers in livestock management were stated recently:

- **Epigenetic Biomarkers Can be Used to Manage Health**: It is possible to develop epigenetic factors that have been demonstrated to significantly correlate with livestock diseases for use in prognostic, predictive, therapeutic, or monitoring purposes. For instance, created epigenetic biomarkers may be crucial to the treatment of silent, chronic livestock diseases like paratuberculosis, subacute ruminal acidosis, chronic mastitis, and metabolic disorders. These diseases also often have no overt clinical symptoms. The US Food and Drug Administration (FDA) has authorized the use of DNMT1 (DNA methyltransferase 1) inhibitors for the management of human health. These drugs, which include 5-azacytidine and 5-aza-2′-

deoxycytidine, have shown excellent efficacy in the treatment of hematological malignancies (Montalvo-Casimiro et al., 2020). An FDA-approved cancer DNA methylation assay (test) based on DNA methylation alteration of SEPT9 (septin 9) shows 71.1% to 95.6% sensitivity and 81.5% to 99% specificity for colorectal cancer diagnosis.

- **Epigenetic Biomarkers Can be Used to Manage Livestock Reproduction, Growth, and Production**: Epigenetic biomarkers linked to health, productivity, and reproduction in livestock can shed light on how these organisms interact with other processes (e.g., biological, molecular, cellular, and immune mechanisms) to influence the outcome of phenotypes and can be utilized for the management of livestock. Particular epigenetic factors are linked to phenotypes related to reproduction, growth and development, and products in response to various stimuli such as nutrition, environmental conditions, disease pathogens, and paternal environment, as indicated in Tables 2.3 and 2.4. For example, sperm epigenetic biomarkers can facilitate the analysis of sperm quality and breeding male selection. Furthermore, placenta and embryo biopsies, along with newborn blood, may be used as predictive biomarkers for later-life phenotypes of interest. Understanding how an individual's epigenetic signature varies throughout life as a result of their diet, surroundings, and other factors may help anticipate how they will react to various environmental stressors and how to take preventative action.
- **Epigenetic Biomarkers Can be Used for Livestock Trait Improvement**: Which is represented by epigenetics and other factors, provides opportunities to further improve the genetic gain because it is well known that genetic data alone do not explain all of the phenotypic variance in livestock traits. For livestock traits, there is a wealth of evidence demonstrating transgenerational epigenetic inheritance. Swartz et al. used somatic cell score (SCS) as a stand-in for mammary gland health. (2021) examined the lactation records of 3,570 daughters from 1,554 herds (second lactation) and 15,992 daughters from 4,366 herds (first lactation), and discovered that the fat yield of daughters born to first lactation cows with elevated SCS declined. According to the authors, a few underlying mechanisms, such as the daughters' susceptibility to mastitis pathogens and epigenome modifications, likely controlled the daughters' production of milk fat. This case illustrates an intergenerational effect on fat

yield by showing that the observation in the daughters was partially caused by maternal stress brought on by disease pathogens that altered the epigenome.

The nature of the epigenetic variation—whether it is intragenerational (transient) or transgenerational/intergenerational (heritable)—the dependence or independence of the epigenetic effect on DNA variant effect, and the causal or consequential effect of the epigenetic effect are some of the considerations that need to be made when analyzing epigenetic information for livestock improvement. Consequently, it's important to precisely define and apply the nature and source of epigenetic variation. Transgenerational effects can be taken into account when calculating GEBV (genomic estimated breeding values), even though transient epigenetic effects can be utilized for routine management. This is consistent with the idea that the assessment of trait heritability should separate and account for the effects of both genetic and nongenetic inheritance.

## 2.6.2. Practical Directions for Incorporating Epigenetic Data for Improving Disease Resistance in Farm Animals

Combining three main approaches will improve animal health: (i) disease elimination and vaccination from the standpoint of treatment, prevention, and immunization; (ii) nutritional needs optimization and environmental sanitation improvement from the standpoint of nutrition and management; and (iii) animal breeding for disease resistance from the standpoint of incorporating genetic and epigenetic data in breeding programs for disease resistance (Figure 2.8).

| Parameter | Breed | Organ | Epigenetic alteration | Production trait | Model and references |
|---|---|---|---|---|---|
| Milk | | | | | |
| High and low milk yield | Holstein cows | Peripheral blood mononuclear cells | 72 DMRs between high and low milk yield, 252 DMRs across herd environments | Milk yield | Dechow and Liu, 2018 |
| High and low milk yield | Holstein cows | Jugular venous blood | DNA methylation rates in the lower-yield cows were significantly higher than those in the higher-yield animals | Milk and protein yield | Wang L. et al., 2019 |
| Experimentally induced mastitis by E. coli and S. aureus | Holstein cows | Liver and mammary gland tissues | Remethylation of upstream promoter of CSN1S1 gene in response to E. coli infection | αS1-casein synthesis | Vanselow et al., 2006 |

| | | | | | |
|---|---|---|---|---|---|
| Once/twice daily milking | Holstein cows | Liver and mammary gland tissue | Altered methylation of four CpG within the distal upstream regulatory region of the CSN1S1 gene | Milk production (yield and milking frequency) | Nguyen et al., 2014 |
| Different lactation stages | Holstein cows | Blood | DNA methylation of EEF1D was lower in the dry period than the early stage of lactation | Milk production | Liu X. et al., 2017 |
| Different lactation stages | Xinong Saanen goats | Mammary gland tissue | Methylation levels of 95 and 54 genes in the lactation period were up- or downregulated, respectively, relative to the dry period | Milk production | Zhang X. et al., 2017 |
| Different lactation stages | Xinong Saanen goats | Mammary gland tissue | Inhibition of miR−145 increased methylation levels of FASN, SCD1, PPARG, and SREBF1 | Milk fat synthesis | Wang et al., 2017b |
| High- and low-concentrate diet feeding | Guanzhong goats | Mammary gland tissue | Increased DNA methylation level in the promoter regions of the ACACA and SCD genes | Milk fat production and composition | Tian et al., 2017 |
| Meat | | | | | |
| Two sheep breeds differing in meat production ability | Small-tailed Han and Dorper × small-tailed Han crossbred sheep | Longissimus dorsi muscles | 808 DMRs and global loss of DNA methylation in the DMRs in the crossbred sheep, 12 potential DMGs | Meat production | Cao et al., 2017 |
| Two cattle breeds exhibiting different meat production ability | Japanese black and Chinese Red Steppes cattle | Longissimus dorsi muscles | 23,150 DMRs identified, 331 DMRs correlated negatively with expression of DE genes, 21 DMRs located in promoter regions | Muscle development and related meat quality traits | Fang et al., 2017 |
| Divergent beef tenderness | Angus beef | Longissimus dorsi muscles | DNA methylation profiles related to beef tenderness, and 7215 DMRs between tender and tough beef | Beef tenderness | Zhao et al., 2020 |
| Three cattle breeds differing in meat production abilities | Simmental, Yunling, and Wenshan cattle | Longissimus dorsi muscles | 18 DM and DE genes between Simmental and Wenshan cattle, 14 DM and DE genes between Simmental and Yunling cattle, 28 DM genes between Wenshan and Yunling cattle | Meat quality | Chen Z. et al., 2019 |

| | | | | | |
|---|---|---|---|---|---|
| Three growth stages | Polled yak | Longissimus dorsi muscles | 1,344, 822, and 420 genes with DM CCGG sites and 2,282, 3,056, and 537 genes with DM CCWGG sites between 6-month-old vs 90-day-old, 6-month-old vs 3-year-old, and 3-year-old vs 90-day-old fetuses, respectively | Muscle development and growth | Ma et al., 2019 |
| Feed restriction | Angus–Simmental crossbred cows | Longissimus dorsi and semitendinosus muscle | One DMR in IGF2 | Muscle function | Paradis et al., 2017 |
| Fetal and adult cattle | Qinchuan cattle | Longissimus dorsi muscles | Three DMCs in the core promoter region of SIX1, histone H4 and E2F2 bind to SLX1 | Muscle development | Wei et al., 2018 |
| Obese, lean, and miniature pig breeds | Tongcheng, Landrace, and Wuzhishan pigs | Blood leukocytes | 2,807, 2,969, and 5,547 DMGs in the Tongcheng vs Landrace, Tongcheng vs Wuzhishan, and Landrace vs Wuzhishan comparisons, respectively | Fat-related phenotype variance | Yang et al., 2016b |
| Obese and lean type pig breeds | Landrace pigs and Rongchang pigs | Backfat | 483 DMRs in the promoter regions | Fat deposition and fatty acid composition | Zhang S. et al., 2016 |
| Castrated and non-castrated pigs | Male Huainan pigs | Liver and adipose tissues | GHR methylation rate in the liver of castrated and non-castrated pigs were 93.33% and 0, respectively | Castration-induced fat deposition | Wang et al., 2017e |
| Pig breeds differing in metabolic characters | Duroc and Pietrain | Longissimus dorsi muscles | More than 2,000 DMCs | Muscle metabolism | Ponsuksili et al., 2019 |
| Three pig breeds differing in fatness traits | Polish Large White, Duroc and Pietrain | Subcutaneous fat, visceral fat, and longissimus dorsi muscle | H3K9ac and H3K4me3 correlated to the expression level of selected genes | Adipose tissue accumulation | Kociucka et al., 2017 |
| Highest and lowest pH among littermates | Duroc | Longissimus dorsi muscle | 3,468 DMRs, including 44 and 21 protein-coding genes with hyper- and hypomethylation regions in their gene bodies | Postmortem energy metabolism and pH | Park et al., 2019 |
| High and low boar taint | Pigs | Testis | 32 DE genes with DMCs | Boar taint | Wang and Kadarmideen, 2019a,c |

| | | | | | |
|---|---|---|---|---|---|
| Different growth stages | Gushi hens | Breast muscle | 2,714 DMRs and 378 DMGs | Intramuscular fat deposition and water-holding capacity | Zhang M. et al., 2017 |
| Different feed conditions and breeds | Daninghe and Qingjiaoma chickens | Breast muscle | 46 CpG sites and 3 CpG islands in UCP3, different methylation levels of UCP and FATP1 between groups | Breast muscle (intramuscular fat content) | Gao et al., 2015, 2017 |
| Egg | | | | | |
| Before and after reproductive maturation | Hy-Line Brown commercial female chickens | Ovaries | Increased methylation of two CpG sites in ERα; increased H3K27ac and decreased H3K36me3 related to increased ERα mRNA transcript | Reproductively mature, egg production | Guo M. et al., 2020 |
| Betaine supplementation | Laying hens | Liver | Hypomethylation of promoter in GR | Egg production | Omer et al., 2018, 2020 |
| Wool | | | | | |
| Different generations of cashmere goats | Cashmere goats | Skin | 336 hyper- and 753 hypomethylated 5mC, corresponding to 214 hyper- and 560 hypomethylated genes | Cashmere traits | Dai et al., 2019 |
| Anagen and telogen stages | Cashmere goats | Skin | 1,311 DMRs corresponding to 493 DMGs (269 hyper- and 224 hypomethylated DMGs) | Hair cycling and cashmere growth | Li et al., 2018 |
| Coarse type and fine type cashmere | Cashmere goats | Skin | 9,085 DM N6-methyladenosine sites | Cashmere fiber growth | Wang et al., 2020f |
| Cashmere goats and other goat species | Cashmere goats | Skin | Altered methylation degree of HOXC8 exon 1 | Cashmere fiber growth | Bai et al., 2017 |
| Anagen and telogen stages | Cashmere goats | Skin | Altered promoter methylation of HOTAIR gene | Cashmere fiber growth | Jiao et al., 2019 |

**Figure 2.8**. *Alteration of epigenetic markers in relation to livestock production traits.*

***Source**: Wang M and Ibeagha-Awemu EM (2021) Impacts of epigenetic processes on the health and productivity of livestock. Front. Genet. 11:613636. doi: 10.3389/fgene.2020.613636.*

*DM, differentially methylated; DE, differentially expressed; DMR, differential methylated region; DMG, differential methylated gene; DMC, differential methylated CpG; H3K9ac, histone H3 lysine 9 acetylation; H3K4me3, trimethylation at the fourth lysine residue of histone H3 protein; H3K27ac, acetylation at the 27th lysine residue of histone H3 protein; H3K36me3, trimethylation at the 36th lysine residue of histone H3 protein.*

The importance of breeding for disease resistance is paramount among these tactics. The two main components of disease resistance breeding are as follows. One viewpoint comes from the field of genetics, which focuses primarily on the genetic variation described by Mendel's genetic law. In other words, if members of one generation who are resistant to a particular disease mate with each other to create the next generation, the population's health will gradually improve. Simultaneously, it is important to properly utilize the non-Mendelian inheritance-related epigenetic regulatory mechanisms to enhance disease resistance. The characteristics and mechanisms of epigenetics can be used to improve the health status of both the current and future generations due to its flexibility and intergenerational and transgenerational inheritance. We can therefore encourage the improvement of animal health in a timely and long-term manner from the current generation and the next generations by taking into account the genetic mechanisms of Mendelian inheritance and the epigenetic mechanisms of non-Mendelian inheritance. High-yielding farm animals' disease resistance can be effectively preserved and enhanced in these two ways.

## REFERENCES

1. Adam-Vizi, V., & Starkov, A. A. (2010). Calcium and mitochondrial reactive oxygen species generation: How to read the facts. *Journal of Alzheimer's Disease*, *20*(Suppl. 2), S413–S426.
2. Adesogan, A. T., & Dahl, G. E. (2020). MILK symposium introduction: Dairy production in developing countries. *Journal of Dairy Science*, *103*, 9677–9680.
3. Ahmed, B., & Averill-Bates, D. A. (2015). Thermotolerance induced at a mild temperature of 40 °C alleviates heat shock-induced ER stress and apoptosis in HeLa cells. *Biochimica et Biophysica Acta: Molecular Cell Research*, *1*, 52–62.
4. Allen, J. L., Clusella-Trullas, S., & Chown, S. L. (2012). The effects of acclimation and rates of temperature change on critical thermal limits in *Tenebrio molitor* (Tenebrionidae) and *Cyrtobagous salviniae* (Curculionidae). *Journal of Insect Physiology*, *58*, 669–678.
5. Amanzadeh, J., Gitomer, W. L., Zerwekh, J. E., Preisig, P. A., Moe, O. W., Pak, C. Y. C., & Levi, M. (2003). Effect of high protein diet on stone-forming propensity and bone loss in rats. *Kidney International*, *64*, 2142–2149.
6. Anderson, O. S., Sant, K. E., & Dolinoy, D. C. (2012). Nutrition and epigenetics: An interplay of dietary methyl donors, one-carbon metabolism, and DNA methylation. *Journal of Nutritional Biochemistry*, *23*, 853–859.
7. Bartels, A., Han, Q., Nair, P., Stacey, L., Gaynier, H., Mosley, M., et al. (2018). Dynamic DNA methylation in plant growth and development. *International Journal of Molecular Sciences*, *19*, Article E2144.
8. Bateson, P., Barker, D., Clutton-Bruck, T., Deb, D., D'Udine, B., Foley, R. A., Gluckner, P., Godfrey, K., Kirkwood, T., Lahr, M. M., McNamara, J., Metcalfe, N. B., Monaghan, P., Spencer, H. G., & Sultan, S. B. (2004). Developmental plasticity and human health. *Nature*, *430*, 419–421.
9. Baumgard, L. H., Collier, R. J., & Bauman, D. E. (2017). A 100-year review: Regulation of nutrient partitioning to support lactation. *Journal of Dairy Science*, *100*, 10353–10366.
10. Belhadj, S. I., Najar, T., Ghram, A., Dabbebi, H., Ben Mrad, M., & Abdrabbah, M. (2014). Reactive oxygen species, heat stress and oxidative-induced mitochondrial damage: A review. *International Journal of Hyperthermia*, *30*, 513–523.
11. Berna-Erro, A., Woodard, G. E., & Rosado, J. A. (2012). The versatility and universality of calcium signalling. *Journal of Cellular and Molecular Medicine*, *1*, 11–21.

12. Bewley, J. M., Robertson, L. M., & Eckelkamp, E. A. (2017). A 100-year review: Lactating dairy cattle housing management. *Journal of Dairy Science*, *100*, 10418–10431.
13. Bowler, K., & Hollingsworth, M. J. (1965). The effect of inbreeding on temperature acclimatization in *Drosophila subobscura. Genetical Research*, *10*, 1–12.
14. Caiafa, P., & Zampieri, M. (2005). DNA methylation and chromatin structure: The puzzling CpG islands. *Journal of Cellular Biochemistry*, *94*, 257–265.
15. Cairns, R. B. (2009). The logic of chromatin architecture and remodelling at promoters. *Nature*, *461*, 193–198.
16. Castro, M. M. D., Albino, R. L., Rodrigues, J. P. P., Sguizzato, A. L. L., Santos, M. M. F., Rotta, P. P., Caton, J. S., Moraes, L. E. F. D., Silva, F. F., & Marcondes, M. I. (2020). Energy and protein requirements of Holstein × Gyr crossbred heifers. *Animal*, *14*, 1857–1866.
17. Cheung, P., & Lau, P. (2005). Epigenetic regulation by histone methylation and histone variants. *Molecular Endocrinology*, *19*, 563–573.
18. Crompton, M., Barksby, E., Johnson, N., & Capano, M. (2002). Mitochondrial intermembrane junctional complexes and their involvement in cell death. *Biochimie*, *84*, 143–152.
19. Deaton, A. M., & Bird, A. (2011). CpG islands and the regulation of transcription. *Genes & Development*, *25*, 1010–1022.
20. Desrosiers, R., & Tanguay, R. M. (1988). Methylation of *Drosophila* histones at proline, lysine, and arginine residues during heat shock. *Journal of Biological Chemistry*, *263*, 4686–4692.
21. Dor, Y., & Cedar, H. (2018). Principles of DNA methylation and their implications for biology and medicine. *Lancet*, *392*, 777–786.
22. Dror, D. K., & Allen, L. H. (2018). Overview of nutrients in human milk. *Advances in Nutrition*, *9*, 278S–294S.
23. Duplessis, M., Pellerin, D., Robichaud, R., Fadul-Pacheco, L., & Girard, C. L. (2019). Impact of diet management and composition on vitamin B12 concentration in milk of Holstein cows. *Animal*, *13*, 2101–2109.
24. Erickson, P. S., & Kalscheur, K. F. (2020). Nutrition and feeding of dairy cattle. *Animal Agriculture*, 157–180.
25. Faulk, C., & Dolinoy, D. C. (2011). Timing is everything: The when and how of environmentally induced changes in the epigenome of animals. *Epigenetics*, *6*, 791–797.
26. Fekete, Á. A., Givens, D. I., & Lovegrove, J. A. (2013). The impact of

milk proteins and peptides on blood pressure and vascular function: A review of evidence from human intervention studies. *Nutrition Research Reviews*, *26*, 177–190.

27. Fodor, I., Abonyi-Tóth, Z., & Ózsvári, L. (2018). Management practices associated with reproductive performance in Holstein cows on large commercial dairy farms. *Animal*, *12*, 2401–2406.
28. Fraga, M. F., Ballestar, E., Paz, M. F., Ropero, S., Setien, F., Ballestar, M. L., Heine-Suner, D., Cigudosa, J. C., Urioste, M., Benitez, J., Boix-Chornet, M., Sanchez-Aguilera, A., Ling, C., Carlsson, E., Poulsen, P., Vaag, A., Stephan, Z., Spector, T. D., Wu, Y. Z., & Esteller, M. (2005). Epigenetic differences arise during the lifetime of monozygotic twins. *Proceedings of the National Academy of Sciences of the United States of America*, *102*, 10604–10609.
29. Gabory, A., Attig, L., & Junien, C. (2011). Developmental programming and epigenetics. *American Journal of Clinical Nutrition*, *94*, 1943S–1952S.
30. Goldberg, L. A. (2003). Protein degradation and protection against misfolded or damaged proteins. *Nature*, *426*, 895–899.
31. Górska-Warsewicz, H., Rejman, K., Laskowski, W., & Czeczotko, M. (2019). Milk and dairy products and their nutritional contribution to the average Polish diet. *Nutrients*, *11*, Article 1771.
32. Gutierrez, A., & Sommer, R. J. (2004). Evolution of *dnmt-2* and *mbd-2*-like genes in the free-living nematodes *Pristionchus pacificus*, *Caenorhabditis elegans*, and *Caenorhabditis briggsae*. *Nucleic Acids Research*, *32*, 6388–6396.
33. Hagan, B. A., Moro-Mendez, J., & Cue, R. I. (2020). Realized genetic selection differentials in Canadian Holstein dairy herds. *Journal of Dairy Science*, *103*, 1651–1666.
34. Hetz, C. (2012). The unfolded protein response: Controlling cell fate decisions under ER stress and beyond. *Nature Reviews Molecular Cell Biology*, *13*, H2410–H2418. https://doi.org/10.1038/nrm3270.
35. Ho, S. M., et al. (2012). Environment epigenetics and its implication on disease risk and health outcomes. *ILAR Journal*, *53*, 289–305.
36. Holliday, R., & Pugh, J. (1975). DNA modification mechanisms and gene activity during development. *Science*, *187*, 226–232.
37. Hristov, A. N., Melgar, A., Wasson, D., & Arndt, C. (2022). Symposium review: Effective nutritional strategies to mitigate enteric methane in dairy cattle. *Journal of Dairy Science*, *105*, 8543–8557.
38. Jaton, C., Schenkel, F. S., Chud, T. C. S., Malchiodi, F., Sargolzaei, M.,

Price, C. A., Canovàs, A., Baes, C., & Miglior, F. (2019). Genetic and genomic analyses of embryo production in dairy cattle. *Reproduction, Fertility and Development*, *32*, 50–55.

39. Ji, Y., Nordgren, K. K. S., Chai, Y., Hebbring, S. J., Jenkins, G. D., Abo, R. P., Peng, Y., Pelleymounter, L. L., Moon, I., Eckloff, B. W., et al. (2012). Human liver methionine cycle: *MAT1A* and *GNMT* gene resequencing, functional genomics, and hepatic genotype-phenotype correlation. *Drug Metabolism and Disposition*, *40*, 1984–1992.
40. Jirtle, R. L., & Skinner, M. K. (2007). Environmental epigenomics and disease susceptibility. *Nature Reviews Genetics*, *8*, 253–262.
41. Kouzarides, T. (2007). Chromatin modifications and their function. *Cell*, *128*, 693–705.
42. Lacasse, P., Vanacker, N., Ollier, S., & Ster, C. (2018). Innovative dairy cow management to improve resistance to metabolic and infectious diseases during the transition period. *Research in Veterinary Science*, *116*, 40–46.
43. Lyons, K. E., Ryan, C. A., Dempsey, E. M., Ross, R. P., & Stanton, C. (2020). Breast milk, a source of beneficial microbes and associated benefits for infant health. *Nutrients*, *12*, Article 1039.
44. Marcondes, M. I., & Silva, A. L. (2021). Determination of energy and protein requirements of preweaned dairy calves: A multistudy approach. *Journal of Dairy Science*, *104*, 11553–11566.
45. Marín-García, P. J., & Llobat, L. (2021). How does protein nutrition affect the epigenetic changes in pig? A review. *Animals*, *11*, Article 544.
46. Martin, P., Barkema, H. W., Brito, L. F., Narayana, S. G., & Miglior, F. (2018). Symposium review: Novel strategies to genetically improve mastitis resistance in dairy cattle. *Journal of Dairy Science*, *101*, 2724–2736.
47. McMullen, C. K., Sargeant, J. M., Kelton, D. F., Churchill, K. J., Cousins, K. S., & Winder, C. B. (2021). Modifiable management practices to improve udder health in dairy cattle during the dry period and early lactation: A scoping review. *Journal of Dairy Science*, *104*, 10143–10157.
48. National Research Council. (2001). *Nutrient requirements of dairy cattle: Seventh revised edition*. Washington, DC: National Academies Press.
49. Palmquist, D. L., Beaulieu, A. D., & Barbano, D. M. (1993). Feed and animal factors influencing milk fat composition. *Journal of Dairy Science*, *76*, 1753–1771.

50. Rizzoli, R. (2014). Dairy products, yogurts, and bone health. *American Journal of Clinical Nutrition*, *99*, 1256S–1262S.

51. Salah, N., Sauvant, D., & Archimède, H. (2014). Nutritional requirements of sheep, goats, and cattle in warm climates: A meta-analysis. *Animal*, *8*, 1439–1447.

52. Sehested, J., Gaillard, C., Lehmann, J. O., Maciel, G. M., Vestergaard, M., Weisbjerg, M. R., Mogensen, L., Larsen, L. B., Poulsen, N. A., & Kristensen, T. (2019). Review: Extended lactation in dairy cattle. *Animal*, *13*, s65–s74.

53. Sun, H. Z., Plastow, G., & Guan, L. L. (2019). Invited review: Advances and challenges in application of feedomics to improve dairy cow production and health. *Journal of Dairy Science*, *102*, 5853–5870.

Chapter 3

# NUTRIENT UTILIZATION, REQUIREMENTS, AND NUTRIGENOMICS IN SHEEP AND GOATS

## Contents

# INTRODUCTION

Global recognition exists for the positive social and economic effects of sheep and goat production on rural communities' ability to reduce poverty and raise living standards. Knowing how sheep and goats use nutrients improves production systems' effectiveness, maximizes resource economy, and reduces carbon footprint. One of the most crucial factors in precision nutrition for the production of sheep and goats may be nutrient requirements. This crucial body of knowledge includes requirements and utilization of lipids, minerals, vitamins, degradable and bypass proteins, and structural and nonstructural carbohydrates. Numerous factors, including individuality, genetics, climate, diet, age, and physiological stages, influence nutrient requirements, so it's critical to keep up with new research. A better understanding of nutrient requirements and utilization has resulted from relatively recent advances in nutrigenomics, which have improved knowledge of the molecular interactions between nutrients and other dietary bioactive components with respect to the genome, the impact of food constituents on gene expression, and the influence of genetic variation on nutrition. This understanding has the potential to greatly improve the accuracy of determining the nutrient requirements of sheep and goats, as well as to optimize and customize nutrition based on a subject's genotype.

## 3.1. NUTRITION OF SHEEP

If there was one fundamental rule for flock management, it would be nutrition. Properly fed sheep have better growth, higher birth rates, and fewer health problems—such as internal parasites—than poorly fed ones. The biggest expense for any producer is usually feed, especially for flocks that are confined for a portion of the year. For this reason, choices about nutrition have a significant financial impact. In nations with limited resources, inadequate nutrition is a prevalent issue that requires management. Overfeeding more frequently leads to illness and decreased productivity in nations with abundant natural resources. Local circumstances have a big impact on nutrition. As such, the details of any feeding schedule ought to be customized to the flock's immediate surroundings. All sheep should, in general, have daily access to water, high-quality feed, and extra vitamins and minerals.

Despite being a vital nutrient, water is frequently disregarded. Under maintenance conditions, a 50-kg sheep should drink 2–4 gallons of water a day. Changes in the environment (such as heat and cold), activity level, and physiological state (such as growth, pregnancy, and lactation) all affect intake. Fresh, clean water must always be readily available. Since water can have significant mineral concentrations, it should be assessed as part of an all-encompassing feeding program. Sheep should be fed a diet primarily consisting of high-quality forage because they are ruminants. Forage can be pasture/range, silage, or hay (grass or legume). The rumen's resident microorganisms, which supply most of the protein needed by a sheep for energy, thrive best on forage. The forage should be reasonably mature, clean, and mold-free. The amount of neutral detergent fiber (NDF) in the forage is influenced by maturity. One factor that may limit intake is forage NDF. According to studies, the majority of sheep consume 1 and a half percent of their body weight in NDF each day. A sheep will therefore eat less fodder the higher the amount of NDF in the forage.

Sheep eat between 1.8% and 2.0% of their body weight in dry matter each day on average. Sheep should eat forage with a total digestible nutrition (TDN) value of 50% and a crude protein concentration of 7–9% for maintenance. Under higher production pressures and in various physiological states, these values rise. Growing lambs, lactating ewes, and ewes in late gestation require a TDN of up to 70% and a crude protein concentration of up to 16%, respectively.

While forages should be the mainstay of any sheep's diet, different physiological states may call for the addition of other energy sources to meet nutritional needs. Cereal grains or pelleted concentrates containing starches or fermentable fiber supplements (like beet pulp or soy hulls) should be fed during these periods. The quality of the forage as a whole and the sheep's level of productivity determine the quantity and frequency. For instance, to meet her energy needs, a ewe in the last few weeks of pregnancy fed a moderate-quality forage may require 1–2 pounds of cereal grains daily. Cereal grains won't usually be necessary for a pet ewe fed moderate-quality forage. You should never feed sugars and starches in place of subpar forage. In small hobby or pet farms, overfeeding or improper starch feeding is a common practice that frequently results in illness.

There are two types of minerals: trace minerals and macrominerals. The main sources of macrominerals are water and forages, both of which need to be considered when creating a feeding plan. For producers with smaller flocks who frequently buy small amounts of forages from multiple sources, this kind of evaluation can be difficult. Macromineral supplements should be adjusted to try to achieve a 2:1 calcium-to-phosphorus ratio and a 4:1 potassium-to-magnesium ratio after concentrations of calcium, phosphorus, potassium, and magnesium are evaluated. Local geography has a significant impact on trace minerals, such as copper, selenium, zinc, molybdenum, and cobalt. Because trace minerals compete with one another for absorption, relative concentrations of the minerals are frequently just as significant as their absolute values. Feedings of trace minerals frequently include commercial supplements. Choosing the supplement that best fits the flock's needs locally is the goal when choosing such supplements. While trace minerals can be added to a concentrate, they are typically provided as a free-choice item in small flocks (NRC, 1945; Umberger, 2009).

Blocks are not preferred to salt and loose minerals. Sodium chloride is combined with free-choice mineral salts to control consumption. Consumption of the mineral mix may be reduced if salt sources are positioned closer to the minerals. Since sheep are highly susceptible to copper toxicity, a trace-mineral mixture fed to them shouldn't contain more than 30 parts per million of copper. As part of some parasite control programs, administering copper boluses is a practice that should only be carried out very carefully.

In contrast to other physiological states like gestation, lactation, or body growth, wool growth does not call for extra nutrition. The amount of energy needed for maintenance is thought to be adequate for wool growth. However, a sheep's nutritional status affects the quality of the wool that it produces. Wool produced on a high nutritional plane has slightly coarser fibers but is produced in greater quantities. A low nutritional plane results in less wool being produced, and the fibers are so thin that they break, significantly lowering the fleece's value. Plenty of strong, fine fibers should be produced by a moderately nutritious plane.

### 3.1.1. Nutritional Requirements of Sheep

To ensure maximum growth and productivity, a sufficient diet should consist of water, proteins, minerals, vitamins, and energy sources such as fats and carbohydrates. It might be necessary to add more nutrients in fields with specific stressors.

#### *3.1.1.1. Water Requirements of Sheep*

There should always be a fresh, clean, and convenient source of water for sheep. In temperate climates, one gallon (3.85 L) of water per day is typically recommended as a minimum requirement, with 1.5 gallons per day for ewes nursing lambs, and 0.5 gallons per day for lambs being finished. Water is often the limiting nutrient in many range areas. Every sheep's water availability should be checked daily, in all weather conditions, to ensure optimal productivity. However, due to the high cost of water supply, it is often more cost-effective to water range sheep every other day. Unless dry feeds like alfalfa hay and pellets are being provided, range sheep do not require additional water when soft snow is present. If ice forms a crust on the snow, it must be broken to allow access. Nevertheless, sheep should have unrestricted access to clean, fresh water whenever possible.

### *3.1.1.2. Energy Requirements of Sheep*

A sheep's diet consists mostly of grass and other forages, which can be scarce or of low quality, so it's critical to provide them with enough energy from their food. Even in large quantities, low-quality fodder may not supply enough readily available energy for upkeep and productivity. The first 8–10 weeks of lactation are when ewes need the most energy. Once the lambs start to forage and milk production starts to decrease, the ewe's need is decreased to a level similar to prelambing.

Assessing energy adequacy in sheep is best done by performing an objective 1–5 scoring system, where 1 represents extremely thin and 5 represents extremely obese, and recording body condition. The amount of fat covering on the spinous processes and transverse processes in the lumbar region is palpated to determine the body condition score. A score of 2–3 points is what most healthy, productive ewes will have. To improve their score, sheep with a score of 1–2 should be examined and fed, while those with a score of more than 3 points should receive less food. It is always best to avoid making sudden cuts in total energy intake, especially in the middle to late stages of pregnancy, and to make dietary adjustments gradually (NRC, 1945; Umberger, 2009).

### *3.1.1.3. Protein Requirements of Sheep*

A mature, nongrowing, nonlactating sheep can usually get enough protein from pasture and good-quality fodder. Most sheep require a minimum of 7% of their diet in crude protein for maintenance. The amount of protein needed depends on the stage of production (growth, gestation, lactation, etc.) as well as the existence of specific illnesses (dental disease, internal nematode parasites, etc.). Protein supplements, such as oilseed meals (cottonseed meal, soybean meal), or commercially blended supplements, should be fed to meet nutrient requirements if the available forages are unable to provide an adequate amount of dietary crude protein. Feeding protein should correspond to needs. Overfeeding on protein can be helpful when there are high internal parasite loads, but it also raises production costs and may lead to an increase in disease incidence (e.g., heat stress, pizzle rot).

In the rumen, sheep can convert nitrogen that is not protein (like urea, ammonium phosphate, and biuret) into protein, though maybe not as effectively as beef cattle. In high-energy diets with a nitrogen: sulfur ratio of 10:1, this source of nitrogen can supply at least some of the additional nitrogen that is required. Alfalfa, authorized growth stimulants, and a source of fermentable carbohydrates (ground corn, ground milo, for example) improve nitrogen utilization in lamb-finishing diets.

### 3.1.1.4. Mineral Requirements of Sheep

The major minerals that sheep need are potassium, sodium, chlorine, calcium, phosphorus, magnesium, sulfur, and trace minerals like zinc, selenium, iron, manganese, copper, iodine, and molybdenum. An affordable method of preventing deficiencies in sodium, chlorine, iodine, manganese, cobalt, copper, iron, and zinc is to use trace mineralized salt. In areas lacking in selenium, rations, mineral mixes, or other supplements should contain selenium. Sheep diets typically include enough iron, magnesium, sulfur, manganese, potassium, and sulfur. The most accurate method for determining an ewe's iodine, cobalt, and copper status among the trace minerals is to analyze liver biopsy tissue. When nonhemolyzed blood is carefully collected and put in collection tubes devoid of trace elements, the amount of zinc in the blood can be evaluated. It is simple to determine the selenium status of a patient by drawing whole blood, ideally heparinized.

#### 3.1.1.4.1. Salt

In the United States, sheep should have free access to salt (sodium chloride), except for some alkaline regions of the western range and the coast. For sheep to continue being frugal, gain weight economically, lactate, and procreate, thcy need salt. A maturc sheep will take in about 0.02 pounds (9 grams) of salt per day; lambs will take in half of this. A typical supply of salt/ewe/mo from range operators is 0.5–0.75 lb (225–350 g). Typically, 0.2–0.5 percent of the dry matter in the diet should be made up of salt.

#### 3.1.1.4.2. Calcium and Phosphorus

When it comes to plants, the leafy sections usually have a higher calcium content and a lower phosphorus content, while the seeds or grain have the opposite effect. Generally speaking, legumes contain more calcium than grasses. Phosphorus is transferred to the seed (grain) as grasses mature. Furthermore, the amount of phosphorus in the soil has a significant impact on the amount of phosphorus in the plant. As a result, poor pasture that lacks legumes and range plants typically has a low phosphorus content by nature, especially as the forage ages and the seeds fall.

Sheep that graze on mature, brown summer fodder and winter range occasionally experience phosphorus shortages. A phosphorus supplement, such as defluorinated rock phosphate, should be given to sheep kept on such forages or fed low-quality hay without grain in addition to a salt-trace mineral mixture. The majority of forages have a relatively high calcium content, especially when mixed with other legumes, so diets typically satisfy the element's maintenance requirements. But if corn silage or other feed made entirely of cereal grains is the only feed given, ground limestone should be given every day at a rate of 0–02–0–03 lb (9–14 g). As long as their diets contain more calcium than phosphorus, sheep appear to be able to tolerate a wide range of calcium:phosphorus ratios. On the other hand, osteodystrophy or the formation of urinary calculi may be facilitated by an excess of phosphorus. For feedlot lambs, a calcium:phosphorus ratio of 1:5 is suitable. ≥0.18 percent should be included in the diet for pregnant ewes and ≥0.27 percent for nursing ewes. Calcium contents between 0.2 and 0.4 percent are deemed sufficient as long as the ratio stays between 1:1 and 2:1.

### *3.1.1.4.3. Iodine*

Sheep occasionally require iodine supplements because their natural diet does not provide enough of it. Many plant species, including Brassica spp., contain goitrogenic substances that prevent the thyroid from using iodine as it should. There are naturally deficient regions in the western United States, the Great Lakes region, and other parts of the globe. Pregnant ewes can avoid iodine deficiencies by feeding stabilized iodized salt to them. Iodine deficiencies can manifest as goiter in adults and lack of wool and/or goiter in lambs. Iodine-deficient ewes may have stillbirths, abortions, or goiters in their young. Iodine levels in diets between 0.2 and 0.8 parts per million (ppm) are typically adequate for animals in various stages of production (growth, lactation, maintenance, etc.) (NRC, 1945; Umberger, 2009).

### *3.1.1.4.4. Cobalt*

Sheep need about 0–1 parts per million of cobalt in their diet. Although they exist in North America, cobalt-deficient soils are uncommon there in comparison to other regions of the world. Legumes typically contain more than grasses do. Feeding trace mineralized salt containing cobalt is a good practice because feedstuff cobalt levels are rarely known.

### *3.1.1.4.5. Copper*

When the forage has ≥5 ppm of copper (Cu), it provides the ~5 mg of copper (Cu) that pregnant ewes need daily. However, other dietary components, particularly molybdenum (Mo), iron, and inorganic sulfate have an impact on the amount of copper required in the diet to prevent copper deficiency. Copper requirements rise with high molybdenum intake when sufficient sulfate is present. Sheep are more vulnerable to copper toxicity than cattle, so it's important to avoid consuming too much of it. Lambs given diets containing 10–20 ppm of copper may develop toxicity, especially if the Cu:Mo ratio is greater than 10:1. It is recommended to maintain the Cu:Mo ratio between 5:1 and 10:1.

### 3.1.1.4.6. Selenium

Nutritional muscular dystrophy can be at least partially controlled by selenium. Selenium levels seem to be low in the northwest of the United States and east of the Mississippi River. It is approximately 0.3 parts per million. If animals are given unrestricted access to a mineral mixture containing selenium, selenium deficiency may be avoided. Toxic levels could be present at or above 7–10 ppm.

### 3.1.1.4.7. Zinc

In dry matter, growing lambs need about 30 parts per million of zinc in their diet. It is a little bit more for normal testicular development. Although it is more common in other small ruminants (goats, llamas, and alpacas), classic zinc deficiency (parakeratosis) can also occur in sheep, especially if they are fed an excessive amount of calcium-rich diet (legumes).

## 3.1.1.5. Vitamin Requirements of Sheep

Sheep diets typically have high levels of provitamin A (vitamin A), D, and E. However, there are situations in which supplements might be required. Under normal circumstances, supplements are not required because the rumen microorganisms synthesize vitamin K and B vitamins. However, polioencephalomalacia is evident and results from abnormalities in the metabolism of thiamine in the rumen, which are secondary to changes in the ruminal pH and/or microflora content. The tissues of sheep are where vitamin C is synthesized. Sheep can store a significant amount of vitamin A in their livers when fed diets high in carotene, such as premium pasture or green hays. This liver storage is frequently adequate to meet the sheep's needs for up to six months.

Sun-cured forage provides vitamin $D_2$, while skin exposure to UV light provides vitamin $D_3$. The amount supplied may not be sufficient if the skin is not exposed to sunlight due to extended cloud cover or confinement rearing, or if the diet is low in vitamin $D_2$. When the diet contains low levels of calcium or phosphorus, or when the ratio between the two is large, the body needs more vitamin D. Dietary changes should be made gradually and carefully to prevent vitamin D toxicity. During the winter months, fast-growing lambs kept out of direct sunlight or fed green forages high in carotene (low irradiation) may exhibit symptoms of vitamin D insufficiency. Sheep on pasture typically don't require vitamin D supplements.

Green feeds and seed germs are the main sources of vitamin E in sheep's natural diet. Vitamin E must be consumed daily because the body does not store it well. Supplemental vitamin E may improve production, lamb weaning weights and colostrum quality in ewes fed subpar hay or forage. If selenium intake is low, vitamin E deficiency in young lambs may contribute to nutritional muscular dystrophy (NRC, 1945; Umberger, 2009).

### 3.1.2. Feeding Practices in Sheep

#### *3.1.2.1. Feeding Farm Sheep*

High-quality roughage stored as hay, low-moisture grass-legume silage, or occasionally chopped green feed is excellent for sheep. Regardless of quantity, low-quality forage is only good for maintenance; high-quality hay or stored forage is a very productive feed. Hay quality is determined primarily by the following:

- its composition, e.g., a mixture of grasses and legumes such as brome/alfalfa or bluegrass/clover
- the stage of maturity when cut, e.g., the grass before heading and alfalfa before one-tenth bloom
- method and speed of harvesting due to loss of leaf, bleaching by sun, and leaching by rain
- spoilage and loss during storage and feeding

Generally speaking, the same elements affect silage quality. A comprehensive examination of chopped and stored forages improves the use of these feed materials and enables the best possible use of added grains and minerals.

#### *3.1.2.2. Feeding Ewes*

If a high twinning rate is desired, the period from ewe weaning to breeding is crucial. During this time, ewes should steadily gain weight without becoming overly obese. At breeding, ewes should weigh approximately 60–70% of their predicted mature weight, and at lambing, they should reach 80–90% of their projected mature weight, with a body condition score of 2.5–3 out of 5, depending on their rate of gain. If pasture production is insufficient, ewes may be confined and fed premium hay along with a small amount of grain. Breeding sheep on pastures with legumes (such as sage and white clovers) may reduce the number of lambs produced and limit their consumption of certain feedstuffs. After mating, ewes can remain on pasture, which saves feed for later in the year. If provided with good-quality pasture during this period, ewes will enter the winter-feeding season in good condition. If pasture is unavailable, a suitable ration must be prepared. The last 6–8 weeks of pregnancy are when the fetus grows at its fastest rate. Nutritionally, this is a crucial time, especially for ewes carrying multiple fetuses. The plane of nutrition should be gradually increased starting 6–8 weeks before lambing and should continue uninterrupted until after lambing. The quantity given is determined by the forage's quality and the ewes' state or fat covering. Ewes in fair to good condition should typically receive 0.5–0.75 lb (225–350 g) per day. All nonlactating ewes should receive all the protein they need from the ration's roughage content. The ewes may be divided into groups for distinct treatments based on the number of fetuses, age, and condition, if necessary.

#### *3.1.2.2.1. Nutrition Requirements of Lactating Ewes*

Ewes and lambs get all the energy, protein, vitamins, and minerals they need from succulent pasture; no additional grain is needed. When confinement rearing is not being utilized, ewes should be fed one of the rations listed in Rations for Pregnant Ewes up to 6 Weeks before Lambing. They should also be fed 1–1.5% lb (450–675 g) of one of the grain mixtures listed in Grain Mixture for Pregnant Ewes. Access to a combination of dicalcium phosphate and trace mineralized salt is necessary for ewes. Separating ewes bearing twins or triplets from those bearing singles is important. The latter group should be fed higher concentrations of grain or superior-quality forages. Compared to ewes caring for single lambs, those nursing twin lambs yield 20–40% more milk. Weaning of lambs at two months of age is possible under confinement rearing or accelerated lambing. After this point, the ewe's milk production rapidly decreases, and feeding creep feed to lambs instead of her results in more efficient weight gain.

### *3.1.2.3. Feeding Lambs*

Lambs should have free access to creep feed starting at about two weeks of age. They should be creep-fed for a period of one to two months in areas with limited pasture until sufficient forages are available. The lambs can be finished in a dry lot if pasture won't be available until they are three to four months old. Grain should be rolled or coarsely ground, but whole grains can be used as the feeding time goes on. The lambs' diet should be gradually expanded to include small amounts of clean, fresh grain. Grain is added progressively until the lambs are receiving their full diet.

In the USA, it has grown more common to feed lambs in a dry lot from birth to market, along with the practice of early weaning at 2–3 months of age. Hay, grain, and a supplement containing vitamins and minerals are ground, combined, and fed either unprocessed or pressed into 3/16- or 3/8-inch pellets. (5–10 mm) in length. These lambs typically attain market weight in 3.5–4 months.

### *3.1.2.3.1. Rearing Lambs on Milk Replacer*

To increase productivity, lambs that are orphaned, extras, triplets, or from ewes that don't milk well enough can be raised on milk replacers. During the first 18 to 24 hours of life, these lambs should be fed 10 to 20% of their body weight in colostrum in several feedings. If ewe colostrum is not available, a frozen mixture of colostrum from multiple cows can be utilized instead. There are milk substitutes available that are made especially for lambs; they have a high concentration of antibiotics, 25% protein, and about 30% fat. It might be wise to give orphaned lambs injections of vitamins A, D, and E as well as selenium in certain situations. While high-quality replacers meant for calves can be fed to lambs, ewe milk replacers are preferred in hand-rearing systems. Making sure the powder and water are properly mixed into a suspension is important when preparing milk substitutes. Feeding small amounts over several feedings lowers the risk of diarrhea and/or bloat. During the first week of life, milk replacers should be given to lambs at a rate of 10–20% of their body weight, split into 4–6 feedings per day. By 3 to 4 weeks of age, the frequency of feedings can be gradually decreased to twice daily.

You can use multiple-nipple pails or containers. Older lambs that nurse more often can use cold milk replacer. If a creep ration is provided, lambs should be given water in addition to milk by the time they are 9–10 days old. At 4–5 weeks of age, they can be abruptly weaned provided they are consuming a reasonable amount of water and creep feed.

### *3.1.2.3.2. Finishing Feeder Lambs*

Before lambs are allowed to leave the producer's property, they should be preconditioned. This covers beginning feeding, immunizing, deworming, and, in certain cases, shearing. If this isn't done, the lambs should be brought to the feedlot, rested for a few days, and fed dry, mediocre hay. For lambs to be finished, there is no ideal technique or diet. They don't need extra grain to finish on good to excellent quality forage, such as wheat or alfalfa. When the forage is depleted, they can be switched to grain-feeding systems from pasture or crop residue. They are typically given free access to feedstuffs when fed in a dry lot. These diets can be high-concentrate, pelleted, ground and mixed, or a combination of grain and ground forage (alfalfa) pellets. Self-feeding typically yields the highest possible feed intake and gain at the lowest possible labor costs. An auger system or self-unloading wagon can automate the hand-feeding process. It entails feeding the lambs at regular intervals so that they finish all of the food before receiving more. Gain and feed consumption are controllable. Corn silage should be fed by hand when in use to reduce spoilage.

Lambs that are fed year-round or heavily by producers typically want to start full feed as soon as feasible (10–14 days). Lambs can be started on diets that contain 60–70% hay safely, either by ground or pelleted feed. If the ration is not pelleted, the amount of hay can be decreased to 30–40% in just two weeks. You can use silage or other roughages like cottonseed hulls in a similar way. With hand-feeding, about half of the hay can be replaced with corn, sorghum, or alfalfa silage; however, finish and yield will be somewhat reduced. When using corn, barley, milo, wheat, or a combination of these, add 0.5 percent salt and 0.5 percent bone meal, or its equivalent, to the grain. When using high-roughage or low-grade roughages in finishing lamb rations, pelleting is advantageous. When feeding large quantities of wheat, caution should be exercised because lambs that are not acclimated to it are more likely to experience acute dyspepsia than if they are fed grains like corn, sorghum, or barley. Salt and other mineral supplements ought to be provided independently of the grain mixture, regardless of their inclusion in it.

### *3.1.2.4. Feeding Mature Breeding Rams*

Mature breeding rams should be fed rations 1, 2, or 3 following the guidelines for rations for pregnant ewes up to six weeks before lambing, or they can graze on pasture when available. It should not be necessary to grain-feed the rams while they are with the ewes if the rams are in frugal condition at breeding time and the ewes are on a good flushing pasture. Rams should be maintained at a good body condition (3–3.5 on a 1–5 scale) before the breeding season.

### *3.1.2.5. Feeding Range Sheep*

The type and quantity of supplement to feed depends on a sheep's body condition, the amount and kind of forage available on the range, and the climate. High-protein pellets, such as cottonseed meal and salt, medium-protein pellets, low-protein pellets, such as corn, alfalfa hay, and minerals, are the typical components of supplements. Proper supplementation of the diets of sheep in the western winter range can increase the production of wool by approximately 1 lb (400–500 g) per ewe and increase the lamb crop by 10%–15%.

Feeding approximately 0. 25 lb (115 g) of high-protein (36%) supplement or 0.33–0.5 lb (150–225 g) of medium-protein (24%) pellets is one suggested practice. This should be done approximately three weeks before and during the breeding season, during extremely cold weather, and for approximately one month before green feed starts in the spring. Furthermore, old ewes with poor teeth, thin ewes, small lambs, and yearling ewes should be fed one of the above supplements after being separated from the main flock, roughly from December 1 (northern hemisphere) until shearing time. For a particular dietary supplement, it is frequently possible to keep older ewes, lambs, and yearlings in a flock. If heavy snowfall or other weather prevents sheep from getting their full ration of forage, they should be fed 1–3 pounds (450–1,350 g) of alfalfa hay and 0.2–0.3 pounds (90–150 g) of corn or a low-protein pellet mixture. When alfalfa hay is unavailable, emergency feeding periods should be used to feed 0.5–1 lb (225–450 g) of a low-protein pellet mixture per head each day (NRC, 1945; Umberger, 2009).

### *3.1.2.5.1. Deficiencies of Range Forages in Sheep*

The dietary components of protein, energy, salt, and phosphorus are linked to deficiencies that are most likely to be observed in range forages. When the forages get closer to maturity or go dormant, these become more common. While these diet-related issues can occur alone, combinations of them are more typical. Range sheep need more energy because they travel greater distances and are exposed to colder climates more frequently. Alfalfa pellets, soybean or cottonseed meal, and other protein supplements improve the digestibility and utilization of subpar forages. Productivity can be significantly increased by adding phosphorus supplements, such as dicalcium phosphate, monocalcium phosphate, or defluorinated rock phosphate, to a salt or trace mineral salt mixture when it is feasible.

The majority of winter grazing areas are deemed sufficient in carotene, as numerous browse species provide an equivalent amount of carotene as sun-dried alfalfa hay. Nevertheless, if sheep are forced to graze dry grass ranges for more than 6 months without regular breaks for 'green' pastures, vitamin A supplements are advised. Supplementing with around 45–50 International Units of vitamin A per kilogram per day can enhance productivity during prolonged consumption (>2 months) of dry or deteriorated forages.

### *3.1.2.5.2. Mineral Mixtures for Sheep*

Sheep find portable mineral boxes handy when they're out on the range. It is recommended to offer a choice of one of these mineral mixtures. If there are no iodine or trace-mineral deficiencies, a salt and dicalcium phosphate or phosphorus supplement (defluorinated rock phosphate) can be used. When there is an iodine deficiency, iodized salt is used instead of regular salt; if there is a trace mineral deficiency, trace mineralized salt is used instead of regular salt.

The type of range forage available, feeding frequency, and pellet composition all affect how much phosphorus supplement should be added to range pellets under winter range conditions. According to recommendations, the phosphorus content of protein pellets should be 36%, 24%, and 12%, respectively. When feeding supplemental protein, 36% protein pellets should be fed at the rate of 0.25 lb (115 g) per head daily, the 24% protein pellets at 0.33–0.5 lb (150–225 g), and the 12% protein pellets at 0.2–0.5 lb (90–225 g), together with alfalfa or clover hay. Care should be taken when adding supplemental phosphorus and magnesium to the diet of rams or whethers, because this is associated with urolithiasis.

### 3.1.3. Nutrition-Related Diseases of Sheep

Diseases related to nutrition are some of the most common and most preventable in a sheep flock:

- **Pregnancy Toxemia**: A negative energy balance causes a condition known as hypoglycemia and ketosis, which affects ewes in their late gestations. Pregnant animals, whether thin or fat, that are fed inadequate or low-quality feed, have had their access to feed restricted, or have gone through a stressful period that has limited their feed intake, are more likely to develop toxemia than ewes carrying multiple fetuses. Reduced appetite, hesitancy to stand, and limping are some of the early symptoms of toxemia. Many ewes have this condition unnoticed until they are anorectic and recumbent. Even with the best veterinary care, ewes and lambs rarely survive when clinical signs get this bad.
- **Hypocalcemia**: When late-gestation or early-lactation ewes are fed a diet deficient in calcium, their bodies are unable to meet the demands of growing fetal bones and producing milk, which leads to hypocalcemia. Lethargy and an inability to stand are examples of clinical signs that can resemble those of pregnancy toxemia. On the other hand, giving calcium can be a simple way to treat hypocalcemia. If this condition is detected and treated in a timely manner, both ewes and fetuses frequently survive.
- **Lactic Acidosis**: The reason behind the suppression of fiber fermentation and cytolytic activity in the rumen is the overabundance of starch-digesting microorganisms due to sudden and significant increases in dietary sugar and starches. The GI tract lining is further harmed by the rapid pH decline of rumen and the accumulation of lactic acid. The clinical manifestations include bloating, diarrhea, isolation, and finally recumbency. The type and amount of concentrated foods consumed have a direct impact on how severe the illness is. Even with intensive care, the prognosis is directly correlated with the severity of the disease.

- ***Clostridium perfringens* Type C Infection**: Very young lambs experience an overgrowth of *C.perfringens* type C as a result of ingesting large milk meals from incorrect bottle-feeding, the ewe producing a lot of milk, or having more access to milk as a result of losing a sibling. These lambs get drowsy, refuse to eat, and frequently pass away. By giving ewes the *C.perfringens* type C and D vaccine beforeparturition, making sure lambs eat enough high-quality colostrum, and following proper bottle-feeding procedures, the infection can be avoided. Injections of *C.perfringens* type C and D antiserum should be part of the treatment.
- ***Clostridium Perfingens* Type D Infection (Enterotoxemia)**: It can cause sudden death in older, fast-growing lambs fed a high-sugar/high-starch diet or transferred to rich pastures because it is caused by an overgrowth of *C. perfringens* type D. Although they are difficult to save, lambs may be discovered alive, particularly if they have received prior vaccinations. As part of a treatment plan, injections of *C.perfringens* type C and D antiserum may be given. Prevention can be aided by giving the late-pregnant ewe and growing lamb the *C. perfringens* type C and D vaccine.
- **Urolithiasis**: In male ruminants, the development of urinary calculi can be life-threatening. Urinary tract calculi can cause painful bladder distention, bladder rupture, and uremic poisoning within days. While early castration and genetic susceptibility are often cited as causes, diet is the most critical risk-reduction strategy. To lower the risk of calculus formation, ram lambs, wethers, and rams should be fed a diet low in magnesium, with a calcium-to-phosphorus ratio of about 2:1. Minimizing grain consumption is recommended to maintain low phosphorus concentrations and dilute the urine. For certain types of calculi, ammonium chloride may help acidify the urine and facilitate dissolution, although not all calculi will respond. In cases of single calculi, removing the urethral process may be effective, but severe obstructions often require surgical procedures such as perineal urethrostomy or tube cystotomy.
- **Polioencephalomalacia**: This is a central nervous system (CNS) illness that can cause abnormal vocalization, ataxia, incoordination, or blindness. Although the exact cause of the illness is unclear, high-sulfur diets have been linked to the disease in some cases. Additionally, metabolites produced during gastrointestinal (GI) disorders may play a role. Both subclinical and clinical episodes of lactic acidosis can also lead to this condition. Treatment often

involves administering thiamine and flunixin meglumine, along with supportive care, which is frequently successful.

- **White Muscle Disease**: This degenerative muscle disease is brought on by a selenium and/or vitamin E deficiency. It frequently affects young, rapidly growing lambs, causing them to become weak, unable to stand up, or stiff, particularly in their hind limbs. It affects both skeletal and cardiac muscle. It is possible for acute heart failure to cause sudden death. Treatment with selenium and vitamin E may not totally resolve clinical signs, so prevention through appropriate mineral and vitamin supplementation is preferred.

### 3.1.4. Breeding, Parturition, and Common Lamb Diseases of Sheep

Pre-breeding ewes and rams should have comprehensive physical examinations. To make sure that every animal has sufficient reserves to meet the demands of breeding and, for ewes, parturition and lactation, body condition scores should be assessed. A body condition score of 3.5/5 is ideal on average. It is important to identify both thin and obese animals for possible diet modification and treatment. Examining the teeth, feet, and range of motion is important. It is not a good idea to breed animals that have trouble eating or walking. It is important to check conjunctival membranes for possible complications from GI parasites as well as clinical indicators of anemia.

Rams' scrotums and ewes' udders should be inspected. It is recommended to palpate both to check for any clinical signs of illness or injury, such as consistency, asymmetry, or inflammation. Because there is a direct correlation between scrotal circumference and reproductive health and capacity, scrotums should be measured. Sheaths should be checked as well, especially in males who are heavily fleeced and consume a lot of protein. When urine gets caught in the wool, it can irritate the preputial orifice and cause scabbing, which can hurt and make breeding difficult. Producers should only choose the best members of their flock to breed; breeding stock selection is an often-overlooked aspect of flock management. Breeding pools should not include animals with poor footwork, udder or scrotal defects, poor conformation, unthrifty phenotypes, or histories of rectal or vaginal prolapses. Retaining animals with proper conformation and robust production traits, like resistance to parasites or adept mothering, is recommended. While it may seem counterintuitive to breed fewer animals, the flock will be healthier and more productive in the long run if care is taken to choose the best stock for reproduction. If necessary, vaccinations against infections caused by Campylobacter and Chlamydia should be given in accordance with the manufacturer's instructions, usually 60–90 days before the introduction of rams. Flushing: Two to four weeks before breeding, ewes should start receiving a high-energy diet to increase their ovulation rate. When ewes were given an extra 5 kg of dry matter per head per day, ovulation rates increased by 25%, according to a study conducted in New Zealand. As long as the dry matter contains noticeably more protein than the maintenance diet, it can be either grain or forage. Flushing is most effective in thin animals or those fed a low-protein diet. In ewes with above-average body condition scores on a medium to high nutritional plane, ovulation rate changes might not be apparent. Sheep are short-day, seasonally polyestrous breeders whose seasonality varies by breed. Because breeding seasons are longer in the equator, geographic location can also have an impact on breeding. Estrus lasts 30 hours on average during the 17-day sheep estrous cycle. Ewes should weigh between 60 and 65% of their mature body weight (or their dam's mature body weight) before breeding, even though they reach puberty at 6 to 9 months of age.

Ewes exhibit extremely mild increases in mucus secretion and vulva enlargement as clinical indicators of estrus. Mixing rams and ewes for several estrous cycles is the most popular breeding technique used by producers because it can be challenging to determine whether an ewe is in heat. A ram-to-ewe ratio of 1:40 on average provides enough coverage in a short amount of time. More seasoned rams might benefit from a higher ratio, while less experienced ram lambs might require a lower one. It has been reported that ratios of 1:150 work well in very large flocks of grazing animals. If more than one ram is being used, visual guides like labeling harnesses can be used to identify the breeding date, the sire, and unsettling ewes. If rams and ewes are kept apart before mingling, ewes can naturally synchronize through the "ram effect." Artificial insemination and estrous cycle manipulation techniques are well-developed for sheep and are commonly employed in flocks under intensive management.

Transabdominal ultrasound and serum testing can both confirm pregnancy. You can use both techniques as soon as 30 days following breeding. Fetuses can be counted using ultrasonography, which can help with more effective dietary management in the later stages of pregnancy. Forty to seventy days after breeding is the ideal window of time to count fetuses.

Sheep have an average gestation period of 147 days. Ewes that are expecting require both exercise and a healthy diet. Ewes carrying more than one fetus require a proportionately higher energy intake than do ewes that are not pregnant or carrying only one fetus. It is best to stay away from both obesity and emaciation. In order to meet the demands of rapid fetal growth, a pregnant ewe's plane of nutrition should increase during the final trimester of her pregnancy. Pregnancy toxemia and hypocalcemia may result from inadequate feeding of ewes during this period. In order to prevent lactic acidosis, ewes that need concentrates in the late stages of pregnancy should be handled carefully and given sugars and starches gradually rather than all at once. The vaccination should be given to ewes 4–6 weeks before parturition to boost the amount of antibodies in the colostrum. The most often given vaccines at this stage of pregnancy are clostridial ones, such as CDT (the vaccination against *C. perfringens* types C and D) and C tetani. In order to provide lambs with easy access to the udder shortly after birth, ewes may choose to undergo partial or total shorning around their udders and hindquarters, a process known as crutching.

The development of the udder, ewe behavior changes (isolation, vocalization, nesting), and vulvar discharge can all be signs that parturition is about to begin. For lambing, ewes must be taken out of the flock and put in a hygienic, dry, warm, and draft-free area. Ewes should be watched over and disturbed as little as possible because they can suppress labor when under stress. Within an hour of the first lamb's contractions, the abdomen should become very tight. Correcting a malpositioned lamb should be done with lots of obstetric lubricant and clean hands, gloves, or sleeves. The lamb should be handled carefully; don't pull it too hard. The placenta should pass during the course of the following twelve hours after every lamb has been delivered.

To reduce the chance of navel illness in newborn lambs, their navels should be dipped in 7% strong iodine (explained in more detail below). After the navel has been quickly dried off, it should be checked for congenital abnormalities such as cryptorchidism, umbilical hernias, and entropion. Drying and warming lambs can prevent many deaths because hypothermia is a major cause of death for neonates, particularly in colder climates. Within thirty minutes of birth, lambs should be able to stand and nurse. Before being incorporated into the larger flock, ewes and lambs should bond on their own for two to three days. It is possible to use individual pens (jugs), but they must be big enough (1.5% square meters) to keep the lambs from being crushed when the ewe lies down. The birth weight of a lamb can be used as a tool to track general health and productivity targets. During the first few weeks of life, any lamb that fails to gain weight or loses weight should have a medical evaluation conducted.

For the first 24 hours after birth, lambs that are rejected or need extra attention should be bottle-fed at least 10% of their body weight in premium colostrum. Insufficient colostrum increases the risk of infection and mortality in lambs. Following the first feeding of colostrum, lambs should receive three feedings of milk or milk replacer, each at a rate of 10% of their body weight per day. Lambs that are bottle-fed should get their milk from as many different sources as possible. Lambs are more susceptible to illness when their GI environment is frequently disrupted. To stop milk-borne diseases from infecting lambs, any fresh colostrum or milk should be heat-treated, pasteurized, or sourced from animals free of disease. Within a few days of birth, young lambs can be given access to hay and grain-based creep feed. When solid food makes up the majority of a lamb's diet, it's time to wean them. Lambs are usually not ready to be weaned before they are six weeks old. A premium diet should be given to weaned lambs to supply them with enough energy for growth (NRC, 1945; Umberger, 2009).

Lambs should receive a clostridial vaccine at birth and thereafter according to the manufacturer's instructions if ewes were not vaccinated before parturition. After the ewe-lamb bond has been established, castration and tail docking should be performed within the first 7–10 days of life. Although anesthetic use is recommended for older lambs, these procedures can still be carried out on them. Tails should be docked at a minimum length that leaves the distal ends of the caudal tail folds intact.

Diseases related to breeding and parturition include nutritionally related diseases and common lamb diseases, such as:

- An ascending umbilicus infection known as "navel ill" is frequently observed in lambs born in unclean, damp, crowded environments. Clinical signs of this disease include a swollen umbilicus, swollen joints, a hesitancy to move, and a failure to gain weight. The risk of navel illness can be reduced by immediately sealing and drying the umbilicus by dipping it in 7% iodine; nevertheless, a clean birthing and jugging area is crucial. This illness can resemble erysipelas arthritis. Antimicrobial therapy is necessary.
- The intestinal protozoa known as coccidia, which are frequently found in the environment and are carried and excreted in the feces of subclinical adults, cause coccidiosis. Coccidia should be tested for in thin lambs, those with diarrhea, and those that pass away unexpectedly at 3 weeks or less. Coccidiosis is especially dangerous in crowded flocks where it is hard to avoid the accumulation of feces. From an early age, lambs consume coccidian oocysts through both the ground and the contaminated teats of their dam. Lambs that are ill tend to gain weight slowly and may even die from thirst. The secret to managing this illness is having clean housing. Drugs like amprolium, toltrazuril, and sulfadimethoxine may be administered as part of the treatment. Lambs can be given scoclidomas, like decoquinate, in milk or feed to manage the illness, as long as their feed intake is adequate to produce therapeutic drug dosages.
- A significant number of lamb deaths occur each year from two management-dependent diseases: hypothermia and starvation. The

secret to keeping young lambs alive in cold, wet environments is to house them in a warm, draft-free, bedded area that is shielded from precipitation. Coats have additional advantages. Always check for starvation in lambs that are shivering, thin, or hunched over. Starvation can be brought on by mismothering, untreated mastitis, or sibling dominance at the teat.

- The ORF virus causes contagious ecthyma, also known as "sore mouth," a skin condition that causes painful lesions on ewes' teats and around lambs' mouths. The virus by itself is not lethal, but the lesions can make it difficult to eat, which can result in starvation and anorexia. The ORF virus can spread through intimate contact as well as through fomites. Although the virus can survive in the soil for years, cases typically resolve on their own within a month or two. Since the sore mouth vaccine is a live vaccine, it shouldn't be given unless the illness is already present on the property.
- There are several reasons why sheep flocks experience abortion storms. Common pathogens include Chlamydia, Listeria, Toxoplasma, and Campylobacter. It can be difficult to stop or slow down a storm once it has started. A precise diagnosis is the foundation for both prevention and treatment plans. The placcnta is frequently the greatest tissue to submit for diagnostic analysis; it is often even better than fetal tissue.

### 3.1.5. Additional Common Diseases Associated with Management Practices of Sheep

#### *3.1.5.1. Internal Parasites (Helminths) of Sheep*

Most parts of the world's pastured and free-ranging sheep suffer from helminthiasis, a clinically significant disease caused by internal parasites called helminths. Lungworms, nematodes, and flukes can all have a negative impact on sheep and result in severe production losses. Even though tapeworms don't usually cause illness, discussing parasitic infections with producers and creating a program for the flock to control them can be facilitated by the fact that tapeworms are easily seen. All the disease complications associated with parasites can affect hobby flocks that are permitted to graze, just like they can large production flocks. Clinical manifestations of helminthiasis that are common include diarrhea, submandibular edema (bottle jaw), anemia, poor growth, and weight loss. Specific clinical symptoms, like coughing or bacterial pneumonia, can be caused by specific parasites, such as lungworms.

In a flock, factors such as stocking density and environmental conditions affect the quantity and severity of parasite-related issues. While flocks grazing in arid or semiarid environments may experience minimal production losses due to helminths, overcrowded flocks in warm, moist conditions frequently struggle with severe helminthiasis.

Drug efficacy has decreased and parasite resistance has increased as a result of overuse of anthelmintic medications. According to a 2008 study, Haemonchus contortus infections on 48% of small ruminant holdings in the Southeast of the United States were found to be resistant to all commercially available anthelmintic medications. The current paradigm for treating helminth infections in sheep flocks is to: (i) treat only when sheep exhibit clinical signs of disease; (ii) assess drug efficacy using fecal egg courts; (iii) reduce exposure risk through pasture and nutrition management; and (iv) select for sheep that are genetically resistant to parasites. This approach is due to the loss of these drugs for treatment. Most of the parasite eggs are frequently shed by a small number of sheep heads. It is also advisable to identify and remove sheep that shed a lot. The severity of an *H. contortus* infection can be determined by utilizing the FAMACHA scoring system to evaluate anemia. It is an inexpensive, proven method of keeping an eye out for certain internal parasites.

### *3.1.5.2. External Parasites of Sheep*

Sheep frequently contract lice, mites, and keds (*Melophagus ovinus*), three common external parasites. These organisms rarely cause serious illness, but if an infestation gets bad enough, it can lead to anemia and debility. Alopecia, pruritus, and the organisms' obvious presence on the skin and in the hair are clinical indicators of infestation.

### *3.1.5.3. Ovine Progressive Pneumonia*

Maedi-visna, also known as ovine progressive pneumonia, is a lentiviral illness that is believed to impact more than 25% of sheep in North America. It is connected to encephalitis and caprine arthritis. Clinical signs include hard, unproductive udders, breathing difficulties, exercise intolerance (lagging behind the flock), and progressive, long-term weight loss. Sheep carry the disease for life and can spread it both horizontally and vertically. In a flock, serum tests can be used to assess the prevalence of ovine progressive pneumonia.

### *3.1.5.4. Paratuberculosis (Johne's Disease) of Sheep*

*M. aviumMycobacteriuM. avium* paratuberculosis is the causative agent of paratuberculosis, a bacterial infection of the intestinal wall. Both the

S and C strains of the bacterium can infect sheep. Sheep contract *M. avium* paratuberculosis through the fecal-oral route, usually when they are very young. The disease is excreted in the feces. The pathogen can also spread through milk and colostrum, as well as in utero. Clinical symptoms may not appear until the animals are much older and include weakness, submandibular edema, and weight loss. Compared to cattle, sheep do not experience diarrhea as frequently. The bacterium that causes Johne's disease can survive in the soil for years, making environmental decontamination challenging. The disease cannot be cured. It is advised to use fecal PCR testing to locate and eliminate diseased animals. Necropsies are advised for animals exhibiting persistent weight loss to check for thicker intestines with noticeable lymph vessels and enlarged mesenteric lymph nodes. Under a microscope, bacteria that stain acid-fast are visible in the infected tissue.

### 3.1.5.5. Caseous Lymphadenitis (Cheesy Gland) of Sheep

A lymph node infection brought on by Corynebacterium pseudotuberculosis is known as caseous lymphadenitis. At any age, the most common way for an infection to spread is through skin breaks that come into direct contact with pus or contaminated objects like shears and feeders. There are two variations of the illness: an exterior form and an internal form. Internal abscesses can cause chronic weight loss and poor production, but they are difficult to diagnose before death. When they burst, external abscesses—which originate in the superficial lymph nodes—pollute the surrounding area with copious amounts of infectious pus. In the soil, organisms can survive for months without treatment, and treatment is not thought to be effective. Animals with visible enlarged lymph nodes or abscesses shouldn't be added to the flock; external abscesses should be cultured to confirm the illness.

### 3.1.5.6. Additional Clostridial Diseases of Sheep

In sheep, castration and tail docking are the most common procedures that result in tetanus due to Clostridium tetani infection, especially if elastic bands are utilized. Tetanus shots should be given to lambs either before or right away following these procedures. In rams, *Clostridium novyi* is the cause of "big head." Head butting causes tissue damage that becomes infected, causing pain and swelling. A vaccination can prevent tetanus, but the CDT vaccines that are frequently given do not contain it (NRC, 1945; Umberger, 2009).

## 3.2. NUTRITION IN GOATS

Goats and sheep are similar animals in many aspects, but they have different nutritional needs. Although they will graze, goats are primarily browsing animals. Nonetheless, a lot of the nutritional and feeding guidelines for sheep also apply to goats. Body condition assessment should be used for nutritional evaluation, especially when it comes to energy intake.

## 3.2.1. Nutritional Requirements of Goats

### *3.2.1.1. Water Requirements of Goats*

Water that is clean, fresh, and easily accessible should be available to goats without restriction. Even though goats are among the most water-efficient domestic animals, only about 10% of body water loss can be fatal. Although they should be housed in areas with shade, they seem to be less susceptible to extreme heat stress than other domestic livestock species. Goats can prevent bodily water losses through reduced urine and fecal losses, in addition to having a reduced requirement for body water evaporation to stay comfortable in hot weather. Goat water intake is influenced by several factors, including lactation, temperature, the amount of water in the forage they eat, exercise level, stage of production (growth, maintenance, lactation, etc.), and the amount of salt and minerals in their diet. Compared to goats fed dry hay, those grazing lush pastures may use significantly less water. Nevertheless, regardless of age, breed, purpose, life cycle stage, or habitat, it is essential to provide unrestricted access to water for all goats.

### *3.2.1.2. Energy Requirements of Goats*

Energy limitations can be caused by inadequate feed intake, poor diet quality, or an excessive amount of water in the feedstuffs. Age, body size, body condition, stage of production (growth, maintenance, pregnancy, and lactation), and coexisting medical conditions (e.g., parasitism, dental disease, arthritis) all have an impact on energy requirements. The environment, hair growth, activity level, and interaction with other nutrients in the diet can all have an impact on energy needs. Elevations in temperature, humidity, sunshine, and wind speed have the potential to reduce energy needs. Shearing cashmere and Angora goats' pashmina and mohair increases energy requirements and reduces insulation.

Goat grazing activity varies widely; it can be light for goats under intensive management, moderate for goats grazing on semiarid land, or high for goats grazing on mountain pastures that require daily long-distance travel and sparsely vegetated grassland. The best indicator of a goat's adequate dietary intake is good body condition, or the amount of fat on the ribs, inner thighs, loin, and brisket. A standardized body condition score (1–5, where 1 is extremely thin and 5 is extremely obese) should be used to track changes in body fat and help make less arbitrary decisions about long-term dietary energy adequacy using herd/individual medical record systems. Goat body condition evaluation is covered in more detail by Michigan State University Extension. Animals that are underconditioned but free of parasites and diseases are typically fed a diet low in energy; the opposite is true for obese animals. The amounts of energy needed for lactation and growth are fairly similar to those needed for sheep and cattle, respectively. Therefore, when it comes to energy, the principles of sheep nutrition will likely be sufficient for all classes of goats, except dairy goats that are nursing.

### 3.2.1.3. Protein Requirements of Goats

The majority of the body's regular processes, such as immune system function, growth, reproduction, lactation, and maintenance, require protein. Lack of protein in the diet causes the blood, liver, and muscles to lose its stores, which puts animals at risk for a number of dangerous and sometimes fatal illnesses. If dietary crude protein is less than 6%, feed intake and dietary digestibility are decreased, which exacerbates an energy-protein deficiency. Therefore, a minimum of 7% crude protein should be included in the diet to maintain mature, healthy animals. There is an increased need for dietary crude protein during growth, gestation, and lactation.

A diet fortified with legumes or protein supplements (such as soybean meal, cottonseed meal, etc.) may be necessary for lactating, growing, sick, or disabled animals, as most forages provide sufficient amounts of dietary protein for maintenance. It seems that feeding sufficient to slightly more protein than needed helps keep internal nematode parasites under control (both in terms of resilience and resistance).

### 3.2.1.4. Mineral Requirements of Goats

Goats' definitive mineral requirements have not been determined, either for maintenance or for production. Goats have been used in studies on the metabolism of minerals, particularly calcium and phosphorus. These data, taken together, provide credence to the hypothesis that goats' mineral needs are largely similar to those of sheep. Satisfying the goat's nutritional requirements will enhance its growth, procreation, and defense mechanisms. Most mineral deficiencies can be avoided and performance can be enhanced by adding specific minerals (phosphorus for dry winter forages, selenium in deficient areas, etc.) to salt (NaCl), preferably in granular form and offered free choice.

- **Calcium:** Under grazing conditions, calcium requirements are usually satisfied by either meat-type or Angora goats; however, high-producing dairy goats should have their levels checked because a deficiency can result in decreased milk production. Goats that are nursing must get enough calcium in their diet to avoid parturient paresis, or "milk fever." The calcium needs of browsing or grain-fed goats are typically met by adding a calcium supplement (calcium phosphate, limestone, etc.) to the feed or to a trace mineral–salt mixture. Legumes, such as kudzu, alfalfa, and clover, are also excellent providers of calcium.

- **Phosphorus Deficiency:** A lack of phosphorus causes stunted growth, an unattractive appearance, and sometimes an insatiable appetite. Goats that are fed a diet low in phosphorus can continue to produce milk for a few weeks by using the phosphorus stored in their bodies, but prolonged phosphorus deficiency has been shown to cause a 60% reduction in milk production. Due to the goats' propensity for urinary calculi, the calcium:phosphorus ratio should be kept between 1:1 and 2:1, preferably 1.2–1.5:1. In grazing goats, phosphorus deficiency is more common than calcium deficiency. The ratio ought to be kept at 2:1 in struvite calculi cases.
- **Magnesium Deficiency:** Grazing goats are less likely than cattle to experience hypomagnesemic tetany, or grass tetany, which is linked to magnesium deficiency. The ability of goats to partially offset low magnesium levels by excreting less magnesium is present. A magnesium deficiency lowers milk production as well as urine excretion.
- **Salt:** Although salt (NaCl) is widely acknowledged as an essential dietary component, it is frequently overlooked. When salt is provided ad libitum, goats may eat more than is necessary; this does not pose a nutritional risk, but it may reduce feed and water intake in some arid regions where the drinking water has a high salt content. Because goats have a strong desire to consume sodium, salt formulations are used as carriers for trace minerals.
- **Potassium:** One of the key components of metabolism is potassium. Nonetheless, potassium deficiency in grazing goats is extremely uncommon because forages are typically fairly rich in this mineral. The consumption of marginal potassium is limited to lactating does feed a diet primarily consisting of cereal grains. In dairy goats, hypocalcemia may be linked to excessive potassium intake, especially during the latter stages of pregnancy. Reduce or monitor

feedstuffs high in potassium, such as alfalfa, if hypocalcemia is a problem for the herd.

- **Iron Deficiency**: It is rare to find mature grazing goats lacking in iron. Little children may exhibit this kind of deficiency due to their low birth weight stores and the low iron level of the mother's milk. This is more frequently observed in children fed in total isolation and in animals with high parasitic load. Having access to pasture or a high-quality trace mineral salt that contains iron can help prevent iron deficiency. Iron dextran injections at 2- to 3-week intervals (150 mg, IM) for the first few months may be curative in severe cases and for children raised in confinement. When administering iron dextrans to patients with mixed iron/selenium deficiencies, care should be taken until the selenium deficiency is also treated.
- **Iodine Deficiency:** In certain parts of the United States, there is an iodine deficiency in the soil and in the crops grown there. Iodine should therefore be supplied in stabilized salt. Goats that eat goitrogenous plants may develop conditional iodine deficiency with normal to marginal iodine intake. A significant iodine deficiency causes an enlarged thyroid, stunted growth, small, feeble newborns, and impaired fertility.
- **Zinc Deficiency:** A zinc deficiency causes smaller testicles, parakeratosis, stiff joints, and decreased libido. Deficiencies are avoided by consuming a trace mineral salt mixture containing 0.5–2 percent zinc or a minimum of 10 ppm of zinc in the diet. Goats that consume a lot of calcium-rich food (alfalfa) may be more likely to experience zinc deficiency.
- **Copper Deficiency:** A copper deficiency can cause diarrhea, some types of metabolic bone disease, poor production, lighter or faded hair color, poor fiber quality, infertility, poor health, slowed growth, and potentially a higher susceptibility to internal parasites. Reduced copper-molybdenum ratios, high dietary sulfur, or insufficient copper intake can all contribute to copper deficiency. Compared to sheep, goats seem to be far more resistant to the toxicity of copper.
- **Selenium Deficiency:** Dietary deficiencies in selenium are typically linked to mastitis, poor growth, weak or premature children, retained placentas, and metritis.

### 3.2.2. Herbage and Browse Utilization in Goats

Goats favor shrubs and tree leaves, whether they are deciduous or evergreen, in contrast to other farm animals—with the exception of llamas. Goats have been used to manage invasive shrub-like growth in pastures due to their preference for this kind of vegetation. Goats and sheep of comparable size eat roughly the same amount of fodder. Goats typically choose the most easily digested parts of most forages, and they prefer to browse in rough patches and along fence lines. Generally speaking, goats thrive on improved pastures and favor browse over grass (consuming more than 80% of their diet in browse).

During their growing season, browse (leaves and twigs of trees and shrubs) typically contain higher levels of phosphorus and crude protein than do grasses. However, one or more inhibitors that may bind to or otherwise prevent the use of the nutrients contained in the plants limit the value of some palatable browse species. The lignification of tree leaves and woody twigs, which physically binds (or encapsulates) the desired nutrients, is one such inhibitor. Some range shrubs contain terpenes, which are plant-based oils that appear to inhibit the growth of rumen bacteria. However, terpenes may also have beneficial effects on internal parasite control.

Some browse plants have high tannin concentrations that impede the digestion of feedstuffs by binding enzymes or by blocking enzymatic activity. Overindulgence in tannins may also raise the need for sulfur, which is especially important for goats that produce hair. Nevertheless, despite these possible issues, goats seem to be able to choose more beneficial and easily digested browse when given the chance. Several species of internal nematode parasites may be controlled by grazing or browsing on plants that contain tannins.

## 3.2.3. Nutritional Diseases of Goats

### *3.2.3.1. Enterotoxemia in Goats*

Enterotoxemia is a feed-related illness that results in nearly instantaneous death because of a toxin generated by type D and occasionally type C *Clostridium perfringens*. The organism seems to be found widely in the natural world. The causative bacteria multiply quickly in the presence of high carbohydrate consumption or high intake of immature succulent forage, resulting in the production of a sigma toxin that increases intestinal permeability. Goats can develop enterotoxemia in certain cases; these animals are typically fed diets high in carbohydrates. Both young goats and adult goats may experience diarrhea, depression, loss of coordination, digestive disturbances, coma, and even death if they are overfed carbohydrates. For stable-fed goats, frequent, small-volume feedings of milk, grain, and forage are the most effective ways to prevent enterotoxemia. It is best to avoid feeding large meals only once a day. Feeding high-quality forages as an energy/protein supplement in place of cereal grains can help lower the intake of fermentable carbohydrates in goat populations that are considered to be "at risk." Acute indigestion and a rumen pH of < 4.8 indicates lactic acidosis, which can lead to the secondary complication of enterotoxemia. Proper vaccination programs for both the dam and kid will also be of value in prevention.

### 3.2.3.2. Polioencephalomalacia in Goats

Polioencephalomalacia manifests clinically as disorientation, dullness, aimless wandering, loss of appetite, circling, extensor spasms, progressive cortical blindness, and sporadically head pressing. Certain animals develop recumbent states and may pass away without medical attention. Diets high in grain and low in forages, or containing certain contaminants (such as sulfate or sulfide) may put ruminants at risk for developing polioencephalomalacia. The rumen may produce thiaminases, thiamine antimetabolites, or reduced amounts of thiamine as a result of these dietary circumstances. Thiamine (200–500 mg, IV, IM, or SC) is typically an effective treatment for affected animals. If there has been significant brain damage, animals rarely recover to a level of production that is satisfactory, despite the dramatic and nearly immediate response. Treatment must be given quickly. The diet needs to be adjusted to decrease excessive intake of sulfate and sulfide, increase the amount of forage, and decrease grain intake when necessary. Incorporating thiamine mononitrate into the diet may help prevent when predisposing diets are unavoidable or during stressful times.

### 3.2.3.3. Pregnancy Toxemia in Goats

Pregnancy toxemia is typically observed in late pregnancy and is significantly more common in mothers carrying multiple fetuses who are extremely thin, live in stressful environments, and/or have other coexisting illnesses. Atypically elevated blood levels of ketone bodies accompanied by hypoglycemia are among the clinical characteristics. Numerous symptoms of enterotoxemia are present in affected animals. The developing fetuses during the late stages of gestation have high glucose requirements, which prompt the dam to start metabolizing adipose tissue (fat) to meet the demands. Ketone bodies are released into the bloodstream at a higher rate as a result of the liver's impaired capacity to metabolize this excess fat load. The symptoms include opisthotonos, depression, dullness, and finally death.

When the disease is still in its early stages and symptoms start to show, the dam can use a 200–300 mL drench of glycerol or propylene glycol as an energy source to stop the body from metabolizing so much fat. Nonetheless, the preferred course of treatment is to inject glucose (5 percent dextrose or 50–120 mL of 23% calcium borogluconate solution into a liter of 5 percent dextrose IV). Reducing the incidence of chronic disease, shearing long-fibered does in late gestation, keeping a proper body condition score, identifying and feeding females with twins and triplets, and potentially including niacin (1 g/day in the diet during late gestation) in the diet are all important preventive measures. The incidence of this condition can be decreased by ensuring sufficient winter feedstuffs and keeping an eye on changes in body condition throughout gestation.

### *3.2.3.4. Urinary Calculi in Goats*

Mineral deposits in the urinary tract cause calculi. Stuffing the feet, kicking at the penis, stooping, and slow urination are signs of painful and difficult urination. Urinary obstruction is typically more common in males who are castrated or intact. The obstruction could burst the bladder, which would cause "waterbelly," a condition that could be fatal. It is typical when feedlot lambs, pet goats, and other animals receive diets high in cereal grains. The urine produced by affected animals is alkaline and contains a lot of phosphorus.

Lowering phosphorus consumption to minimal levels and maintaining calcium:phosphorus ratio >2:1 (with the phosphorus content < 0.45 percent of the diet) can help reduce the incidence of urinary struvite calculi. The application of anionic salts, such as ammonium chloride (0.5% of the total diet), dietary tetracycline, sufficient intake of vitamin A (or beta-carotene), increased dietary intake of NaCl, decreased grain intake, decreased use of pelleted cereal grains, and guaranteeing a sufficient supply of clean, fresh, and appetizing water free choice have all shown positive results. When exposed to ammonium chloride (7–14 g/day) for 3–5 days, affected animals may respond well. The disease in range sheep and goats is linked to the ingestion of forages high in silica.

### *3.2.3.5. White Muscle Disease in Goats*

Low selenium and possibly vitamin E levels are the cause of white muscle disease. Compared to sheep, goats appear to experience less development of it. The symptoms include arched backs, tucked-up rear flanks, pneumonia, and acute death, particularly in the hindquarters. White striations are observed on necropsy in the skeletal, diaphragmatic, and cardiac muscles. Elevated levels of AST and lactic dehydrogenase signify damage to the muscles. The selenium-containing glutathione peroxidase is found in lower concentrations in the blood (Lu et al., 1989).

## 3.3. LIMITATIONS IN ESTABLISHING NUTRITIONAL STANDARDS

Nutrient requirements can be determined by a variety of factors, and it is clear that there are relative rather than absolute nutritional standards. Unless it is clearly defined at the start of the experiment, it is directed towards a population rather than a particular individual or subset of animals. Deduction or empirical methods can be used to determine nutritional standards. Once established, they are not necessarily corroborated by more animal tests. These standards represent the expert opinion or professional judgment of a group of people at a specific time. To effectively implement these standards, users need to identify the primary sources of variation, which include individual, genetic, climatic, dietary, age, and physiological stages.

Factorial and empirical methods are part of the fundamental methodology used to assess nutrient requirements. The empirical method is verifiable and based on observation or experience. A factor or essential components of a factor are related to the factorial method. Nutritional requirements in the empirical method are the minimum amounts of nutrients required to maximize or minimize population responses for one or more performance criteria, like growth, over a specified time period. The factorial method combines the estimated needs for maintenance and production, such as growth and milk production, to determine the daily requirements for a single animal at a given time. The factorial approach determines the requirements of a reference animal over a very brief time, while the empirical method determines optimal nutrient allowances from a population perspective. The disparity between factorial and empirical approaches' dietary requirements must always be reconciled. Advanced techniques for assessing nutrient requirements can include calorimetry for both direct and indirect measurements of energy intake and expenditure, whole-body dynamic non-invasive body composition detection using methods like computed tomography, positron emission tomography, single photon emission tomography, and magnetic resonance imaging, system biology using statistical and mathematical modeling, integrating large data sets to interpret and comprehend complex physiological systems, and nutrigenomics. Predicting an animal's nutritional needs is one of the problems that artificial intelligence (AI) has helped solve. AI uses strong machine learning and deep learning techniques to analyze data. Tedeschi talked about the benefits and drawbacks of using mathematical modeling to forecast ruminant nutrient needs.

## 3.4. NUTRIGENOMICS AND NUTRIENT REQUIREMENTS

Nutrient requirements can be significantly impacted by developments in molecular nutrition, particularly in the field of nutrigenomics. The field of molecular nutrition studies how nutrients and food ingredients affect the physiology and state of health of the entire body at the molecular and cellular levels. Advanced biostatistics, imaging, calorimetry, cell biology, challenge tests (meals, exercise, etc.), single nucleotide polymorphisms (SNPs) and epigenomics, transcriptomics, proteomics, metabolomics, and state-of-the-art epidemiology are all used in these descriptive and mechanistic studies), and systems biology's integration of all data will yield knowledge in the field of molecular nutrition research that is far superior to what it is today.

### 3.4.1. The Concept of Molecular Nutrition Research is Broader Than Nutrigenomics

Accurately identifying the molecular mechanisms underlying animal health and disease presents a significant opportunity to improve health, reduce mortality and morbidity, and advance the field of nutrigenomics. The relationship between nutrients, diet, and gene expression is known as nutrigenomics, and it is the study of how food ingredients affect gene expression. It is able to detect the molecular interactions that occur between the genome and various dietary bioactives and nutrients. Additionally, it focuses on how genetic variation affects nutrition and links gene expression or SNPs—a DNA sequence variation that frequently occurs within a population and causes differences in a single nucleotide, such as A, T, C, or G—between members of the same species or paired chromosomes—with the biological effects, metabolism, absorption, and elimination of nutrients. It could tailor and optimize nutrition based on the genotype of the subject. It shows promise and may help with anti-aging, health promotion, and disease prevention. Despite being in its early stages, nutrigenomics in sheep and goats has seen some recent advances.

The fields of nutrigenomics, omics (genomics, epigenomics, transcriptomics, proteomics, and metabolomics), and bioinformatics have the potential to significantly advance our comprehension of the various levels of regulation that dietary nutrients induce in small ruminants when they are being used for milk, meat, wool, or reproduction. According to preliminary data, the nutrigenomics approach could eventually result in more accurate management of sheep and goats, which would enhance feed resource utilization in an optimal manner and have an impact on nutrient requirement estimation. Bacteria are home to nature's largest gene bank. In sheep and goats, the microbes in the rumen play a crucial role in the digestion and utilization of nutrients. Knowledge of the microorganisms in the rumen can improve knowledge of how nutrients are used, which will then increase the accuracy with which nutrient requirements in sheep and goats are determined. Rumen microorganisms are made up of two types of genomes: core and accessory. Understanding how genes from different strains interact with nutrients will undoubtedly improve with the separation of genes that are shared by all strains (accessory genome) from genes that are unique to that strain (core genome). Another area where nutrients can be precisely delivered to each animal is precision nutrition, which can meet the nutrient requirements for the upkeep and production of meat, milk, and fiber in sheep and goats. It is possible to meet each animal's unique needs through the use of software (feed formulation, livestock administration, reproductive optimization, and quality management) and EIDs (ear tags that can be read by a handheld reader). Readers have internet access so they can process the data for each individual anime right away. The system maximizes animal productivity by identifying unique characteristics in each animal and utilizing cutting-edge technologies. It makes it possible to tailor each animal's dietary requirements and makes the application of nutrient requirements efficient (Gonzalez et al., 2018).

# REFERENCES


1. Chadwick, R. (2004). Nutrigenomics, individualism and public health. *Proceedings of the Nutrition Society*, *63*, 161–166.
2. Ferreira, L. M. M., Hervas, G., Belenguer, A., Celaya, R., Rodrigues, M. A. M., Garcia, U., Frutos, P., & Osoro, K. (2017). Comparison of feed intake, digestion and rumen function among domestic ruminant species grazing in upland vegetation communities. *Journal of Animal Physiology and Animal Nutrition (Berl)*, *101*, 846–856.
3. Gonzalez, L. A., Kyriazakis, I., & Tedeschi, L. O. (2018). Review: Precision nutrition of ruminants: Approaches, challenges and potential gains. *Animal*, *12*(s2), s246–s261.
4. Howell, S. B., Burke, J. M., Miller, J. E., et al. (2008). Prevalence of anthelmintic resistance on sheep and goat farms in the southeastern United States. *Journal of the American Veterinary Medical Association*, *233*(12), 1913–1919.
5. Lu, C. D. (1989). Energy and protein nutrition in lactating dairy goats. In *Proceedings of 24th Pacific Northwest Animal Nutrition Conference* (pp. 133–142). Boise, Idaho, USA.
6. Lu, C. D. (2023). The role of goats in the world: Society, science, and sustainability. *Small Ruminant Research*, *227*, Article 107056.
7. Lu, C. D., Sahlu, T., & Fernandez, J. M. (1987). Assessment of energy and protein requirements for growth and lactation in goats. In *Proceedings of the 4th International Conference on Goats* (Vol. 2, pp. 1229–1248). Brasilia, Brazil, March 8–13.
8. Luo, J., Goetsch, A. L., Nsahlai, I. V., Sahlu, T., Ferrell, C. L., Owens, F. N., Galyean, M. L., Moore, J. E., & Johnson, Z. B. (2004). Prediction of metabolizable energy and protein requirements for maintenance, gain and fiber growth of Angora goats. *Small Ruminant Research*, *53*, 339–356.
9. National Research Council (NRC). (1945). *Recommended nutrient allowances for sheep*. National Academy Press.
10. National Research Council (NRC). (1981). *Nutrient requirements of goats: Angora, dairy, and meat goats in temperate and tropical countries*. National Academy Press.
11. Neary, M. (2009). The basics of feeding sheep. Purdue University. http://ag.ansc.purdue.edu/sheep/articles/basics.html (accessed on 7th October 2024).
12. Norheim, F., Gjelstad, I. M. F., Hjorth, M., Vinknes, K. J., Langleite, T. M., Holen, T., Jensen, J., Dalen, K. T., Karlsen, A. S., Kielland, A., Rustan, A.

C., & Drevon, C. A. (2012). Molecular nutrition research—The modern way of performing nutritional science. *Nutrients*, *4*, 1898–1944.

13. Osorio, J. S., Vailati-Riboni, M., Palladino, A., Luo, J., & Loor, J. J. (2017). Application of nutrigenomics in small ruminants: Lactation, growth, and beyond. *Small Ruminant Research*, *154*, 29–44.
14. Qi, K., Lu, C. D., & Owens, F. N. (1992). Sulfate supplementation of Alpine goats: Effects on milk yield and composition, metabolites, nutrient digestibilities, and acid-base balance. *Journal of Animal Science*, *70*, 3551–3560.
15. Sahlu, T., Fernandez, J. M., & Lu, C. D. (1992). Dietary protein level and ruminal degradability for mohair production in Angora goats. *Journal of Animal Science*, *70*, 1526–1533.
16. Sahlu, T., Fernandez, J. M., Lu, C. D., & Potchoiba, M. J. (1992). Influence of dietary protein on performance of dairy goats during pregnancy. *Journal of Dairy Science*, *75*, 220–227.
17. Schlecht, E., Richter, H., Fernandez-Rivera, S., & Becker, K. (2007). Gastrointestinal passage of Sahelian roughages in cattle, sheep, and goats, and implications for livestock-mediated nutrient transfers. *Animal Feed Science and Technology*, *137*, 93–114. https://doi.org/10.1016/j.anifeedsci.2006.11.001.
18. Schoenian, S. (2010). What sheep eat. sheep101.info. http://www.sheep101.info/whatsheepeat.html (accessed on 7th October 2024).
19. Smith, J. F., Jagusch, K. T., & Farquhar, P. A. (1983). The effects of the duration and timing of flushing on ovulation rate in ewes. *Proceedings of the New Zealand Society of Animal Production*, *43*, 13–16.
20. Tedeschi, L. O. (2019). Mathematical modeling in ruminant nutrition: Approaches and paradigms, extant models, and thoughts for upcoming predictive analytics. *Journal of Animal Science*, *97*, 1921–1944.
21. Umberger, S. H. (2009). Feeding sheep. Virginia Cooperative Extension. http://pubs.ext.vt.edu/410/410-853/410-853.html (accessed on 7th October 2024).

*Chapter* **4**

# NUTRIGENOMICS IN LAMBS

## Contents

# INTRODUCTION

The study of the relationship between diet and gene expression in lambs is a developing field called "nutrigenomics," which aims to maximize productivity, growth, and health. This field of research is essential to raising the quality of meat and increasing the productivity of lamb production. By comprehending how distinct nutrients can impact gene expression in lambs, scientists can create customized feeding plans that encourage desired characteristics like muscle development, fat accumulation, and general well-being. Finding particular genes that react to different dietary components is one of the main goals of nutrigenomics. For example, research has demonstrated that specific fatty acids, vitamins, and amino acids can alter the expression of genes related to fat metabolism and muscle growth. It is possible to improve feed efficiency and muscle growth in lambs by adjusting these dietary components. Furthermore, by knowing the genetic variations among the various lamb breeds, nutritional programs can be specifically tailored to each breed's unique genetic makeup, which will ultimately improve production results.

Nutrigenomics not only influences growth performance but also disease resistance and general health in lambs. Enhancing immune function through nutrition can help lower disease incidence, which lowers the need for antibiotics and improves animal welfare. For instance, it has been demonstrated that adding particular bioactive substances to the diet, such as probiotics and prebiotics, can positively affect the expression of genes linked to immune responses. Nutrigenomics in lambs presents a promising new direction in livestock production advancement. Producers can increase meat quality, encourage sustainable farming practices, and increase the efficiency of lamb production by fusing genomic technologies with nutritional science. The knowledge gained from nutrigenomics will be crucial for addressing the difficulties of feeding a growing world population while maintaining the health and welfare of livestock as research continues to advance (Banchero et al., 2016; Ahn and Lee, 2018; Alim et al., 2020).

## 4.1. ROLE OF SPECIFIC GENES IN DETERMINING THE NUTRITIONAL NEEDS OF LAMBS

As ruminant animals, lambs have intricate dietary needs that are impacted by their genetic composition. Knowing how particular genes contribute to these nutritional requirements can help with diet formulation optimization, health outcomes improvement, and lamb productivity enhancement. This paper explores the major genes that influence lambs' nutritional requirements, emphasizing their roles and implications for nutrition management.

### 4.1.1. The Role of the DGAT1 Gene in Fat Metabolism

Among ruminants, including lambs, the DGAT1 (Diacylglycerol O-Acyltransferase 1) gene is one of the most researched. This gene plays a major role in the metabolism of fat, especially in the synthesis of triglycerides, which are the main lipid stored in animals. The DGAT1 gene is a major regulator of fat deposition in lambs because it encodes an enzyme that catalyzes the last stage of triglyceride synthesis. The DGAT1 gene has been linked to variations in lamb meat's fat content, which affects the meat's nutritional value. Lambs with specific DGAT1 gene alleles might need to have their diets changed to maximize fat metabolism and enhance the quality of their meat.

### 4.1.2. MC4R Gene and Energy Balance

In animals, the MC4R (melanocortin 4 receptor) gene is essential for controlling appetite and energy balance. The MC4R gene affects body composition, growth rate, and feed intake in lambs. The leptin and insulin signaling pathways rely on the receptor that the gene encodes to play a crucial role in the central regulation of energy homeostasis. Because variations in the MC4R gene have been associated with variations in energy expenditure and appetite control, this gene is crucial for figuring out what lambs need to eat. In order to maximize growth and health, lambs with particular MC4R gene variants may show variations in feed efficiency. This calls for the use of specialized nutritional strategies.

### 4.1.3. LEP and LEPR Genes in Fat Regulation

In lambs, the regulation of fat deposition and energy balance is influenced by the genes for the hormones leptin (LEP) and leptin receptor (LEPR). Adipose tissue secretes the hormone leptin, which is essential for controlling appetite, energy expenditure, and body weight. The leptin receptor, which mediates the effects of leptin on target tissues, is encoded by the LEPR gene, whereas the leptin protein is encoded by the LEP gene. Variations in the sensitivity to leptin caused by mutations or polymorphisms in these genes can impact the way lambs store fat and manage their energy levels overall. Comprehending the genetic differences in the LEP and LEPR genes can assist in developing diet plans that control fat accumulation and enhance feed efficiency in lambs.

### 4.1.4. FASN Gene and Fatty Acid Synthesis

In lambs, the synthesis of fatty acids depends on the FASN (Fatty Acid Synthase) gene. An enzyme that is responsible for catalyzing the synthesis of long-chain fatty acids—essential building blocks of energy-storing molecules and cell membranes—is encoded by this gene. The nutritional state of the lamb affects the FASN gene's activity; animals fed high-energy diets have higher expression levels of this gene. The fatty acid composition of lamb meat can vary due to variations in the FASN gene, which can impact the meat's nutritional value. Through an understanding of the genetic regulation of the FASN gene, dietary strategies that maximize fatty acid synthesis and enhance lamb meat quality can be developed.

### 4.1.5. The Influence of SCD Gene on Fatty Acid Composition

To produce monounsaturated fatty acids from saturated fatty acids, the SCD (Stearoyl-CoA Desaturase) gene is essential. This gene produces an enzyme that changes stearic acid into the monounsaturated fatty acid oleic acid by adding a double bond. Lamb meat's fatty acid composition is influenced by the activity of the SCD gene, specifically the ratio of saturated to unsaturated fatty acids. The nutritional profile of lamb meat can vary depending on variations in the SCD gene, which can impact the meat's suitability for consumption. It is possible to create nutritional strategies that improve the production of healthier meat with a desirable fatty acid profile by knowing how the SCD gene is genetically regulated.

### 4.1.6. GHR Gene and Growth Regulation

One important gene that controls growth and development in lambs is the GHR (Growth Hormone Receptor) gene. Growth hormone, which is necessary for promoting muscle growth, overall body composition, and growth, is encoded by this gene. The GHR gene affects feed efficiency, growth rate, and carcass composition by influencing how the lamb responds to growth hormone. GHR genotype variations can cause variations in growth performance, requiring dietary formulations to be adjusted to meet the unique growth requirements of lambs with various GHR genotypes. Understanding the genetic variations in the GHR gene allows for the optimization of nutrition to improve lamb productivity and growth.

### 4.1.7. CAPN1 Gene and Meat Tenderness

Lamb meat tenderness and muscle development are influenced by the CAPN1 (Calpain 1) gene. This gene produces an enzyme that is essential for the breakdown of muscle proteins, which affects the softness and texture of lamb meat. Meat quality can vary depending on variations in the CAPN1 gene, where certain alleles are linked to higher tenderness. Gaining knowledge of the genetic regulation of the CAPN1 gene can aid in the development of nutritional approaches that improve the quality of meat, especially by maximizing the ratio of nutrients supporting the growth and development of muscle. Lambs' unique genetic requirements can be catered for in their diets to produce meat that is more tender and appealing to consumers.

### 4.1.8. MSTN Gene and Muscle Development

One important modulator of muscle growth in lambs is the growth differentiation factor 8 (GDF8) gene, commonly referred to as the MSTN (myostatin) gene. This gene produces a protein that functions as a negative regulator of muscle growth by preventing the growth of new muscle. Differences in muscle mass and composition can result from variations in the MSTN gene, with some mutations linked to increased muscle growth. To support their increased muscle growth and make sure their nutritional needs are satisfied for optimal development, lambs with mutations in the MSTN gene might need to consume particular dietary formulations. Comprehending the function of the MSTN gene in the development of muscles can aid in creating diet plans that encourage the growth of lean muscle mass and enhance the quality of the carcass in lambs.

### 4.1.9. ACACA Gene and Lipid Metabolism

Lipid metabolism in lambs is regulated by the ACACA (Acetyl-CoA Carboxylase Alpha) gene. An enzyme that catalyzes the carboxylation of acetyl-CoA to malonyl-CoA, an essential step in the biosynthesis of fatty acids, is encoded by this gene. Lambs' total lipid content and the rate at which fatty acids are synthesized are both influenced by the activity of the ACACA gene. Lamb meat's fat content and composition can be impacted by variations in the ACACA gene, which can cause variations in lipid metabolism. It is feasible to create dietary plans that maximize lipid metabolism and raise the nutritional value of lamb meat by comprehending the genetic regulation of the ACACA gene.

### 4.1.10. SLC2A2 Gene and Glucose Metabolism

One important modulator of lambs' glucose metabolism is the SLC2A2 (Solute Carrier Family 2 Member 2) gene. This gene produces a glucose transporter protein, which is essential for preserving glucose homeostasis because it makes glucose easier for cells to absorb. The effectiveness of glucose transport can be impacted by variations in the SLC2A2 gene, which can have an impact on the lamb's energy metabolism and general health. Lambs will have enough energy for growth and development if diets that optimize glucose metabolism are developed with an understanding of the genetic regulation of the SLC2A2 gene. Nutritional strategies can be adapted to meet the unique genetic requirements of lambs, improving their productivity and overall health (Banchero et al., 2016; Ahn and Lee, 2018; Alim et al., 2020).

## 4.2. DIFFERENT NUTRIENTS CAN INFLUENCE GENE EXPRESSION IN LAMBS

The study of the complex relationship between nutrients and gene expression, or "nutrigenomics," is a fast-developing field that sheds light on how diet affects an animal's genetic composition, including lambs. The interplay of nutrients and genes is a major factor in determining the growth, development, and general health of ruminant animals such as lambs. This chapter examines the ways in which various nutrients—such as proteins, fats, carbohydrates, vitamins, and minerals—can affect the way genes are expressed in lambs. It also discusses the mechanisms at play and how these findings may affect livestock production and nutrition management.

### 4.2.1. Protein and Amino Acids in Gene Expression

Dietary staples like proteins and amino acids are essential for proper gene expression. Amino acids, the building blocks of cellular structures and enzymes involved in a variety of metabolic processes, make up proteins. Certain amino acid availability in lambs can have a direct impact on the expression of genes related to immune system, muscle growth, and growth.

The mammalian target of rapamycin (mTOR) signaling pathway, for example, is known to be activated by the essential amino acid leucine and to control both protein synthesis and cell growth. Lambs' increased expression of genes involved in muscle protein synthesis is caused by the activation of the mTOR pathway, which occurs when they eat diets high in leucine. Lambs' overall body composition and muscle growth may both benefit from this. Likewise, methionine—another necessary amino acid—is important for the methylation of DNA, a process that can modify the expression of genes by turning them on or off. Lamb health and growth may be impacted by the expression of genes involved in different metabolic pathways, which can be influenced by adequate levels of methionine in the diet.

### 4.2.2. Fats and Fatty Acids in Gene Regulation

Fatty acids and fats are essential dietary components that have a big impact on lambs' gene expression. Fatty acids function as sources of energy, building blocks of cell membranes, and messenger molecules that control the expression of certain genes. Gene expression related to inflammation, immune system function, and lipid metabolism can be influenced by the kinds and quantities of fatty acids consumed in the diet. Polyunsaturated fatty acids (PUFAs) with omega-3 and omega-6 ratios are especially crucial for controlling gene expression. It is well known that omega-3 fatty acids, like docosahexaenoic acid (DHA) and eicosapentaenoic acid (EPA), alter the expression of genes related to immunological response and inflammation. Omega-3 fatty acid-rich diets can lower the expression of genes that promote inflammation and raise the expression of genes that reduce it in lambs, strengthening their immune systems and lowering their risk of inflammatory illnesses.

Saturated fats, on the other hand, can affect how genes related to energy balance and lipid metabolism are expressed. For instance, the peroxisome proliferator-activated receptor alpha (PPARα) pathway, which controls the expression of genes involved in fatty acid oxidation, can be activated by the common saturated fatty acid stearic acid. Saturated fatty acids can improve feed efficiency and growth performance by helping the lamb better utilize fats as an energy source by influencing the PPARα pathway.

### 4.2.3. Carbohydrates and Gene Expression

For lambs, carbohydrates are the main source of energy and are important for controlling gene expression. The expression of genes involved in energy metabolism, insulin signaling, and lipid synthesis can be influenced by the availability of glucose, which is derived from carbohydrates. One important modulator of gene expression in a variety of tissues is insulin, a hormone released in reaction to glucose consumption. High-carb diets can cause lambs to secrete more insulin, which triggers the insulin signaling pathway. This pathway affects the expression of genes related to lipid metabolism, glycogen synthesis, and glucose uptake. Insulin, for instance, can enhance the expression of the SLC2A4 gene, encoding the GLUT4 glucose transporter protein, thereby promoting the uptake of glucose into cells. Furthermore, insulin can promote the synthesis and storage of fat by upregulating the expression of genes involved in lipogenesis, such as the FASN (Fatty Acid Synthase) gene.

Nevertheless, consuming too many carbohydrates can cause these pathways to become dysregulated, which can lead to metabolic diseases like obesity and insulin resistance. Thus, it is essential to comprehend how carbohydrates affect gene expression in order to create diets that maximize energy metabolism and shield lambs from metabolic disorders.

### 4.2.4. Vitamins and Their Role in Gene Expression

Vitamins are necessary micronutrients that are important for many biological functions, including the expression of genes. A number of vitamins influence lambs' gene expression by functioning as cofactors for enzymes involved in DNA synthesis, repair, and methylation. Retinoic acid, a form of vitamin A, is a powerful modulator of gene expression. Nuclear receptors called retinoic acid receptors (RARs) and retinoid X receptors (RXRs), which control the expression of genes related to immune system, cell division, and growth, are bound by retinoic acid. Sufficient amounts of vitamin A in lambs can boost the expression of immune response genes, improving the animal's defenses against infections and general well-being.

Another crucial modulator of gene expression, especially in the metabolism of calcium and phosphorus, is vitamin D. The vitamin D receptor (VDR), which is bound by the active form of vitamin D, calcitriol, controls the expression of genes related to calcium absorption and bone mineralization. Adequate levels of vitamin D are essential for healthy bone development and the avoidance of illnesses like rickets in lambs. Folate and vitamin B12 are examples of B vitamins that are involved in DNA methylation, a process that has the ability to turn on or off genes. Vitamin B12 is a cofactor for the enzymes involved in DNA methylation, whereas folate supplies methyl groups for the process. Their growth, development, and general health can be impacted by the expression of genes involved in different metabolic pathways when these vitamins are present in adequate amounts in their diet.

## 4.2.5. Minerals and Their Impact on Gene Expression

Minerals are inorganic nutrients that are necessary for many physiological functions, including the expression of genes. A number of minerals directly affect lamb gene expression by functioning as cofactors for enzymes involved in DNA synthesis, repair, and transcription. One of the most crucial minerals for controlling gene expression is zinc. More than 300 enzymes, including those involved in DNA synthesis and repair, depend on it as a cofactor. Zinc finger proteins are transcription factors that bind to DNA and control the expression of genes. Zinc is also involved in the structure of these proteins. Enough zinc levels are necessary for the correct expression of genes related to immune response, development, and reproduction in lambs. A zinc deficiency can cause altered gene expression, which can lead to immune system weakness, growth retardation, and problems with reproduction.

Another essential mineral that affects gene expression is iron, especially when it comes to the control of genes related to energy metabolism and oxygen transport. Heme, a component of hemoglobin, the protein in the blood that carries oxygen, contains iron. Hypoxia-inducible factor (HIF), a transcription factor that reacts to low oxygen levels, is another protein that iron regulates. For lambs to grow and develop to their full potential, appropriate iron levels are essential for sustaining healthy oxygen transport and energy metabolism. A trace mineral called selenium is essential for the control of antioxidant defense mechanisms. It is a part of the glutathione peroxidase enzyme, which guards against oxidative damage to cells. Additionally, genes related to thyroid hormone metabolism and immune system performance are influenced by selenium. Enough selenium in lambs is essential for preserving oxidative balance and warding off oxidative stress-related illnesses like white muscle disease.

### 4.2.6. Nutrient-Gene Interactions and Epigenetics

The term "epigenetics" describes modifications to gene expression caused by environmental factors, such as nutrition, rather than changes in DNA sequence. Through epigenetic processes like DNA methylation, histone modification, and non-coding RNA expression, nutrients can alter the expression of genes. One of the most researched epigenetic processes impacted by nutrition is DNA methylation. A methyl group is added to the DNA molecule during methylation, usually at cytosine residues within CpG islands. This process inhibits transcription factors from binding to DNA, which can repress the expression of genes. Methionine, choline, and folate are among the nutrients that supply methyl groups for DNA methylation, which affects the expression of genes related to a number of physiological processes.

Another epigenetic mechanism through which nutrition can affect gene expression is histone modification. Histones are proteins that form the chromatin structure, which is encircled by DNA. Histones can be modified by acetylation, methylation, or phosphorylation to change DNA accessibility to transcription factors, which in turn controls gene expression. The expression of genes involved in immunological response, metabolism, and development can be influenced by histone modifications, which can be influenced by nutrients like vitamins A and D and specific fatty acids.

Non-coding RNAs, like microRNAs (miRNAs), are short RNA molecules that bind to messenger RNA (mRNA) and stop it from being translated into protein, thereby controlling gene expression post-transcriptionally. Target gene expression can be modulated by nutrients through their effect on miRNA expression. For instance, it has been demonstrated that omega-3 fatty acids affect the expression of miRNAs related to lipid metabolism and inflammation, which in turn affects the general health and metabolic state of lambs.

### 4.2.7. Practical Implications for Lamb Nutrition

For the management of nutrition and the production of livestock, it is crucial to comprehend how various nutrients affect gene expression in lambs. Optimizing growth, health, and productivity of lambs can be achieved by customizing diets to meet their unique genetic requirements. For instance, feeding diets high in particular amino acids, like methionine and leucine, can promote feed efficiency and muscle growth. In a similar vein, increasing the intake of omega-3 fatty acids in the diet can improve immune function by regulating the expression of genes linked to inflammation (Banchero et al., 2016; Ahn and Lee, 2018; Alim et al., 2020).

## 4.3. THE ROLE OF NUTRITION IN THE EARLY GROWTH AND DEVELOPMENT STAGES OF LAMBS

Lambs' early growth and development are greatly influenced by nutrition, which also establishes the basis for their general health, well-being, and productivity. A lamb's first few weeks and months of life are crucial because they are when its immune system is developing, it grows quickly, and metabolic processes that will affect the animal's long-term performance are set in motion. The importance of maternal nutrition, colostrum intake, the switch to solid feed, and the effects of particular nutrients on growth, immunological response, and metabolic programming are all covered in this chapter's analysis of the role nutrition plays during these formative years.

### 4.3.1. Maternal Nutrition and Fetal Development

The foundation for a lamb's early growth and development is laid during the gestational period, which is even before birth. Fetal development is greatly impacted by maternal nutrition, which can have an impact on birth weight, muscle development, organ development, and general vitality. Pregnant ewes who eat well have a higher chance of producing robust, healthy lambs with good growth potential and robust immune systems.

The ewe's nutritional requirements rise during gestation, especially in the later stages when fetal growth is at its fastest. To support this growth, a diet high in protein, energy, vitamins, and minerals is necessary. Protein is required for the synthesis of new tissues, such as muscle and bone, while energy is essential for the development of the fetal tissues and organs. For healthy fetal development, micronutrients like vitamins A, D, and E as well as minerals like selenium, zinc, and copper are essential. Inadequate intake of these nutrients may result in underwhelming lambs at birth, low birth weights, and heightened disease susceptibility.

Studies have indicated that intrauterine growth restriction (IUGR) resulting from undernutrition during pregnancy may have long-term effects on the productivity and health of the lamb. Low birth weight lambs can have stunted growth, make it harder for them to compete for milk, and increase their risk of illness, all of which can lower their lifetime performance. For this reason, it's crucial to give ewes a balanced diet during pregnancy in order to maximize fetal growth and set up lambs for a healthy start in life.

### 4.3.2. The Importance of Colostrum Intake

Consuming colostrum, the first milk produced by the ewe right after birth, is a crucial phase in a lamb's life. Immunoglobulins, which are abundant in colostrum, offer passive immunity to a newborn lamb, shielding it from infections in its early days of life. In addition, it is packed full of protein, energy, vitamins, and minerals—all of which are critical for boosting the lamb's metabolism and promoting quick growth.

Immunoglobulins from colostrum are best absorbed during the first twenty-four hours of life; after that, the lamb's capacity to absorb these antibodies quickly declines. For this reason, it is essential that lambs get a sufficient amount of colostrum during this period. Insufficient colostrum exposes lambs to a markedly elevated risk of neonatal infections, potentially resulting in elevated mortality rates.

Apart from fortifying the immune system, colostrum supplies the energy required to sustain body temperature, especially in chilly climates. Lambs that are born without much time to warm up can quickly develop hypothermia if they do not get enough energy from colostrum. Making sure lambs have enough colostrum can make a big difference in their chances of survival, growth, and general health.

### 4.3.3. Transition to Solid Feed: Creep Feeding and Weaning

Lambs gradually switch from a milk-based diet to solid feed as their nutritional needs increase with growth. A key component of this transition is creep feeding, which is the practice of giving lambs additional feed while they are still nursing. Generally speaking, creep feed is a highly digestible, nutrient-dense meal that gives young animals the energy, protein, and vital nutrients they need to support their rapid growth.

Weaning-related stress can be minimized, feed efficiency can be increased, and lamb growth rates can be increased with creep feeding. Lambs that are introduced to solid feed at an early age acquire the rumen function and digestive enzymes required to effectively process and digest increasingly complex diets. As a result of their early exposure to solid feed, lambs may perform better after weaning because they will be better able to meet the nutritional needs of growing once they are not as reliant on milk.

Weaning is a crucial stage in a lamb's life because it means that its diet will change from being mostly made of milk to only solid feed. In order to reduce stress and preserve growth rates during the weaning process, proper nutrition is crucial. Sufficient energy, protein, vitamins, and minerals should be included in a well-planned post-weaning diet to promote ongoing growth and development. In order to promote feed intake and lower the risk of digestive disturbances, the diet should also be appetizing and simple to digest.

## 4.3.4. The Role of Specific Nutrients in Early Growth

Lambs' early growth and development are greatly impacted by the nutrients that are available to them in their diet. Since protein is needed for the synthesis of enzymes, muscle tissue, and other essential bodily components, it may be the most significant nutrient at this point. A sufficient intake of protein is necessary to attain ideal growth rates and enhance feed efficiency. Another essential nutrient is energy, which serves as the fuel for various metabolic functions such as growth and thermoregulation. To maintain their fast growth rates, lambs need a steady supply of energy, especially during times of cold stress when their energy needs are higher. The main sources of energy in the diet are fats and carbohydrates, and the lamb needs a balance of these nutrients to meet its energy needs.

Minerals and vitamins are also crucial for lambs' early growth and development. For instance, selenium and vitamin E are essential for immune system development and the defense against white muscle disease, which can afflict lambs that grow quickly. Zinc contributes to healthy skin, wound healing, and general growth, while calcium and phosphorus are necessary for the development of bones. Lambs must receive enough of these micronutrients in order for their development to be optimal and for them to grow to their full potential.

## 4.3.5. Immune Function and Disease Resistance

Lambs' immune system development and operation are closely related to their diet. Early life is a crucial time for the development of the immune system, and building robust immune defenses requires a healthy diet. Adequate nutrition helps lambs fight off infections and heal from illnesses, increasing their chances of surviving and being healthy in general.

Immune system function is especially dependent on protein, energy, and certain micronutrients like zinc, selenium, and copper in addition to vitamins A, D, and E. The immune system's cell, antibody, and other component production is aided by these nutrients. For instance, selenium and vitamin E cooperate to shield cells from oxidative stress, which can impair immunity. White blood cells are vital for battling infections, and zinc is involved in their formation.

Adequate nutrition can not only boost immunity but also hasten the recuperation of sick lambs. Well-fed lambs have the energy reserves required to fight off infections and are better able to mount an efficient immune response. This enhances flock health generally and lessens the need for medical interventions.

### 4.3.6. Metabolic Programming and Long-Term Implications

According to the theory of metabolic programming, a lamb's growth, productivity, and metabolism may be impacted in the long run by the food it is given in its early years. Animals' metabolic pathways can be programmed by nutrition during key developmental stages, which will impact how the animal uses nutrients and adapts to dietary changes as it gets older.

For instance, lambs that are fed well in their early life are more likely to have effective feed conversion ratios later on, which can result in higher growth rates and increased productivity all around. On the other hand, lambs that don't get enough nutrition during this crucial stage might be more likely to develop metabolic problems, grow slower, and be less productive as adults.

Early nutrition can affect the rumen's development, which is the main location of microbial fermentation in ruminants, according to research. Effective digestion and nutrient absorption in the rumen depend on the development of a diverse and healthy microbial population. The development of the rumen can be aided by early exposure to solid feed, such as through creep feeding, and this will enhance the lamb's capacity to use a variety of feeds during its lifetime.

### 4.3.7. Practical Implications for Lamb Nutrition Management

Comprehending the function of nutrition in the initial growth and development of lambs holds significant pragmatic consequences for the management of livestock. Producers can maximize growth, health, and productivity by feeding lambs a balanced diet that fulfills their nutritional needs at every developmental stage.

It's critical for the health of ewes and their lambs during pregnancy to make sure they eat enough. This entails giving the ewe enough protein, energy, and micronutrients to support fetal development and get her ready to give birth. Problems like low birth weight and inadequate milk production can be avoided by keeping an eye on body condition and making necessary dietary adjustments.

Lambs must have access to high-quality colostrum immediately after birth in order to develop passive immunity and promote early growth. Within the first few hours of life, producers should make sure lambs are given colostrum, and they should keep an eye out for symptoms of insufficient nutrition, such as weakness or failure to thrive. Crawl feeding is a useful method for supplying the nutrients lambs need to support their continued growth and get ready for weaning as they move from liquid to solid food. A balance of energy, protein, vitamins, and minerals should be present in the highly digestible and nutrient-dense creep feed. Lambs should also have access to clean water, and producers should keep an eye on their feed intake to spot any problems with the health of their digestive systems or feed acceptance.

To reduce stress and promote continued growth during weaning, it's critical to maintain a steady, nutrient-rich diet. The post-weaning diet should support the growth of the rumen and the general health of the lambs' digestive systems in addition to offering enough energy and protein to meet their needs. A smooth transition to an entirely solid diet can be facilitated by making gradual dietary changes, which also lowers the risk of digestive disturbances (Banchero et al., 2016; Ahn and Lee, 2018; Alim et al., 2020).

## 4.4. METABOLIC PATHWAYS IN LAMBS INFLUENCED BY DIET

Lambs' metabolic pathways are greatly influenced by their diet, which affects not only their growth and development but also their general health and productivity. A complex web of biochemical reactions known as the metabolism of nutrients builds and repairs tissues, produces energy from food, and controls several physiological functions. Several dietary components, such as carbohydrates, proteins, fats, vitamins, and minerals, can have a big impact on these metabolic pathways in lambs. With an emphasis on energy metabolism, protein synthesis, fat metabolism, and the function of micronutrients in promoting metabolic processes, this chapter investigates how diet affects important metabolic pathways in lambs.

### 4.4.1. Energy Metabolism

Lambs use a process called energy metabolism to turn the nutrients in their food into energy that can be used for different physiological functions. Proteins are largely unimportant in the metabolism of energy, but fats, carbs, and oils are the main nutrients. These macronutrients are reduced to simpler molecules, which then enter different metabolic pathways. These pathways include glycolysis, the citric acid cycle (also called the Krebs cycle), and oxidative phosphorylation. The final step produces adenosine triphosphate (ATP), which is the cell's energy currency.

#### 4.4.1.1. Carbohydrate Metabolism

Lambs primarily get their energy from carbohydrates, particularly during times when they are growing quickly. The main product of dietary carbs is glucose, which is subsequently used by cells to make ATP. Glycolysis, a process that produces pyruvate in the cytoplasm of cells, is the first step in the metabolism of glucose. The citric acid cycle can then be used to further metabolize pyruvate in the mitochondria, producing ATP through oxidative phosphorylation.

Lambs that are growing depend on the availability of glucose and the effectiveness of its metabolism to meet their energy needs. Eating a diet high in easily digested carbohydrates, like grains, can improve glucose availability, which will support rapid growth and optimal energy metabolism. But consuming too many carbohydrates can result in metabolic diseases like acidosis, a condition in which the rumen produces too much lactic acid due to rapid fermentation, which lowers blood pH and causes health issues.

#### 4.4.1.2. Fat Metabolism

For lambs, fats constitute an additional vital energy source, offering over twice the energy content per gram in contrast to proteins and carbs. Cells can use the fatty acids and glycerol that are produced when dietary fats are broken down to produce energy. In the mitochondria, fatty acids undergo β-oxidation, resulting in the production of acetyl-CoA, which subsequently enters the citric acid cycle to produce ATP.

When there is a low availability of carbohydrates, like during fasting or when lambs are fed a low-carb diet, fat metabolism is especially crucial. In these circumstances, the body switches from using glucose as its main energy source to using fatty acids. Because of their flexible metabolism, lambs can continue to grow and produce energy even in situations where their diet is low in carbohydrates.

Furthermore, certain fatty acids—polyunsaturated fatty acids like omega-3 and omega-6—have a significant impact on immune system and inflammatory metabolic pathways. These fats have the ability to affect how genes are expressed, which can result in the synthesis of signaling molecules that control inflammatory reactions. This emphasizes how crucial it is to include essential fatty acids in the diet to support not only healthy overall functioning and disease resistance, but also energy metabolism.

#### 4.4.1.3. Protein Metabolism

Proteins are important macronutrients that the body uses to make muscle, hormones, enzymes, and other essential parts. The process of breaking down dietary proteins into amino acids, which are subsequently utilized for protein synthesis or transformed into energy, is known as protein metabolism. Depending on the requirements of the body, amino acids enter different metabolic pathways. They can be used, for instance, to create new proteins needed for tissue maintenance, growth, and repair. When lambs are growing quickly and there is a high demand for protein synthesis, this process is especially crucial. When the amount of carbohydrates consumed is insufficient, certain amino acids, referred to as glucogenic amino acids, can also be transformed into glucose through the process of gluconeogenesis, offering an extra energy source.

The amount and quality of protein in the diet have a significant impact on how lambs metabolize protein. High-quality proteins are better used for protein synthesis because they have a balanced profile of essential amino acids. Reduced growth rates, muscle atrophy, and compromised immune system can result from consuming insufficient amounts of protein. In contrast, as excess amino acids are deaminated and eliminated as urea, eating too much protein can raise the nitrogen burden on the liver and kidneys.

## 4.4.2. The Role of Micronutrients in Metabolic Pathways

Micronutrients, such as vitamins and minerals, are essential for a variety of lamb metabolic processes. These nutrients support the synthesis of proteins, the regulation of other metabolic processes, and the conversion of macronutrients into energy by functioning as cofactors and coenzymes in enzymatic reactions.

### *4.4.2.1. Vitamin Metabolism*

Vitamins are organic compounds that are required in small amounts for various metabolic functions. Different vitamins play distinct roles in metabolic pathways:

- **Vitamin B Complex**: The B vitamins are necessary cofactors in energy metabolism. These include thiamine (B1), riboflavin (B2), niacin (B3), pyridoxine (B6), and cobalamin (B12). They participate in the process that turns proteins, fats, and carbs into ATP. For example, riboflavin, niacin, and thiamine are components of flavoproteins and nicotinamide adenine dinucleotide (NAD), which are essential for oxidative phosphorylation. Niacin is also a cofactor for enzymes in the citric acid cycle and glycolysis.
- **Vitamin A**: Vitamin A is involved in the expression of genes and the differentiation of cells, especially in the immune system's development and the upkeep of epithelial tissues. Moreover, it

contributes to the synthesis of rhodopsin, a protein required for vision.

- **Vitamin D**: In order to support bone mineralization and development, vitamin D is essential for the metabolism of calcium and phosphorus. It has also been demonstrated to affect gene expression linked to calcium transport and bone formation, and it modulates immune function.
- **Vitamin E**: As an antioxidant, vitamin E shields cells from the oxidative damage that comes from free radicals. It is especially crucial for preserving the integrity and functionality of cells by safeguarding the polyunsaturated fatty acids found in cell membranes.
- Vitamin deficiencies can cause metabolic disturbances that impair immune system function, stunted growth, and other health problems. For instance, a deficiency in vitamin E can cause oxidative stress and increased susceptibility to infections, while a deficiency in vitamin B12 can lead to poor energy metabolism and slower growth rates.

### *4.4.2.2. Mineral Metabolism*

Minerals are inorganic elements that are essential for various metabolic processes. Key minerals involved in lamb metabolism include:

- **Calcium and Phosphorus**: The growth and maintenance of bones depend on these minerals. While phosphorus is a component of ATP and contributes to energy metabolism, calcium is also involved in blood clotting, nerve transmission, and muscle contraction.
- **Magnesium**: Many enzymatic reactions, including those involving the synthesis of ATP and the operation of muscles, require magnesium as a cofactor. It is also necessary to preserve the health of your neuromuscular system and electrolyte balance.
- **Iron**: Hemoglobin and myoglobin, the proteins in the blood and muscles that carry oxygen, depend on iron to function. In addition, it contributes to the synthesis of ATP by participating in the transport of electrons during oxidative phosphorylation.
- **Zinc**: More than 300 enzymes, including those involved in protein metabolism, DNA synthesis, and immune function, depend on zinc for their activity. By promoting the activity of superoxide dismutase, an enzyme that neutralizes free radicals, it also contributes to antioxidant defense.
- **Selenium**: A component of selenoproteins, which have antioxidant qualities and are involved in the metabolism of thyroid hormone,

is selenium. To shield cells from oxidative damage, it functions in concert with vitamin E.

Sustaining these metabolic processes and guaranteeing ideal growth and development in lambs depend on a sufficient intake of minerals. Skeletal abnormalities, stunted growth, and metabolic disorders can result from deficiencies in minerals like calcium, phosphorus, or magnesium. Conversely, consuming too much iron or selenium can be toxic and have negative health impacts.

### 4.4.3. Metabolic Pathways in the Rumen

The digestion and fermentation of dietary fiber is a major function of the rumen in lambs' nutrition metabolism. A wide variety of microorganisms, such as fungi, bacteria, and protozoa, live in the rumen and collaborate to convert complex carbohydrates like cellulose and hemicellulose into volatile fatty acids (VFAs). These volatile fat acids (VFAs), which are the main byproducts of rumen fermentation and a significant source of energy for lambs, include acetate, propionate, and butyrate.

- **Acetate**: The most common VFA made in the rumen, acetate is mostly utilized as a building block for the synthesis of fatty acids. It is crucial for the deposition of fat and the production of fat in milk in nursing ewes.
- **Propionate**: In ruminants, propionate is the primary precursor of glucose because the liver's gluconeogenesis process can transform it into glucose. Because ruminants absorb less glucose from their diets than other animals, this is especially crucial for keeping blood glucose levels in lambs stable.
- **Butyrate**: Butyrate is used as an energy source by the cells lining the rumen wall and plays a role in promoting rumen development and function.

The diet's composition can have an impact on the balance of VFAs generated in the rumen. Diets high in readily fermentable carbohydrates, like grains, can increase the production of propionate, whereas diets high in fiber tend to favor the production of acetate. This emphasizes how diet composition affects rumen fermentation patterns and, in turn, the metabolic pathways that assist lamb growth and energy production (Banchero et al., 2016; Ahn and Lee, 2018; Alim et al., 2020).

## 4.5. CHALLENGES AND OPPORTUNITIES IN TRANSLATING NUTRIGENOMIC RESEARCH INTO PRACTICAL FEEDING STRATEGIES

The knowledge gathered from nutrigenomic research has the potential to completely change feeding practices in the lamb industry, improving productivity and sustainability. However, there are many opportunities and challenges in putting these scientific discoveries into real-world applications. This chapter examines the main obstacles to incorporating nutrigenomic research into lamb feeding plans as well as the opportunities to maximize production systems.

### 4.5.1. The Promise of Nutrigenomics in Lamb Feeding Strategies

A greater knowledge of how particular nutrients interact with the genome to affect gene expression and metabolic pathways is made possible by nutrigenomics. Using this information, feeding plans that are specifically tailored to each animal's genetic requirements can be created in the context of lamb production. Nutrigenomics holds the potential to boost the productivity and sustainability of lamb farming by customizing diets to optimize growth, improve feed efficiency, and enhance health.

1. **Improved Growth and Feed Efficiency:** Better growth rates and feed efficiency in lambs is one of the main advantages of nutrigenomics. Through the identification of genetic markers linked to growth traits, farmers can devise customized feeding plans that maximize the uptake of nutrients. For example, studies may show that particular fatty or amino acids have a major impact on the expression of genes linked to muscle growth. Farmers can minimize feed costs and optimize growth performance by creating diets high in these nutrients.
2. **Disease Resistance and Health Improvement:** Additionally, nutrigenomic research can help lambs become healthier and more resistant to disease. Producers can create diets that improve the expression of genes involved in immunity by knowing the genetic basis of immune responses. To support immune function and lessen the need for antibiotics and other drugs, this may entail consuming particular vitamins, minerals, and antioxidants.

### 4.5.2. Challenges in Translating Nutrigenomic Research

Despite the potential benefits, several challenges must be addressed when translating nutrigenomic research into practical feeding strategies for lambs.

#### 4.5.2.1. Knowledge Gaps and Complexity of Nutrigenomics

Lack of knowledge about the precise mechanisms by which nutrients affect gene expression is one of the main obstacles to applying nutrigenomic research. The relationships between nutrients and genes are still not completely understood in the complicated field of nutrigenomics. Many studies are carried out in controlled laboratory environments or *in vitro*, which may not adequately represent real-world lamb production conditions.

For instance, although some studies may pinpoint particular nutrients that affect muscle growth, the way these nutrients interact with other nutrients in the diet and the surrounding conditions can make it difficult to translate these findings into effective feeding plans. To fully understand the precise mechanisms underlying these interactions and to decide how to apply this knowledge in a farming context, more research is required.

#### 4.5.2.2. Genetic Variation Among Lamb Populations

Lambs show genetic variation both within and between breeds, just like any other livestock. Based on nutrigenomic principles, this variation can have a substantial impact on an individual animal's response to dietary interventions. Due to variations in gene expression patterns, a nutrient that stimulates growth in one genetic line of lambs might not have the same effect in another. The development of feeding plans that are universally effective is hampered by this genetic diversity. Producers may need to use more complex techniques, like genomic selection and customized feeding programs, which can be resource-intensive and call for specialized knowledge and cutting-edge technology, to optimize lamb nutrition based on genetic variations.

#### 4.5.2.3. Cost and Economic Viability

The implementation of nutrigenomic strategies may entail substantial financial outlays for genetic testing, food formulation, and monitoring systems, among other things. Adopting these advanced feeding strategies can come with a financial cost that is prohibitive for many lamb producers, especially small-scale farmers. Furthermore, careful consideration needs to be given to the financial feasibility of implementing nutrigenomic feeding strategies. Producers must determine if the costs of putting these strategies into practice outweigh the possible gains in growth performance and health. Making sure that these innovations provide a quantifiable return on investment is essential for their broad adoption in a competitive market.

#### *4.5.2.4. Practical Application and Integration into Existing Systems*

Integrating nutrigenomic research into current production systems is necessary to translate it into workable feeding strategies. Numerous lamb farmers have developed feeding procedures based on accumulated wisdom and experience from the past. Long-standing practices may need to change to implement nutrigenomic principles, which may encounter resistance. Moreover, producers might not have the resources or expertise necessary to successfully apply these novel tactics. To give farmers the knowledge and abilities to comprehend and incorporate nutrigenomic principles into their feeding practices, educational programs and training are required. The translation of research findings into practical applications can be aided by extension services and partnerships with nutritionists and researchers.

### 4.5.3. Opportunities in Translating Nutrigenomic Research

Despite the challenges, there are numerous opportunities to harness nutrigenomic research for practical feeding strategies in lamb production.

#### *4.5.3.1. Advances in Genomic Technologies*

Genome-wide association studies (GWAS) and next-generation sequencing are two recent genomic technology developments that have significantly improved our knowledge of the genetic basis of lamb trait variation. With the aid of these technologies, genetic markers linked to growth, feed efficiency, and disease resistance can be more easily identified.

Through the amalgamation of genomic data and nutritional information, farmers can formulate more precise feeding tactics that maximize nutrient uptake according to each lamb's unique genetic composition. More sustainable production methods and increased performance are possible outcomes of this customized approach.

#### *4.5.3.2. Precision Nutrition*

In animal husbandry, the idea of precision nutrition is becoming more popular, and nutrigenomics offers the basis for this strategy. Precision nutrition entails customizing diets to each animal's unique requirements taking into account factors such as age, weight, health, and genetic composition. A thorough knowledge of how lambs metabolize nutrients can be obtained by combining nutrigenomics with techniques like metabolomics and microbiome analysis. Through the analysis of individual lambs' responses to distinct dietary formulations, producers can enhance their feeding strategies to achieve superior growth and health results.

#### *4.5.3.3. Collaborative Research and Extension Programs*

Translating nutrigenomic research into useful applications requires cooperation between researchers, extension agencies, and lamb producers. In order to facilitate knowledge transfer and help producers comprehend and incorporate nutrigenomic principles into their feeding strategies, partnerships are needed. To assist farmers in navigating the complexities of nutrigenomics, extension programs can offer helpful training and resources. The creation of feeding recommendations unique to a given area that take into account the genetic diversity, climate, and farming practices of that area can also result from this collaboration.

#### *4.5.3.4. Consumer Demand for Sustainable Practices*

Nutrigenomics has great potential to be applied in lamb production due to rising consumer awareness and demand for sustainably produced food. Customers are growing increasingly concerned about the welfare and health of animals as well as the process used to produce their food. Lamb producers can match their practices to consumer expectations for sustainability by implementing nutrigenomic feeding strategies that enhance feed efficiency, minimize the use of antibiotics, and promote animal health. Improved marketability and customer trust may result from this alignment, which will ultimately be profitable for producers.

### 4.5.4. Case Studies and Success Stories

Several case studies illustrate the successful application of nutrigenomic principles in lamb production.

1. **Genetic Selection for Feed Efficiency:** Genetic selection has the potential to increase lambs' feed efficiency, according to an Australian study. Particular genetic markers linked to feed intake and growth rates have been found by researchers. Producers were able to create a flock with higher feed efficiency and lower feed costs and increased productivity by choosing breeding stock based on these markers.
2. **Nutritional Intervention in Immune Function:** Another study looked into how lambs' immune systems were affected when particular vitamins and minerals were added to their diet. Lambs-fed diets supplemented with vitamin E and selenium showed improved

immune responses and decreased susceptibility to infections, according to research. The present study underscores the significance of focused nutritional interventions in enhancing the well-being and output of lambs (Banchero et al., 2016; Ahn and Lee, 2018; Alim et al., 2020).

# REFERENCES


1. Ahn, J., & Lee, K. (2018). The effects of nutrition on lamb growth and gene expression. *Animal Science Journal*, *89*(5), 847–855.
2. Alim, U., et al. (2020). Nutritional genomics and growth performance in lambs. *Journal of Animal Science*, *98*(12), 123–135.
3. Banchero, G. E., et al. (2016). Genomic responses to dietary interventions in lambs. *Molecular Biology Reports*, *43*(6), 519–526.
4. Barboza, S., et al. (2017). Nutrigenomic effects of omega-3 fatty acids in lambs. *Small Ruminant Research*, *146*, 1–10.
5. Bhatia, R., et al. (2019). Nutrigenomics and health in lambs: A review. *Veterinary Sciences*, *6*(1), 15.
6. Bodnar, G., et al. (2021). Epigenetic modifications in lambs fed different diets. *Nutrients*, *13*(4), 1289.
7. Borghese, A., & G. S. (2020). Genomics of lamb meat quality: Nutrigenomic approaches. *Meat Science*, *168*, Article 108195.
8. Bousquet-Mélou, A., et al. (2018). Nutrigenomic approaches in lambs: Challenges and opportunities. *Animal*, *12*(10), 2020–2033.
9. Dario, R., et al. (2022). The impact of diet on the expression of growth-related genes in lambs. *Livestock Science*, *263*, Article 104850.
10. Doornbos, R. F., et al. (2020). Nutritional modulation of gene expression in lambs. *Agricultural Sciences*, *11*(2), 138–152.
11. Hwang, H. K., et al. (2017). Nutritional effects on gene expression in lamb muscle. *Journal of Agricultural and Food Chemistry*, *65*(1), 85–92.
12. Iqbal, M. A., et al. (2018). Nutrigenomic interventions to enhance lamb performance and health. *Frontiers in Veterinary Science*, *5*, Article 290.
13. Jia, Y., et al. (2021). Transcriptomic analysis of lambs under different dietary conditions. *BMC Genomics*, *22*, Article 789.
14. Karamati, A., et al. (2020). Nutritional strategies to improve lamb growth: Insights from genomics. *Molecular Biology Reports*, *47*(9), 6893–6902.
15. Kearney, J. F., et al. (2019). Dietary influences on the lamb immune response: A nutrigenomic perspective. *Veterinary Immunology and Immunopathology*, *207*, 40–50.
16. Kim, H. J., et al. (2018). Nutrigenomic mechanisms of dietary fat in lambs. *Journal of Animal Physiology and Animal Nutrition*, *102*(4), 983–991.
17. Koenig, K. M., et al. (2021). Nutritional genomics of lambs: Implications for production and health. *Animal Production Science*, *61*(4), 567–578.
18. Liu, Z., et al. (2020). The role of nutrition in the development of lamb

muscle: A nutrigenomic perspective. *Frontiers in Genetics*, *11*, Article 1–14.

19. Loor, J. J., et al. (2016). Nutrigenomics in lambs: Influence of dietary fatty acids on gene expression. *Animal Feed Science and Technology*, *219*, 123–135.
20. Ma, Y., et al. (2018). Functional genomics of lambs: Nutritional modulation of gene expression. *Journal of Animal Science and Biotechnology*, *9*, Article 24.
21. Ma, Z., et al. (2019). Epigenetics and lamb nutrition: Implications for growth and health. *International Journal of Molecular Sciences*, *20*(10), Article 2456.
22. Mazzocchi, A., et al. (2021). Gene expression profiling in lambs fed different forage diets. *Journal of Nutritional Science*, *10*, Article e48.
23. Mohammadi, M., et al. (2020). Nutrigenomic effects of probiotics on lamb growth and health. *Animal Nutrition*, *6*(2), 118–125.
24. Mohan, J., et al. (2021). Role of dietary fiber in lamb growth and development: A nutrigenomic approach. *Animal Science Journal*, *92*(7), Article e13538.
25. Montaño, M., et al. (2020). Nutritional modulation of immune responses in lambs: A nutrigenomic perspective. *Veterinary Research*, *51*(1), Article 49.
26. Naranjo, M. A., et al. (2019). Genomic strategies to improve lamb quality through nutrition. *Meat Science*, *155*, 40–46.
27. Nazari, R., et al. (2020). Impact of dietary protein levels on gene expression in lambs. *Animal Feed Science and Technology*, *267*, Article 114519.
28. Orozco, R. A., et al. (2019). Nutritional strategies to enhance lamb meat quality: Insights from nutrigenomics. *Molecular Nutrition & Food Research*, *63*(14), Article e1900376.
29. Osorio, J. S., et al. (2020). The effects of dietary polyunsaturated fatty acids on lamb muscle gene expression. *BMC Genomics*, *21*, Article 431.
30. Pérez, J. A., et al. (2021). Nutritional modulation of lamb gene expression: Implications for growth and health. *Agricultural Sciences*, *12*(2), 128–140.
31. Pinho, R. S., et al. (2021). Nutrigenomics of lamb meat quality: A review. *Current Opinion in Food Science*, *39*, 115–120.
32. Radhakrishnan, G., et al. (2018). Dietary influences on gene expression in lamb muscle: Implications for meat quality. *Animal*, *12*(8), 1724–1731.

33. Rauscher, A. R., et al. (2019). Nutrigenomics of lambs: Molecular effects of diets on growth. *Animal Production Science*, *59*(5), 813–824.
34. Ribeiro, A. D., et al. (2020). Epigenetic regulation of growth in lambs: Dietary influences. *Molecular Biology Reports*, *47*(10), 8239–8249.
35. Santos, M. A., et al. (2017). Nutritional strategies to modulate the immune response in lambs. *Frontiers in Veterinary Science*, *4*, Article 134.
36. Schlegel, P., et al. (2021). Nutritional genomics in lambs: Implications for growth and performance. *Journal of Animal Science*, *99*(5), Article skab130.
37. Sutherland, M. A., et al. (2021). Nutrigenomic impacts of dietary fatty acids on lamb meat quality. *Animal Science Journal*, *92*(8), Article e13615.
38. Szczerba, A., et al. (2022). Nutritional modulation of gene expression related to growth in lambs. *Animal Biochemistry and Physiology*, *200*(1), 73–83.
39. Tavakkoli, J., et al. (2019). Nutritional interventions and gene expression in lambs: A review. *Animal Nutrition*, *5*(3), 329–336.
40. Tzeng, C. L., et al. (2017). Nutrigenomics of lamb meat quality: Dietary influences on gene expression. *Journal of Animal Science*, *95*(11), 5001–5014.
41. Vasconcelos, J. T., et al. (2020). Nutritional genomics in lambs: Dietary effects on growth performance. *Journal of Animal Physiology and Animal Nutrition*, *104*(3), 648–658.
42. Wang, F., et al. (2021). Nutritional modulation of muscle gene expression in lambs. *Livestock Science*, *245*, Article 104423.
43. Wang, J., et al. (2018). Nutrigenomic approaches to enhance lamb meat quality: A review. *Meat Science*, *145*, 122–130.
44. Wang, Y., et al. (2019). The role of diet in modulating gene expression related to growth in lambs. *Frontiers in Genetics*, *10*, Article 646.
45. Wilkins, B., et al. (2020). Genomic analysis of lambs: Dietary influences on health and performance. *Animal Production Science*, *60*(9), 1133–1144.
46. Xu, J., et al. (2019). Nutritional regulation of muscle growth in lambs: A nutrigenomic perspective. *International Journal of Molecular Sciences*, *20*(10), Article 2457.
47. Yan, Y., et al. (2018). Epigenetic effects of nutrition on lamb development and growth. *BMC Genomics*, *19*, Article 511.
48. Zhang, G., et al. (2021). Nutrigenomic regulation of lamb growth performance: Implications for sustainable production. *Agriculture*, *11*(4), Article 368.

49. Zhao, S., et al. (2022). Nutritional modulation of immune responses in lambs: A nutrigenomic approach. *Veterinary Research*, *53*, Article 15.
50. Zúñiga, J. E., et al. (2020). Nutritional interventions and their effects on gene expression in lambs. *Journal of Nutritional Science*, *9*, Article e3.

Chapter 5

# NUTRIGENOMICS IN LIVESTOCK

## Contents

# INTRODUCTION

One global challenge is producing enough nutrient-dense animal protein to feed the world's expanding population. The world's population is predicted to reach 9.4 billion people by 2050, and in order to meet the growing demand of this population, the world's food supply is projected to grow at a rate of 2.3 percent annually. Worldwide, animal protein is a vital component of the human diet, but as the GDP of developed nations rises, there will likely be a significant increase in the demand for animal protein, particularly in these nations. By 2030, a 14% rise in meat consumption is predicted to be supported by an 11% increase in the world's population. Antibiotic-resistant bacteria and antimicrobial resistance genes have emerged as a result of the livestock industry's indiscriminate use of antibiotics to boost productivity, which poses a threat to global health. Consequently, the only sensible solution is to boost plant and animal productivity per unit, which will enable us to meet our population's expanding nutrient needs (Barros-Rodríguez et al., 2021).

Nutrigenetics is the study of population diversity, whereas nutrigenomics is the short-term evaluation of genome activity in response to dietary changes. Interactions with different genome sequences include metabolomics, epigenomics, proteomics, and microarray (transcriptomics) (Figure 5.1) (Pal, 2022).

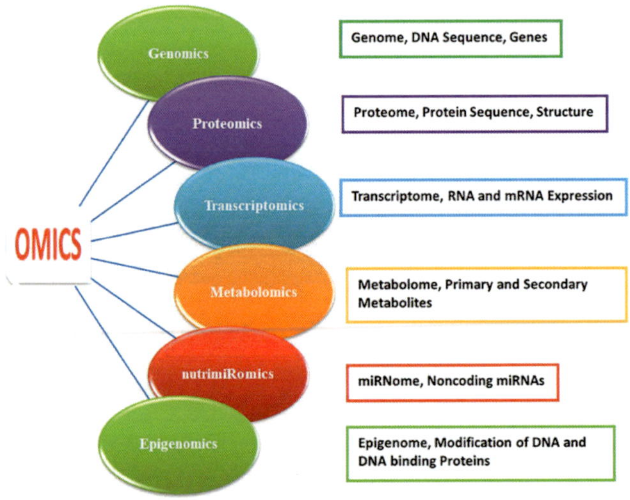


**Figure 5.1.** *OMIC tools used in nutrigenomic research.*

***Source:*** *Pandita, D., and Pandita, A., (2022) Omics technology for the promotion of nutraceuticals and functional foods. Front. Physiol. 13:817247. doi: 10.3389/fphys.2022.817247.*

To improve animal husbandry, however, contemporary production technology must be introduced. One example of this is precision animal husbandry, which uses molecular methods. The animal kingdom is made up of gene expression pathways, as is common knowledge. All of these processes—including the genetic component—are fueled by nutrition. Thus, the goal of research on animal nutrition is to fully understand how nutrients affect an animal's genetic makeup over the course of its lifetime. The study of animal nutrition at the molecular level is known as "nutrigenomics," or "molecular animal nutrition" (Pal, 2022). The objective is to investigate animal biological processes at the level of gene expression or the interactions between nutrients and genes by using modern molecular biological sciences and technology. Ultimately, the key to understanding the basic molecular mechanisms that control the expression of nutrients in an animal's diet, the biochemical reaction, and the physiological and phenotypic processes is to investigate the intricate, bidirectional process interactions between nutrients and genes.

The application of nutrigenomics to food components can alter epigenetic modifications like DNA methylation and histone modifications as well as patterns of gene expression. They rely on epigenetic pathways that are highly responsive to dietary modifications, much like genomic imprinting. For instance, pigs are printed with developmental traits that encourage increased muscle mass and fat storage, such as the genes for insulin-like growth factor 2 (IGF2). According to Ibeagha-Awemu and Zhao (2015), there have been three times as many microRNA precursors discovered in cattle miRBase over the previous ten years.

To comprehend the molecular causes of diseases like obesity, cancer, and cardiovascular disease through the interactions between genes and diet, nutrigenomics studies have been carried out in both humans and mice. Traditionally, studies investigating how diet affects genome function have been conducted on rodent models. Mice and rodents have also been used for measurements of particular life stages (like pregnancy) and cross-generational research. Recent studies in mice have explored the theory that the mother's diet will influence the offspring's genome structure. Nutrigenomics research is booming, especially in the meat industry, for health reasons, since meat products directly affect human health, even though similar farm animal methods are still in their infancy.

Although the field of nutritionomics is not new, a technology with such great potential for the future demands regular reviews and updates to stimulate creative thinking and new methods. This review seeks to give a thorough overview of the core ideas, most recent developments, and research in the field of nutrigenomics, with a focus on how diet affects gene expression, how nutritional intervention can alter an animal's tissue composition, and how all of this relates to human health and the interphase between humans and animals.

## 5.1. NUTRIGENOMICS IN THE LIVESTOCK SECTOR

One of the main sources of income for Indian farmers is the livestock industry. Animal feed makes up about 60–70% of all inputs used in the production of livestock. As a result, the economics of the livestock business are primarily dependent on how well the animals use their feed. The intricate and diverse microbial ecosystem found in the rumen works in harmony and symbiosis to facilitate effective bio-fermentation. Effective anaerobic processes break down lignocellulosic feeds in the rumen. Nevertheless, the amount of potential energy that rumen microbes can extract from lignocellulosic feeds is limited. The livestock industry suffers large financial losses when feed energy is lost. Nutritionists and rumen microbiologists have been working to reduce the losses and understand the intricate microbial ecology. The molecular era and sophisticated molecular methods are utilized to address unexplained nutrition-related disputes at the gene level. Clinical management and prevention of nutritional diseases and metabolic disorders are made possible by the field of nutrigenomics. Research on differential gene expression will make it possible to identify potential genes and the metabolic pathways underlying traits with significant economic value. Nutrigenomics studies how diet affects gene expression or regulatory systems that may be connected to a range of biological processes influencing animal health and productivity in the context of changing nutrition (Asmelash, Mahlet, and Brhane, 2018).

Functional genomic techniques, such as transcriptomics, proteomics, and metabolomics, have long been heralded as having enormous potential to advance nutritional science through the discovery of novel nutritional biomarkers and a fresh perspective on diet-gene interactions. One could argue that transcriptomics is concerned with the pattern of gene expression, whereas genomics summarizes the instructions provided by DNA. Proteomics is the study of complex protein products and their interactions; metabolomics is a step toward understanding an organism's whole metabolism. Practical genomic testing technologies are always developing and getting better. This has led to an ongoing expansion in the field of nutrigenomic research. Real-time quantitative polymerase chain reaction (PCR), DNA sequencing, and DNA microarray techniques can all be used to evaluate the genetic relationship with nutrition. Many genes can now be sequenced thanks to chip technology, which also makes it possible to accurately comprehend how patterns of gene expression change over time. Nutrigenomics is a viable and affordable alternative to costly transgenesis for boosting animal productivity, growth rate, milk production, feed utilization, disease resistance, and fertility (Figure 5.2). By bridging the genetic map of animal husbandry and feeding with the aid of genomics, productivity and transformation efficiency can be increased. Previously, Differential Display PCR and Real-Time Quantitative PCR were used in nutrigenomics studies to track the expression of specific genes. Numerous biological issues can now be solved more successfully thanks to developments in genome analysis. Large-scale gene expression analysis is now feasible thanks to the development of microarrays and massive transcriptome sequencing (RNAseq).

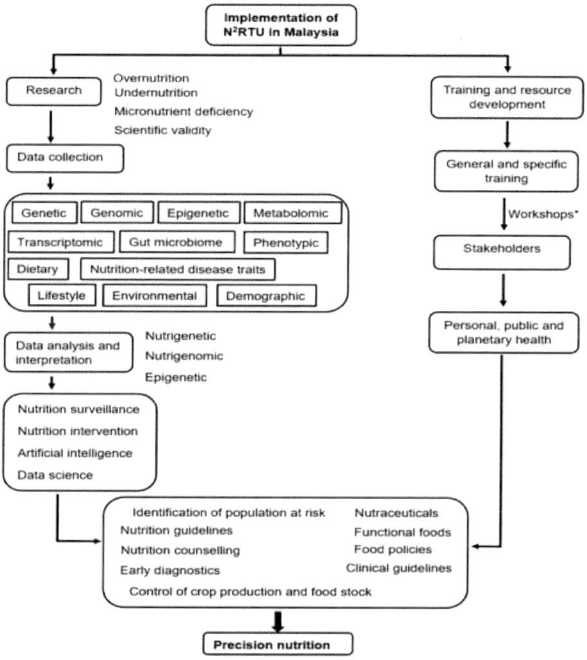


**Figure 5.2.** *Integration of OMIC technology in animal nutrition research.*

*Source: Dhanapal, A.C.T.A.; Wuni, R.; Ventura, E.F.; Chiet, T.K.; Cheah, E.S.G.; Loganathan, A.; Quen, P.L.; Appukutty, M.; Noh, M.F.M.; Givens, I.; et al. (2022). Implementation of nutrigenetics and nutrigenomics research and training activities for developing precision nutrition strategies in Malaysia. Nutrients, 14, 5108. https://doi.org/10.3390/nu14235108.*

These cutting-edge technologies were predicated on the notion of automating and parallelizing DNA/RNA and protein synthesis in order to gather large amounts of data. The most widely used omics techniques for obtaining transcriptomic, metabolomic, and proteomics data have been developed. But now, with the development of bioinformatic tools, scientists can use existing systemic methodologies to study interactions within living animals in conjunction with biological data from transcriptomics and genomics. Using large-scale gene expression techniques, nutrigenomics examines the interactions and effects of nutrition on gene expression (Hasan et al., 2019). By examining gene expression in animals exposed to varying nutritional conditions and the metabolic pathways directly involved in regulating tissue structure, which in turn contributes to tissue quality, researchers can gain more insight into the molecular underpinnings of observed phenotypic differences within organisms.

### 5.1.1. Transcriptomics

The transcriptome was created when it became possible to quantify the expression of almost all transcribed genes in a sample thanks to developments in RNA sequencing and microarray technology. As the next-generation high-throughput sequencing technology, transcriptome research has enhanced our understanding of the RNA-based gene regulatory network. A transcriptome is a snapshot of all the transcripts present in a cell at a specific point in time. In 2007, feed-deprived goats underwent a nutrigenomic analysis of their transcriptome responses, a first for small ruminants. According to this study, the early stages of mammalian involution are influenced by genes that lower milk fat, lactose, and protein levels, inhibit the growth and differentiation of mammalian cells, and increase programmed cell death. In omics science, RNA sequencing has recently taken the place of microarray systems, having been in use since 2011. The transcriptome and micro RNAome of small ruminants, like goats and sheep, were studied using this technology.

Animal nutritionists now prefer to use high-throughput RNA-seq to establish a framework for creating feeds or feed additives that support animal growth, health, and productivity. Using this technique, researchers are looking into global gene expression patterns in tissues associated with economically important characteristics such as feed effectiveness or the quantitative identification of genes or transcripts that may serve as future production trait biomarkers. In fact, recent developments in the sequencing of farm animals' genomes—including those of pigs, cattle, and sheep—have provided crucial reference genomes for the alignment of RNA-Seq short reads, which are used to evaluate alterations in gene expression in response to dietary nutrients.

### 5.1.2. Proteomics

The term proteomics was first used in 1995 to refer to the comprehensive examination of cell lines, tissues, and organisms' whole protein complement. In short, it is the study of proteins, primarily their structure and function. Nutrigenomics, the study of how our genome is expressed in response to nutrition overall, includes proteomics as a key component. Proteomics aims to provide a more comprehensive and integrated understanding of biology by examining every protein in a cell at once. Proteomics has not been widely used in cattle previously because of high cost and a lack of well-optimized protocols. On the other hand, improved knowledge of animal health, productivity, and reproduction efficiency has resulted from the development of new computational tools and analytical techniques for the analysis of proteomic data.

Numerous diagnostic targets, vaccine candidates, pathogenicity factors, functioning protein pathways in different disorders, and changes in expression patterns in response to different signals are all identified by proteomics-based approaches in a range of research settings. Since the sequencing of the human genome, the proteomic scope has expanded from protein recognition and characterization to include protein-protein interactions, protein structure, and protein function. Proteomic research employs various technologies such as mass spectrometry (MALDI-MS, ESI-MS, LC-MS/MS, MALDI-TOF MS, etc.). The ionized peptide mass to charge (m:z ratio) for protein identification; yeast two-hybrid screens; one-dimensional (ID) and two-dimensional (2D) gel electrophoresis; gas and liquid chromatography; and computer prediction programs.

Proteomic approaches have become increasingly popular in the last ten years as a means of studying reproductive issues. Targeting strong protein markers is helpful for surveillance, disease monitoring, overall health monitoring, and the assessment of the animal's response to medications. Proteomics is used to find genetic variants with desirable traits for breeding and selection for different post-harvest animal products, such as milk, meat, cheese, etc. (Almedia et al., 2015).

In addition to clinical management and the treatment of human diseases, proteome approaches to farm animals have been used to enhance animal welfare and avoid a detrimental impact on animal production and human health. The majority of farm animals' serum proteome analysis and characterization are still in the early stages because their genomes have not been fully sequenced like those of humans. Measuring the expression levels of proteins and metabolites offers new insights and methods for assessing and controlling the nutritional balance in farm animals. Such studies' hypotheses can be extrapolated from comparable human studies that support the optimization of diet composition and dietary nutrient supply in specific livestock species populations or individual animals.

Proteomics may be able to assess zoonotic disease risk in farm animals, boost fertility and productivity, and enhance the safety and caliber of animal products—all of which could lower costs. Over the past ten years, the proteomic investigation has been applied to veterinary and animal sciences;

however, its full potential in animal health and production has not yet been realized. Using quantitative proteomic analysis, biofluids and tissue samples from animals such as sheep, cattle, pigs, chickens, cats, and dogs have been used to find novel biomarkers that may be present in an experimental model of *Streptococcus uberis*-induced dairy cow mastitis. Certain milk proteins, such as haptoglobin, cathelicidin, and mammary-associated serum amyloid A3, may be used as biomarkers for diseases. These indicators aid in the diagnosis of both clinical and subclinical mastitis. Proteomic analysis of chicken plasma after *E. coli* stimulation of the inflammatory response is consistent with this. Major alterations in the plasma proteome are revealed by coli LPS (Eckersall, 2020).

In a recent study, LC-MS/MS analysis revealed the overexpression of eight fertility-related proteins in Tibetan pigs that had hereditary adaptation to hypoxic environments. Proteins and fertility were found to significantly correlate, according to LC-MS/MS analysis of seminal plasma proteins. Proteomics can be used to find bulls that are fertile, which will boost output and lower the non-return rate (NRR). Proteomic tools are used to identify superior genetic variants for selective breeding. Mullins et al. (2021) have identified protein markers by proteomics which will help in the selection of genetic variants, accelerate genetic gain and increase profitability.

Over the past ten years, proteomics has emerged as the primary platform for investigating plasma and serum proteomics to uncover next-generation biomarkers in both the veterinary and human sectors. This has been facilitated by the introduction of standard pre-analytical sample therapies, MS-based systems, and data processing systems for data management. The proteomic approach is faster than ELISA and PCR because it allows for the simultaneous investigation of a large number of serum proteins. Nevertheless, a major obstacle to the technique's wider application is the expensive initial cost of equipment acquisition. But the expenses incurred have rapidly dropped, making their purchase more desirable. New proteomic technologies and the identification of next-generation biomarkers that aid in the diagnosis of a variety of diseases in humans and farm animals have the potential to greatly enhance surveillance.

### 5.1.3. Metabolomics

Metabolomics is the study of metabolites, which are small molecules, in biological samples, such as microbes, plants, and animals. Through a comparative analysis of metabolome profiles, also known as metabolic phenotypes or "metabotypes," patterns of variation between various groups can be identified i.e., healthy versus sick, control versus treated. In order to incorporate prognostic techniques in animal health and improve prediction accuracy, metabolomics in livestock has made significant advancements in animal welfare, reproduction, and development. It has also helped identify biomarkers for transformation diseases and production traits in dairy and beef cattle. Furthermore, a multitude of bio-fluids have been examined for metabolomics research, encompassing metabolic profiling of plasma, milk, urine, and serum, thereby mitigating concerns regarding animal welfare (Gowda et al., 2008).

There are currently two basic metabolomics methods in use: i.e., mass spectrometry (MS) and nuclear magnetic resonance (NMR) spectroscopy. But mass spectrometry is frequently used in conjunction with a separation method, like capillary electrophoresis (CE), liquid chromatography (LC), or gas chromatography (GC). In terms of analysis time and sample cost, nuclear magnetic resonance (NMR) has surpassed mass spectrometry (MS) for high-throughput analysis. Typically, the LC analysis takes five to thirty minutes, and the NMR spectrum takes five to eight minutes Wishart, Feunang, and Marcu (2018). Mass spectrometry (MS) covers a wider area since of its high sensitivity. Because of its potential performance advantages in terms of metabolites per minute, more cutting-edge technologies have been developed in this field than NMR. Recent developments in technology have led to a shift in the field of metabolomics toward more rapid, thorough processing with fewer samples at a lower cost per sample (more for less). It will be simpler to implement fully automated (open source) data analysis systems to process the massive amounts of data gathered from omics studies in order to produce novel biological insights if data exchange initiatives and standard data reporting are increased.

In metabolomics research, integrating data with phenotypic and other omics data is the most difficult task. Expanding the types and number of animal breeds studied, as well as examining a larger range of samples like saliva, urine, amniotic fluid, and semen, would be a more thoughtful approach to the design and execution of metabolome studies.

### 5.1.4. Lipidomics

Global meat production has recently focused on raising the marbling and fatty acid profile of meat. Gaining more insight into the mechanisms controlling fat synthesis is necessary to achieve induced marbling. It has recently been found that vitamin A has a positive effect on practically every stage of animal adipogenesis. The active metabolite of vitamin A, retinoic acid receptor (RAR) and retinoid X receptor (RXR), can be stimulated by retinoic acid, and this can result in epigenetic mutations in the central regulatory genes of adipogenesis. Receptors for retinoic acid, vitamin D, and folate control adipogenesis and prevent the buildup of fat. The development and differentiation of adipose tissue are altered by vitamin D through its interaction with retinoic acid receptor signals. Furthermore, during particular stages of pregnancy, such as the prenatal period, nutrients are manipulated in such a way that they affect not only the developing fetus but also the primary germ cells that give rise to the next generation. Giving particular nutrients will therefore have both immediate and long-term effects.

These recently developed techniques will aid functional lipidomics in characterizing and comprehending the distinct roles of every lipid subtype. By directly applying lipidomics to human plasma, new and predictive disease biomarkers, such as the course of coronary disease and treatment response, may be found. Lipidomics offers a promising technique for examining the connection between dietary lipids and cardiometabolic risk, enabling the identification of lipid biomarkers, since many distinct lipid species are synthesized *in vivo* from dietary precursors.

### 5.1.5. Impact of the Relationship Between Nutrients and Genome in Animal Health and Production

In addition to adjusting to environmental stresses, there has been a remarkable increase in the amount of effort put into meeting the significant challenges posed by the sharply anticipated rise in the world's demand for food and premium animal proteins. The goal of nutrigenomics research is to uncover the molecular signatures required to create successful nutritional strategies and to explain the molecular interactions between genes and diet. The majority of nutrigenomics research has focused on how human nutrition affects the genesis of disease. Because specific gene transcripts are necessary for certain biochemical parameters, studies of gene expression analysis are conducted in response to dietary regimens. The experiments in dairy cattle concentrate on the composition and yield of milk.

On the other hand, research on beef cattle focuses on determining the fatty acid composition of muscle tissue and subsequently adjusting feed regimens. Pigs and poultry alike depend on animal health, which is why prebiotics, vitamins, and amino acids are used as supplements. These substances influence gene transcription and are critical for enhancing immune system performance.

Most studies on how diet affects genome functioning have used experimental models such as mice, rats, and pigs. Rats and mice have been used in studies of particular life stages, such as pregnancy, and rodents have also been used to study the phenomenon of fetal programming, which shows how mother nutrition influences the functioning of the progeny's genome. The use of similar strategies for farm animals is still in its infancy. Nonetheless, the ideas of nutrigenomics have become increasingly significant, particularly in the meat industry and in enhancing animal health. According to a recent study by Kang, Madkour, and

Kuenzel (2017), birds' early nutritional stress and psychological stress affected key DNA methylation regulators. Another study found that under conditions that are challenging for acidosis, there is a relationship between nutrients, bovine genotype, and the enteric biome. For metabolites such as butyrate and lactate, acidosis, and host-microbiome interactions, several putative QTL were found. It has been reported that feeding broilers a diet high in *Morinda citrifolia* (Noni) reduces heat stress by modifying the hypothalamic orexin-AMPK-mTOR pathways. These studies demonstrated that by identifying important molecular markers for personalized nutrition approaches, the interaction between nutrition and genome may be able to modulate and alter future dietary guidelines.

## 5.1.6. Epigenetic Control on Livestock Production

Gene-diet interactions are the subject of the study of nutritional epigenetics. We are still in the early stages of studying nutritional epigenetics in livestock production. Nonetheless, a basis for research on nutritional epigenetics has been established and verified. Diet may influence the first-generation consumer's epigenetic composition, but it also has an impact on the performance of their offspring. Given that pigs and poultry have quick reproductive cycles and produce a lot of kids, this could be a historic year for the sector (Roberts et al., 2019).

When taking into account the phenotypic variation, there are numerous gene and environment interactions. Nutrient quantity and quality are important environmental factors in these phenotypes. For instance, honeybees and mice have been shown to exhibit nutrient-induced epigenetic events (Kucharski et al., 2008).

The perspectives of growth, feed intake, and efficiency are the main ones used to study nutritional phenotypes. The relationship between epigenetic modifications and nutritional phenotypes has been well-documented in studies on genome-wide nutritional epigenomics in livestock and other animals. In livestock, correlations between genetic variation, epigenetic mechanisms, and consequently economically significant phenotypes have been found. Within the livestock genomics community, one of the main objectives is to look into the economically important traits and find out how much of the variation in phenotype can be explained by genetic variation. Nevertheless, instruments that enable us to examine the epigenetic modifications must be developed in order to understand a mechanistic relationship between genotype and phenotype. Previously, next-generation sequencing techniques based on DNA methylation have been applied to epigenetic studies in livestock animals to explore the function of methylation in the phenotypic variation of their economically significant trait. In humans, a high-resolution DNA methylation array has been utilized. When the same technology is applied to animals, it will allow for EWAS and eventually allow for epigenetic modifications for breeding and livestock production (Rolf et al., 2014).

### 5.1.6.1. Epigenetic Control of Adipose Tissue

Progenitor cells and stem cells can maintain pluripotency or multipotency while promoting gene expression linked to stem cell self-renewal through reversible inhibition of lineage-specific genes. Pluripotency and multipotency genes are inhibited during separation, but lineage-specific genes are released. When a central development gene is activated, progenitor cells can commit to a specific lineage by activating a transcription factor and a gene specific to that lineage.

One of the most important CpG-rich genes in development, Zfp423 is mostly regulated by epigenetic changes. Other examples of epigenetic modifications are DNA methylation and histone modifications. Histone methylation is catalyzed by the Polycomb Inhibition Complex (PRC), which is in charge of reversible gene suppression. PRCs 1 and 2 are the two that have undergone extensive research. A key element that mediates the trimethylation of the histone 3-lysine 27 (H3K27me3) gene silencing marker is the PRC2senhancerZeste 2 (EZH2). PRC2 has been shown to bind to promoters with lots of CpG sites in order to attract PRC1 binding, but its precise DNA binding abilities have not been established. These promoters are typically found in DNA that is methylated when PRC release stimulation is not present.

The complexity of these epigenetic regulatory mechanisms in important developing genes is influenced by both genetic and environmental factors. Gene polymorphisms in critical developed genes alter how epigenetic modulation complexes bind and affect how the progenitor's cell lineage interacts with itself during development. Furthermore, environmental cues and indicators, like nutrition, affect animal development by altering cellular signage pathologies or by attracting epigenetic transcription factors, like adipogenesis.

### 5.1.6.2. Role of B-Complex Vitamins and Methyl Donors

Histone and DNA methylation are examples of epigenetic modifications; however, methylation is unlikely to occur without the assistance of methyl donors. DNA's methyl methionine (SAM) adds a methyl group to DNA by means of S-adenosyl methionine (SAM). SAM is transformed into S-adenosylhomocysteine (SAH) after being transferred to DNA or histones. Folic acid, vitamin B12, choline, and concrete nutrients support this process. Gene methylation is reliably promoted by food methylation, folic acid, vitamin B12, choline, and betaine. Wingless and Int (Wnt) 10b are modulated by folic acid and vitamin B12 therapy to prevent adipogenic development in 3T3-L1 cells. Burton et al., 2019 suggested the influence of methyl donors on adipogenesis through epigenetic alterations in processing genes.

### 5.1.7. Adipogenesis and the Role of Vitamin A, Retinoic Acid, and Vitamin D Receptors

Intracellular nuclear receptors are activated by a variety of lipid signaling components, including thyroid hormones, retinoids, steroid hormones, and vitamin D metabolites. To promote the expression of a gene, ligand-activated transcription factors bind to the appropriate DNA sequences. The effects of vitamin D and retinoids are discussed in the section that follows as a summary of adipogenesis. When dietary vitamin A is extracted and absorbed, all-trans retinoic acid is created. For retinoic acid receptors (RAR), retinoic acid functions as a ligand. Through retinoid X receptors (RXR), they will attach to retinoic acid reaction elements (RAREs) on target genes. Retinoic acid stimulates lipid oxidation, lipid breakdown, and cell proliferation through the orphan receptor PPAR. The biological effects of retinoic acid will depend on how it is divided between RAR and PPAR. Cellular retinoic acid-binding proteins CRABPII and fatty acid-binding protein type 5 (FABP5) control retinoic acid partitioning; CRABPII supplies retinoic acid to RAR, while FABP5 supplies retinoic acid to PPAR. Adipogenic progenitor cells express a high level of CRABP-II/FABP5, which results in the dominance of RAR signaling. During adipogenic differentiation, CRABP-II and RAR down-regulate, but FABP5 and PPAR/up-regulate, triggering PPAR/signaling in mature adipocytes. On progenitor cells and mature adipocytes, retinoic acid has different effects due to the stage-specific expression of associated transcription factors. Adipogenesis is thought to be regulated by vitamin D, which also plays a part in calcium homeostasis and bone metabolism. It has been documented that vitamin D in obesity influences tissue sensitivity to insulin, insulin secretion, and ultimately systemic inflammation. Local 1,25(OH)2D synthesis, CYP27B1 expression, and VDR activation in pancreatic beta-cells were all caused by the direct and paracrine effects of vitamin D. Polymorphisms in genes related to vitamin D synthesis and signaling decrease obesity and diabetes. It has been demonstrated that vitamin D inhibits adipogenesis in both *in vitro* and *in vivo* studies. In a dose-dependent manner, the active metabolite of vitamin D, 1,25-dihydroxy vitamin D, inhibits the development of 3T3-L1 preadipocytes and downregulates the expression of adipogenic genes. Adipogenesis is regulated by the 1,25-dihydroxy vitamin D receptor, which has been shown to be activated by 1,25-dihydroxy vitamin D (VDR). By downregulating C/EBP expression in the presence of 1,25-dihydroxy vitamin D, VDR inhibits adipogenesis.

### 5.1.8. Summing up the Role of Nutrigenomic in Nutrients

Adipocyte formation during the fattening phase is triggered by vitamin A during the early stages of adipocyte development, which increases marbling. On the other hand, metabolites of vitamin D limit the formation of adipocytes. By binding to the RXR through VDR, vitamin D displaces PPARG ligands and their receptors, which also bind to the RXR and are necessary for adipogenesis. Methyl-donors that control epigenetic modifications during the early stages of adipogenesis may also have an impact on adipogenesis. Conversely, in feedstock beef cattle, vitamin A supplementation promotes lipid degradation, which reduces adipocyte hypertrophy and marbling. Assume a newborn receives vitamin A supplements until they are weaned. In that scenario, there might be an increase in intramuscular adipocyte production, which would lead to an increase in intramuscular pre-adipocytes that could store fat (hypertrophy) during the fattening stage and promote marbling without making the carcass fatter.

Adequate balanced feeding and extensive research on the short- and long-term effects of alternative nutrients are the first steps towards producing healthy meat. Because the specific gene transcript will determine any changes in the biochemical parameter, studies on gene expression are conducted with reference to dietary regimens. Even though studies on transcriptomes have been done on farm animals, more in-depth research on protein levels needs to be done. Moreover, it is still unclear how diet affects epigenetic processes in farm animals (Nowacka-Woszuk, 2020). Thus, a comprehensive approach encompassing genome functioning, breeding objectives, and physical traits ought to be employed in nutrigenomic studies. Cutting edge technology is already being heavily utilized in nutrigenomic studies on humans and rodents, and it will most likely be used more frequently in studies on other species.

## 5.2. NUTRIGENOMICS IN CATTLE

In order to comprehend how dietary components affect gene expression and function, nutrition, genomics, and molecular biology are combined in the field of nutrigenomics. It focuses on determining which genes and pathways react to different nutrients and how these reactions impact the phenotypic (growth, health, reproduction, and production traits) of the animal. The goal of nutrigenomics in cattle is to improve performance and welfare by customizing nutrition based on genetic predispositions.

### 5.2.1. Importance of Nutrigenomics in Cattle Production

The agricultural industry prioritizes the efficient and sustainable production of cattle because they are an essential source of meat, milk, and other products. The goal of traditional nutrition has been to provide cattle with balanced diets that meet their overall needs. Nutrigenomics, on the other hand, enables more focused nutritional interventions that may increase feed efficiency, improve the quality of milk and meat, strengthen immunity, and have a smaller environmental impact.

### 5.2.2. Mechanisms of Nutrigenomic Action in Cattle

Nutrigenomics operates through several key mechanisms in cattle:

- **Epigenetic Modifications:** Without changing the DNA sequence, nutrients can induce epigenetic modifications that control gene expression, such as histone modification and DNA methylation. For instance, the necessary amino acid methionine participates in methylation processes that affect the expression of genes linked to the growth and development of cattle.
- **Regulation of Transcription Factors:** Transcription factors that regulate the expression of several genes can be activated or repressed by nutrients. For example, fatty acids have an impact on the PPAR (peroxisome proliferator-activated receptor) family, which regulates genes related to lipid metabolism and impacts fat deposition and milk composition in dairy cattle.
- **Nutrient Sensing Pathways:** Protein synthesis and growth are regulated by pathways like the mTOR (mechanistic target of rapamycin) pathway, which reacts to nutrients like amino acids. Leucine-mediated mTOR activation in cattle can promote muscle growth and accelerate growth rates.

### 5.2.3. Applications of Nutrigenomics in Cattle

Nutrigenomics offers several practical applications in cattle production, including:

- **Precision Feeding:** Nutrigenomics, the study of the genetic composition of cattle, can direct the development of diets customized for individual animals or breeds. This method contributes to more sustainable cattle farming by optimizing nutrient use, enhancing growth and milk production, and minimizing waste.
- **Improving Milk Quality and Yield:** The impact of various dietary

components on the expression of genes involved in the composition and production of milk can be determined through nutrigenomics. For instance, feeding techniques that modify the diet's fatty acid composition can affect the expression of genes that regulate the synthesis of milk fat, enhancing the milk's quality for use in dairy products.

- **Enhancing Growth and Meat Quality:** The impact of dietary nutrients on genes linked to muscle growth and fat deposition can be uncovered through nutrigenomic studies. For example, adjusting the dietary levels of particular amino acids can change the expression of genes linked to muscle growth, improving the quantity and quality of meat produced.
- **Boosting Immune Function and Disease Resistance:** Cattle's resistance to disease can be increased by specific nutrients, including vitamins and trace minerals, which can improve the expression of genes related to immune response. For example, it is known that zinc and selenium affect the expression of antioxidant enzymes, which helps cattle fight infections and oxidative stress.

## 5.2.4. Case Studies in Cattle Nutrigenomics

Several studies have demonstrated the potential of nutrigenomics in enhancing cattle production:

- **Methionine Supplementation in Dairy Cows:** Studies have demonstrated that adding methionine to the diets of dairy cows can affect the expression of genes related to the synthesis of milk proteins, increasing the amount of protein and yield of milk produced. This is accomplished by controlling the transcription of important genes involved in milk production through epigenetic mechanisms.
- **Omega-3 Fatty Acids and Milk Composition:** It has been demonstrated that adding omega-3 fatty acids, like those in flaxseed or fish oil, to cattle diets can improve the fatty acid profiles of their milk by changing the expression of genes related to lipid metabolism. This can improve milk's nutritional value by increasing its concentration of healthy fatty acids like DHA (docosahexaenoic acid).
- **Amino Acid Regulation of Muscle Growth:** Studies have investigated how different levels of dietary amino acids, particularly leucine and lysine, affect the expression of genes related to muscle protein synthesis in beef cattle. These findings have been used to

optimize feed formulations that promote lean muscle growth, improving the efficiency of meat production.

### 5.2.5. Challenges and Limitations of Nutrigenomics in Cattle

Despite its potential, the application of nutrigenomics in cattle production faces several challenges:

- **Complexity of Gene-Nutrient Interactions:** Numerous factors, such as age, breed, and environmental conditions, can impact the intricate interactions between genes and nutrients. Because of this complexity, it is challenging to forecast with precision how a given dietary component will affect gene expression and animal performance.
- **High Costs and Technological Barriers:** Advanced technologies, which can be costly and not always available, are needed for nutrigenomics research. Examples of these technologies include high-throughput sequencing, bioinformatics analysis, and gene expression profiling. Additionally, not all cattle breeds have access to the comprehensive genetic information needed to implement precision feeding strategies based on nutrigenomics.
- **Ethical and Regulatory Considerations:** Animal welfare and genetic modification are two ethical issues that are brought up by the application of nutrigenomics in livestock production. Although improving productivity and health is the aim, care must be taken to make sure that these methods don't endanger animal welfare or have unforeseen repercussions.

### 5.2.6. Nutrigenomic Roles in Dairy Cows of Ligand-Dependent Nuclear Receptors

The nuclear receptor superfamily includes the LdNR. Of the 48 known LdNR in humans, only 27 have endogenous ligands that are known to be found in lipophilic compounds in general, such as fatty acids, hormones, and xenobiotics. The LdNR with nutrigenomic significance are those that can bind to and activate in response to molecules derived directly or indirectly from food, like vitamins and fatty acids (Table 5.1). On the other hand, as PPAR has amply shown, phosphorylation of the LdNR can also activate it independently of the agonist or agonists. LdNR binds DNA as a dimer, either homo- or heterodimer, once it is activated.

**Table 5.1.** Studies of Transcriptome Expression Analyses by Using DNA Microarray Technology

| Organism | Experiment | Consequences | Reference/s |
|---|---|---|---|
| MCF-7 breast cancer cells | Influence of genistein on gene expression at global levels at physiologic (1 or 5 mu M) and pharmacologic (25 mu M) concentrations | Genistein altered the expression of genes of various pathways, plus estrogen- and p53-mediated pathways. At physiologic concentration (1 or 5 μM), genistein elicited an elevated expression and mitogenic activity, while at the pharmacologic concentration (25 μM), genistein escalated apoptosis, reduced proliferation, and total cell number | Lavigne et al. (2008) |
| Rat model of an alcohol-induced fatty liver | Analysis of hepatic gene expression | The five genes (β-glucuronidase, UDP-glycosyl transferase 1, UDP glucose dehydrogenase, apoC-III, gonadotropin-releasing hormone receptor) involved in immune response, signal transduction, transcription, and protein and amino acid metabolism were controlled by chronic ethanol intake | Park et al. (2008) |
| Blood cell RNA of eight healthy men prior and post 2 h diet ingestion | Effect of high-carbohydrate (HC) or high-protein (HP) breakfast on the transcriptome of human blood cells | Genes (317) for HC breakfast and genes (919) for HP breakfast showed differential expression. HC breakfast ingestion showed differential gene expression of mainly glycogen metabolism and HP breakfast showed differential expression of protein synthesis genes | van Erk et al. (2006) |
| Lymphocytes from 30 post-menopausal women | Effects of dietary soy isoflavones on changes in expression of genes | Isoflavones had a robust effect on some putative estrogen-responsive genes in equol producers than non-producers due to enlarged cell differentiation, cAMP signaling, G-protein–coupled protein metabolism and steroid hormone receptor activity | Niculescu et al. (2007) |

| | | | |
|---|---|---|---|
| Subcutaneous adipose tissue (SAT) in 47 persons with metabolic syndrome | Gene expression after consumption of two carbohydrate modifications (rye–pasta diet with low postprandial insulin response and oat–wheat–potato diet with high postprandial insulin response) | The rye–pasta diet downregulated 71 genes related to insulin signaling and apoptosis. The oat–wheat–potato diet upregulated 62 genes connected to cytokine-chemokine-mediated immunity, stress, and interleukin pathway | Kallio et al. (2007) |
| Adipose tissue from 131 moderately overweight men | Identification of molecular pathways responsive to caloric restriction and dietary composition | Above 1,000 transcripts showed downregulated expression after acute weight loss. The expression of stearoyl-coenzyme A desaturase (SCD) in adipose tissue is autonomously controlled by weight loss and by ingestion of carbohydrates and saturated fat. The expression of SCD and diacylglycerol transferase 2 (DGAT2) may be implicated in the dietary regulation of triacylglycerol metabolism | Mangravite et al. (2007) |
| Rectal mucosa in randomized double-blind crossover trial on 19 healthy volunteers | Effect of daily intake of low-digestible and prebiotic isomalt and digestible sucrose on gene expression for 4 weeks of feeding | No influence on gene expression in lining rectal mucosa after dietary intervention while gene expression of the rectal mucosa can be measured in biopsy material | Schauber et al. (2006) |

Source: Pandita, D., and Pandita, A., (2022). omics technology for the promotion of nutraceuticals and functional foods. Front. Physiol. 13:817247. doi: 10.3389/fphys.2022.817247.

The LdNR is nearly exclusively found in the nucleus, though some has also been seen to be in the cytosol. However, in the absence of an agonist, the majority of LdNR are trapped in a complex made up of inactive LdNR linked to other proteins such as heat shock protein 70 and 90. The LdNR breaks away from the complex when it is activated by the agonist. A nuclear localization signal found in several LdNR enables their transfer into the nucleus by the nuclear shuttling protein importin α. The PPAR, RXR, VDR, and RAR nuclear hormone receptors are among the type II nuclear hormone receptors that are typically found in the nucleus but can also be found in the cytoplasm. Certain agonists and, more powerfully, RXR activation cause the PPAR to localize to the nucleus. It has been shown more recently that distinct agonists can have varying effects on the import and export of PPAR from the nucleus. Thus, the shuttling of the LdNR between the cytoplasm and nucleus determines the activity of the LdNR in addition to the agonist's presence and the abundance of the LdNR and its coactivators.

### *5.2.6.1. Peroxisome Proliferator-Activated Receptors*

#### *5.2.6.1.1. Biological Roles of Peroxisome Proliferator-Activated Receptors in Dairy Cows*

The LdNR PPARs function as heterodimers with RXR and are important in regulating the expression of genes related to inflammation and lipid metabolism. Recently, a comprehensive review was published on PPAR and their potential nutrigenomic role in dairy cows. The review covered the functions of PPARα in regulating hepatic inflammation and lipid metabolism, the possible functions of PPARβ/δ in regulating glucose uptake in breast tissue, and the potential functions of PPARγ in regulating mastitis and milk fat synthesis. Recent overexpression of transcription factor specific protein 1 (Sp1) in goat mammary epithelial cells increased the expression of several genes related to milk fat synthesis, including PARAG (but not SREBF1 or NR1H3), providing additional evidence of the important role of PPARγ in controlling milk fat synthesis. Data from a study with lactating mice showed that rosiglitazone, a PPARγ agonist, was unable to reverse the milk fat depression caused by trans-10,cis-12 CLA (t10,c12 CLA), suggesting that PPARγ is not involved in the synthesis of milk fat. This was expected since it had previously been shown that PPARγ plays no function in murine mammary epithelial cells.

An *in vivo* study using lactating dairy goats as a model for dairy cows produced preliminary evidence that PPARγ activation influences the response to mastitis. To induce subclinical mastitis, goats were injected with 2,4-thiazolidinedione (2,4-TZD), a synthetic PPARγ agonist, for one week before receiving an intramammary infusion of Streptococcus uberis. The total number of somatic cells in the milk of the goats given a 2,4-TZD injection was decreased. Subsequent investigations showed that 2,4-TZD improved the response of neutrophils and the liver to inflammation. Remarkably, there were no overall effects of 2,4-TZD found in adipose or mammary tissue on the expression of putative PPARγ target genes. It was also observed that the expression of potential PPARγ target genes was not affected by 2,4-TZD in the adipose tissue of dairy cows. Researchers have employed a luciferase gene reporter assay in Madin-Darby bovine kidney (MDBK) and bovine mammary epithelial (MacT) cells to determine whether 2,4-TZD is a PPARγ agonist (Figure 5.3). It was evident from the data that 2,4-TZD does not cause PPAR activation. Nonetheless, the simultaneous activation of RXR by its particular agonist, 9-cis-retinoic acid derived from vitamin A, and the activation of PPAR are closely linked. We saw a significant activation of PPAR by 2,4-TZD when cells were treated with it in the presence of 9-cis retinoic acid in the media. Therefore, it appears that *in vivo* vitamin A supplementation may be required as 2,4-TZD alone is not an effective treatment for activating PPARγ.

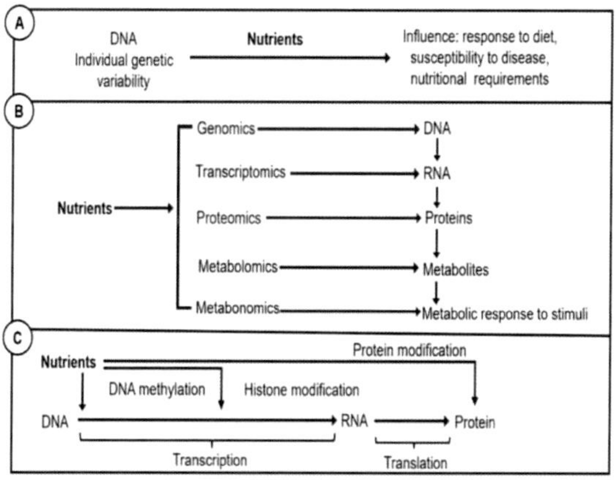


**Figure 5.3.** *Representation of: (A), Nutrigenetics; (B), Nutrigenomics; and (C), Nutri-epigenetics.*

*Source: Dhanapal, A.C.T.A.; Wuni, R.; Ventura, E.F.; Chiet, T.K.; Cheah, E.S.G.; Loganathan, A.; Quen, P.L.; Appukutty, M.; Noh, M.F.M.; Givens, I.; et al. (2022). Implementation of nutrigenetics and nutrigenomics research and training activities for developing precision nutrition strategies in Malaysia. Nutrients, 14, 5108. https://doi.org/10.3390/nu14235108.*

An experiment conducted in our laboratory with bovine mammary alveolar (epithelial) cells only partially supported the previously proposed role of PPARβ/δ in regulating glucose uptake in mammary tissue. Furthermore, it has been recently shown that PPARβ/δ regulates the expression of a carnitine transporter in MDBK cells, which implies that it can have an impact on mitochondrial lipid peroxidation.

### 5.2.6.1.2. Peroxisome Proliferator-Activated Receptors and Long-Chain Fatty Acids

The ability of the PPAR to bind to and be activated by LCFA in both monogastrics and ruminants makes them extremely interesting in the field of nutrigenomics. PPAR receptor sensitivity to LCFA is species-specific, according to research done on LCFA in dairy cows. When compared to saturated LCFA, the PPAR in monogastrics are more sensitive to PUFA. Particularly, CLA and linoleic acid are strong PPAR activators, though there have been conflicting findings for PPARγ. For example, in humans, mice, and rats, PPARγ responds to these LCFA at low to null levels, and in human preadipocytes, t10, c12 CLA inhibits PPARγ activity. According to reports from others, CLA can bind to and activate PPARγ in both human cells and mice living *in vivo*.

Studies using target gene expression in dairy cows for both PPARα and PPARγ revealed a definite "preference" for saturated LCFA, specifically palmitate and stearate. Unsaturated LCFA generally only slightly activate PPAR in bovine cells, and the effect gets smaller the more unsaturated the LCFA is. However, *in vitro*, higher concentrations of unsaturated LCFA were associated with increased expression of CPT1A, a PPARα target gene. This could suggest that, as has been seen in monogastrics in the past, there are differences in PPAR isotypes in the response to LCFA.

Recently, it was shown that t10,c12 CLA is a weak activator of PPAR in MacT cells using a luciferase gene-reporter assay. The activation of PPAR in MacT cells and nonpurified primary goat mammary cells transiently transfected with 3x PPAR response element (PPRE) luciferase was tested in our lab using palmitate, t10, c12 CLA, and the PPARγ agonist rosiglitazone. In MacT cells, rosiglitazone strongly activated PPARγ, confirming earlier findings based on putative target gene expression. On the other hand, rosiglitazone only slightly (only numerically) activated PPARγ in goat mammary cells, which partially validates the low activation that was recently found in goat primary mammary cells. Researchers found that palmitate strongly activated PPAR in both cell types, but not with t10, c12 CLA. Researchers also found that, in comparison to the particular PPARγ agonist rosiglitazone, palmitate has a similar or stronger activation of PPAR, as previously shown by gene expression analysis. This method cannot be used to determine whether palmitate activates particular PPAR isotypes. On the other hand, we have previously shown that, in goat and cow cells, saturated LCFA frequently has a greater effect than artificial agonists at increasing the expression of both PPARα and PPARγ target genes. Monogastrics do not appear to be more sensitive to natural agonists than synthetic ones when it comes to PPAR sensitivity. Given that dietary unsaturated LCFA are extensively biohydrogenated in the rumen, saturated LCFA are the main

LCFA absorbed from the small intestine, which explains the marked response of PPAR to saturated LCFA in ruminants (Figure 5.4).

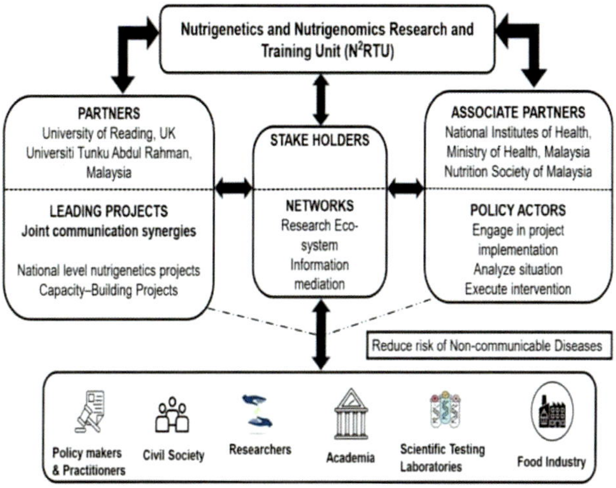


**Figure 5.4.** *Nutrigenomics-an example.*

***Source:*** *Dhanapal, A.C.T.A.; Wuni, R.; Ventura, E.F.; Chiet, T.K.; Cheah, E.S.G.; Loganathan, A.; Quen, P.L.; Appukutty, M.; Noh, M.F.M.; Givens, I.; et al. (2022). Implementation of nutrigenetics and nutrigenomics research and training activities for developing precision nutrition strategies in Malaysia. Nutrients, 14, 5108. https://doi.org/10.3390/nu14235108.*

## *5.2.6.2. Liver X Receptor and Control of Milk Fat Synthesis*

The LXR are LdNR and play a major role in regulating cholesterol synthesis in nonruminant animals. LXRα (gene symbol NR1H3) and LXRβ (gene symbol NRH2) are the two isoforms of LXR. Fatty acids and other derivatives of cholesterol metabolism, as well as oxysterols, are the primary activators of the LXR. In dairy cows and monogastrics, NR1H2 is expressed everywhere while NR1H3 is highly expressed in the liver. Both LXR isoforms are expressed at comparatively low levels in the mammary tissue of cows. It has been shown by multiple independent studies that LXRα, even in low abundance, regulates SREBF1 expression in the mammary epithelial cells of goats and cows. In addition to regulating SREBF1 expression, LXRα can enhance lipogenesis in mammary cells and regulate the expression of lipogenic genes in a SREBP1-independent manner.

The LXR's ability to be activated by LCFA in monogastrics is comparatively low, and there is debate over PUFA's ability to inhibit it. Because LXRα in bovine mammary cells is not responsive to t10,c12 CLA, bovine LXR isotypes are not involved in the transcription inhibition brought on by t10,c12. The nutrigenomic significance of LXR during lactation stems from its numerous documented roles in nonruminants, such as cholesterol metabolism, glucose uptake and lipogenesis in adipose tissue, lipogenesis and very low-density lipoprotein formation in the liver, and reducing expression of inflammatory genes in macrophages. This is in addition to its possible role in regulating milk fat synthesis in dairy cows. Numerous investigations in the rumen epithelium and mammary tissue have offered indirect evidence regarding the involvement of LXR in the inflammatory response in dairy cows.

All things considered, the nutrigenomic potential of LXR is modest but encouraging. Only the nutrigenomic role of unsaturated LCFA was examined because all of the research done on dairy cows to date has concentrated on its function in milk fat depression. Determining whether saturated LCFA activates LXR in dairy cows would be interesting.

### *5.2.6.3. Other Ligand-Dependent Nuclear Receptors with Potential Nutrigenomic Roles in Dairy Cows*

#### *5.2.6.3.1. Vitamin D Receptor*

The nutrigenomic interplay between vitamin D and vitamin A may be made possible by the fact that VDR, an LdNR, forms a heterodimer with RXR, the target for the vitamin A metabolite 9-cis-retinoic acid. VDR is essential for calcium metabolism in dairy cows, particularly in the early postpartum period when milk fever is most common. In addition to its involvement in calcium metabolism, VDR has a long-term "nutriepigenomics" function that can be crucial for the general health of animals. VDR has a very high specificity for binding vitamin D in monogastrics, but it can also be activated by other naturally occurring substances such as ω6 and ω3 PUFA. Thus, nutrigenomic intervention with LCFA may be able to alleviate milk fever in dairy cows.

#### *5.2.6.3.2. The Pregnane X Receptor*

This LdNR regulates the expression of genes linked to detoxification, specifically those encoding cytochrome P450. In addition to xenobiotics, vitamin K2, several herbal extracts, and the metabolites of vitamin E also activate it. Because pregnane X receptor (PXR) is involved in gluconeogenesis, triglyceride synthesis, and bone mineral homeostasis in monogastric animals, research on PXR in dairy cows may be significant.

#### *5.2.6.3.3. The Hepatic Nuclear Factor 4 Alpha*

Dairy cows have also shown that this LdNR, which is highly expressed in the liver, is involved in the transactivation of the GH receptor. Early postpartum and GH treatment both result in an increase in its expression in the liver of cows. The acyl-CoA thioesters of myristic acid (C14:0) and palmitic acid (C16:0) are agonists of HNF4α in monogastrics, while the'H3 and ω6 PUFA and stearic acid (C18:0) are antagonists. To date, no research has been published examining the effects of LCFA on HNF4α activation in cow cells.

## 5.3. NUTRIGENOMICS IN PIGS

Nutrigenomics in pigs focuses on how diet influences gene expression and impacts pig health, growth, and overall productivity. This field combines nutrition, genomics, and molecular biology to understand how different nutrients affect genes related to growth, immune function, fat deposition, and meat quality in pigs.

### 5.3.1. Protein-Restricted Diet and Protein Sources

One of the main sources of meat is the pig farming sector. The majority of swine nutritionogenomic research has focused on the genes that control the function of the liver, intestine, muscle, and adipose tissues (Loor et al., 2015). A study involving a low protein diet with different sources of fat (palm kernel oil, soybean oil, and palm oil) was undertaken by Doran et al. (2006) with a hybrid Duroc × Large White × Landrace pig line. They investigated the expression and activity of SCD, a crucial lipogenic gene, in subcutaneous adipose and muscle tissues. Additionally, they measured the protein levels of FAS and ACACA, two important enzymes in the synthesis of new fatty acids. In response to protein restriction, levels of total fatty acids and monounsaturated fatty acids (MUFAs) were increased in muscle tissue. There was a positive correlation found between the SCD protein level and the intramuscular fatty acid level. The expression of FAS and ACACA proteins was also upregulated by the protein-reduced diet; however, the oil type solely had an impact on ACACA, which was upregulated in animals given palm oil supplements. To ascertain the transcriptome of muscle tissue in finished Duroc gilts, protein restriction was also employed. The expression microarray results demonstrated that by changing the expression of genes involved in lipid biosynthesis and degradation, a protein shortage during the most recent feeding period improved very important traits, such as IMF, in muscle tissue. Adipokines such as leptin (LEP), hypoxia inducible factor 1 subunit alpha (HIF1α), and tumor necrosis factor-alpha (TNFα) all had higher transcript levels. On the other hand, because the genes involved in the protein synthesis pathway were changed, the tested diet may also have decreased the ability for efficient growth by suppressing cell cycle and cellular growth. Nevertheless, this conclusion was solely supported by the prediction and the identification of the investigated genes as belonging to this specific biological pathway.

Pigs have already been used in studies on the sources of protein in food. For example, Schwerin et al. (2002) investigated the impact of diets based on isolates of soy or casein protein and discovered that a diet high in soy changed the transcript level of genes related to the liver's oxidative stress response. A soy-rich diet induced changes in protein biosynthesis at the same loci where transcriptional changes were already found, according to later studies by the same research group that examined the hepatic proteome. This suggests that testing transcript and protein levels simultaneously is necessary to provide a chance for conclusive results. Roh et al. (2015) also looked into soybeans as the main protein source for pigs. They postulated that using fermented soybeans rather than dry soybeans could affect how the immune system responded in piglets given a lipopolysaccharide injection as a kind of immune challenge. The oligonucleotide microarray analysis was utilized to determine the transcriptional profile in leukocytes. Forty genes were discovered to exhibit differential expression, with seventeen upregulated and twenty-three downregulated. These genes were primarily involved in lowering the growing piglets' inflammatory response following the lipopolysaccharide challenge, indicating possible advantages for growth rate in animals fed fermented soybeans (Roh et al., 2015).

### 5.3.2. Fat Sources in Pig Nutrition

Numerous pig studies on fat sources have been conducted. One study examined the effects of supplementing finishing pigs with CLA in relation to muscle tissue proteome changes and IMF content (Zhong et al., 2011). It was demonstrated that consuming CLA through diet increased IMF and raised the amount of proteins involved in the synthesis and oxidation of fatty acids, the metabolism of amino acids, and energy metabolism. Óvilo et al. (2014) studied the effects on fatty acid composition, lipid metabolism, and gene expression patterns in Iberian pigs of different MUFA energy sources, such as high-oleic sunflower oil as an exogenous source of MUFA (HO diet) and carbohydrate diet providing substrates for endogenous MUFA synthesis (CH diets).

They observed that the dietary treatment was reflected in the fatty acid profiles of the adipose, liver, and muscle tissues, with the HO diet having lower saturated fatty acids (SFA) and higher monounsaturated fatty acids (MUFA). The more precise qPCR method only confirmed significant differences for three genes: LEP, malic enzyme 1 (ME1), and retinoid X receptor gamma

(RXRG). Microarray technology revealed 37 differentially expressed genes in the adipose tissue. It was discovered that the chosen gene expression profile for liver and muscle tissues was unaffected by the diet. The same research group subsequently compared the HO diet to the CH diet while taking into account the breed effect (Iberian and Duroc pigs) and feeding versus fasting status. In the adipose tissue, the transcript levels of a few genes involved in lipogenesis, adipogenesis, and lipolysis were examined. The HO diet raised the amounts of MUFA and reduced the amounts of SFA in the adipose tissue in both breeds. Compared to the Duroc breed, the Iberian breed exhibited lower levels of PUFA and higher levels of SFA. Breed-specific variations in the transcript levels of LEP, ME1, SCD, and ELOVL6 were observed.

In the adipose tissue, feeding type HO versus CH affected the transcription of only one perilipin 1 (PLIN1) gene, while the feeding status changed the mRNA of the LEP, ME1, SCD, ACACA, ELOVL6, RXRG, PPARG, FASN, and G0/G1 switch 2 (G0S2) genes (Benítez et al., 2017). Recent work by this group demonstrated the strong breed effect on the transcript level in terms of growth, extracellular matrix formation, carbohydrate and lipid metabolism, and immune response. RNA-seq was used for global transcriptomic analysis of pig subcutaneous adipose tissue in Iberian and Duroc breeds. In Iberian breeds, the dietary effect was more pronounced, altering the transcription of genes linked to immune response, lipid metabolism, inflammation, and fat storage. It was proposed by the authors that structural variants (e.g., glucose response elements) that are nutrient-sensitive and have the ability to control gene expression and are situated close to a gene promoter. Thus, it is important to think about looking for long noncoding RNAs that control it (Benítez et al., 2019).

Diets with different fat sources (rapeseed oil, beef tallow, or coconut oil) and with the addition of dried distillers' grains with solubles as a source of protein were studied in adult pigs by Oczkowicz et al. (2016). By using RNA-seq to analyze the entire liver transcriptome, they discovered that the expression of several genes linked to lipid metabolism, including apolipoprotein A4 (APOA4), acyl-CoA synthetase long chain family member 5 (ACSL5), cytochrome P450 2B22 (CYP2B22), and glutathione S-transferase omega 1 (GSTO1), varied depending on the type of fat. The group that received coconut oil had the highest expression levels of APOA4, CYP2C49, and CYP2B, while the GST1 gene's transcript level was most affected by beef tallow. Subsequently, the liver transcriptome was examined in a study on PUFA supplementation (rich in α-linolenic acid and linoleic acid, respectively) in adult pigs. More than 3500 genes were found to have altered expression profiles. Genes involved in fatty acid activation, fatty acid β-oxidation, lipid transport activity, and lipid binding showed higher levels of expression; genes showing lower levels of expression were primarily related to prostaglandin biosynthesis or peroxidase activity. Furthermore, the experimental group supplemented with PUFAs experienced alterations in their hepatic lipid profile. The ratio of n-6 to n-3 fatty acids decreased as the level of n-3 fatty acids rose and the content of n-6 fatty acids decreased (Szostak et al., 2016).

### 5.3.3. Supplementation with Vitamins and Minerals

A balanced growth requires the right macro- and microelement composition of the diet. Energy homeostasis, which controls the amount of acetyl-CoA in the cytosol and mitochondria, depends on carnitine. It increases muscle mass and protein synthesis. After three weeks of feeding the developing piglets a diet enhanced with L-carnitine, the muscle transcriptome was ascertained. The researchers discovered that, out of the 211 genes that were differentially expressed, many were upregulated in the pathways linked to the binding of insulin receptors and insulin-like growth factor (IGF), while genes associated with atrophy and proapoptopic processes were silenced. This finding provides compelling evidence that L-carnitine may benefit growing piglets' skeletal muscle mass. Pigs have also been used to test diets supplemented with other amino acids, such as glutamine. Because it is necessary for the synthesis of purines and pyrimidines and is a crucial substrate for the synthesis of other endogenous amino acids like arginine, glutamine is necessary for healthy cell division. Within a Wang et al. (2008) investigated the potential of a glutamine-rich diet to prevent intestinal dysfunction and atrophy in weaned piglets. These conditions are frequently seen when the feeding type shifts from sucking to chow. After analyzing the transcriptome of the intestinal tissue, it was discovered that the piglets in the experimental group had higher expression levels of genes related to oxidant elimination and cell growth. Additionally, there was a downregulation of the immune-activating genes. Dietary supplementation with glutamine had a positive impact on intestinal oxidative defense capacity, body weight gain, and small intestine growth (Wang et al., 2008). Threonine is another amino acid crucial in gut physiology, and a study employing a threonine-deficient diet (30% reduction) was carried out in young piglets for 2 weeks by Hamard et al. (2010). A threonine deficiency was found to modify the expression of 324 genes related to energy metabolism, protein synthesis, immune and defense responses, and other processes, according to an analysis of the transcriptome of ileal tissue. Additionally, it was discovered that these animals were more capable of absorbing glucose and had higher paracellular permeability, suggesting that threonine may be essential for preserving intestinal integrity (Hamard et al., 2010). However, it must be pointed out that the mentioned above diet regimens were usually tested for a short time and the consequences for longer supplementation were not considered. Contrary to previous reports, long-term supplementation with selenium in adult pigs was carried out by Song et al. (2013) with the aim of determining transcriptome changes in leukocytes. Selenium is a dietary microelement that is necessary for proper thyroid hormone metabolism, as well as the functioning of the immune system and neural cells. The results obtained indicated that the transcriptional level of genes linked to both acquired and innate immunity increased, indicating that pig immunity may be enhanced by long-term selenium supplementation (Song et al., 2013). Recent research of a similar nature was also carried out by Sirri et al. (2018) in adult pigs, three distinct experimental diets were tested: (a) a diet rich in n-3 PUFA from extruded linseeds; (b) a diet supplemented with vitamin E and selenium as antioxidants; and (c) a diet containing plant extract and extruded linseeds as an additional source of polyphenols. The muscle transcriptome under these three feeding schedules was determined by RNA-seq. The findings showed that genes involved in muscle contraction and development were downregulated in a diet high in polyunsaturated fats. The transcription of genes involved in oxidative phosphorylation was decreased when vitamin E and selenium supplements were taken, indicating that the antioxidant properties of these nutrients helped to eliminate the reactive oxygen species (ROS) brought on by n-3 PUFA. Supplementing the diet with extra polyphenols appeared to have a significant impact on genes related to muscle growth, since it also increased the expression of genes related to immune system and lipid metabolism. The same research team later verified these findings in another study. Piglets were recently used to test the mannan-rich fraction derived from yeast (Fouhse et al., 2019); mannan is regarded as an antiproliferative and antioxidant polysaccharide that guards against gastrointestinal infections. This additive was tested in young piglets to determine how it affected gene expression, jejunal morphology, cecal microbial profiles, and growth performance. The diet changed the microbial structure in cecal digesta and raised the height of the jejunal villus, but it had no effect on performance traits. 1378 genes that were examined in jejunal tissue showed variations in their transcript levels, according to the expression microarray. A diet high in mannan may help maintain interstitial homeostasis because the changed genes were grouped in pathways related to immune modulation, protein synthesis, and cellular homeostasis and development (Fouhse et al., 2019).

## 5.4. NUTRIGENOMICS IN POULTRY

Similar to how nutrition and genetics are correlated, nutrition genomics is now crucial to the raising of chickens. It helps to improve meat quality, bird health, feed effectiveness, and bird performance. Let us first examine molecular biology before delving into the field of nutrigenomics. Each and every animal possesses a genome, which is its entire genetic makeup and serves as the basic blueprint for life. These masterfully designed molecules represent ideal animal lifestyles. However, external factors such as diet can have a significant impact on how this blueprint is read or how this genetic information is expressed. Nutritional genomics is a field that combines nutrition genomics and nutrition genomics to investigate the relationships between genes and nutrients. The study of how food ingredients and physiologically active foods affect an animal's or human's gene expression is known as nutrigenomics.

Based on heredity, "nutrigenetics" controls how the body reacts to food. In other words, nutrigenetics posits that an animal's reaction to food ingredients that are bioactive. The genetic makeup of the animal affects this because it modifies how those ingredients are absorbed, processed, etc.

It examines single nucleotide polymorphisms, or polymorphisms, which are genetic variations in DNA. A single nucleotide, the fundamental building block of DNA, is varied in each SNP. Next-generation sequencing technologies (NGS) provide a means to measure the variance of millions of SNPs.

Nutritional signatures are patterns of gene expression, protein expression, and metabolite synthesis that show up in response to particular diets. Understanding how dietary components impact the genome, transcriptome, proteome, and metabolome is the goal of this emerging field of research. Nutrients are now thought to be signals that a cellular sensor detects and controls the expression of genes and proteins to create metabolites. Long hailed for their immense potential to further nutritional science, functional genomic approaches reveal hitherto undiscovered diet-gene interactions and pave the way for the development of novel nutritional indicators. Precise genetic testing techniques are continuously evolving so that the process of producing poultry can enhance the expression of numerous genes related to growth, metabolism, immunity, and antioxidants (Alagawany et al.,2022).

### 5.4.1. World Is Hustling on Achieving Sustainable and Healthy Food Supply Goals

The poultry industry is positioned to be a growth driver and must make significant progress as we work to achieve the Sustainable Development Goals of the United Nations, which include no poverty, zero hunger, good health and well-being, quality education, climate action, and ecosystem management. Better times for the sector, for society, and for the earth will follow from this. In terms of both structure and operation, the sector in India has undergone a paradigm shift during the last 20 years. It has developed from a modest backyard hobby into a sizable commercial industry with the emergence of major food processing companies and the use of contract poultry raising. The sector is important to India because it generates meat and eggs for human consumption in addition to supporting sustainable development. Future advancements in nutrition genomics hold great promise as they have opened the door to new avenues for disease understanding and prevention through diet adjustments.

#### *5.4.1.1. Diet–Gene Interactions*

The expression of several genes linked to immunity, growth, metabolism, health, and antioxidants may be targeted when nutrients and nutraceuticals—food supplements with extra health benefits—are employed in the production of chickens. It has been demonstrated that a variety of food items influence gene expression. They could function as ligands for transcription factor receptors, alter signal transduction pathways, or influence the amounts of substrates or intermediates. For example, vitamin E supplementation of broilers has been shown to have an antioxidant effect in addition to increasing meat quality and shelf life. Heat stress weakens laying hens' immune response and raises oxidative stress in liver tissue, which can lead to a variety of illnesses.

The addition of vitamin E to diets high in selenium significantly increases levels of antioxidant enzymes, as evidenced by the upregulation of CAT, SOD, and GPx genes (antioxidant-related genes). Antioxidants aid the body's defense against dangerous free radicals. Shakeri et al. (2020) showed that adding 200 mg/kg of vitamin C and 100 mg/kg of vitamin E to a staple diet significantly reduced the expression of interleukin (IL)-1, IL-6, and interferon (IFN), as well as Toll-like receptor (TLR)-4 and HSP70 in the liver, all of which are connected to the body's reaction to stress.

Additionally, it has been found that heat-stressed chickens given dietary curcumin have stronger immune systems and are more resilient to adverse environmental conditions. The effects of egg parameters on the expression of genes have been determined. High egg weight producing birds have lower levels of EAAT3 gene expression, but their intestinal tracts have higher levels of EAAT2 gene expression.

Products derived from capsicum species, such as oleoresin, cinnamaldehyde, and carvacrol, are easily identified in intestinal mucosa leukocytes or tissue following transcriptomic analysis. Garlic metabolites, anethole, and turmeric are also useful for enhancing the immune response and protecting the gastrointestinal tract. These substances can alter the expression of genes that regulate immunity and physiology (e.g., energy and protein metabolism), demonstrating the immune-stimulating properties of plant-derived phytochemicals in chickens. Kimball and Jefferson (2004) state that amino acids are critical for modulating mRNA, which regulates gene expression and other cellular functions. The known maximum amount of lysine that broiler birds can ingest without harming them is well known. A comparison of the high and low lysine treatments reveals 67 upregulated genes and 143 downregulated genes (Khwatenge et al., 2020). Whereas arginine and threonine boost the expression of genes linked to growth, methionine, cystine, and threonine influence the expression of immune genes in broiler chickens.

Choline, betaine, folic acid, vitamins B2, B6, and B12, and zinc supplements have all demonstrated some promise in modifying epigenetic patterns. Recent research has shown that a wide range of bioactive substances, such as vitamins, carotenoids, coenzyme Q, organic acids, essential fatty acids, amino acids, and many more, affect how genes are expressed in various domestic animal species, such as chickens and birds. This demonstrates how the nutrients in their diets interact with their genomes.

The results of their investigation also demonstrate a relationship between the hepatic gene expression patterns of apolipoprotein A-I (ApoA-I) and apolipoprotein B (ApoB) and the concentration of nicotinic acid in feed.

#### *5.4.1.1.1. Importance of Feed Optimization in Poultry Nutrition*

The livestock industry is one of the main sources of income for farmers in India. Since feed accounts for a sizable portion of total production costs, improvements in feed utilization efficiency will reduce the amount of feed needed for growth, which will directly lower production costs, increase profitability, and eventually lead to less manure being produced. There is still genetic heterogeneity in feed efficiency among broiler hens, which needs to be exploited for genetic advancement. Dietary supplementation with three distinct probiotic species also increased the height of the villus in the duodenum (*Bacillus subtilis*, *Saccharomyces cerevisiae*, and *Lactobacillus acidophilus*). Research indicates that probiotic supplementation significantly affects the immunological response and growth performance of broiler chickens. When probiotics are eaten, they flood the digestive tract with lactic acid bacteria. These microbes are well known for changing the environment within the intestines and bringing in healthy substances like enzymes. Thus, the main benefits of using probiotics are increased growth, decreased mortality, and increased feed conversion efficiency. Probiotic bacteria that are added to poultry meals after bacterial challenge have been shown to decrease the expression of some genes associated with intestinal infections. Researchers can now examine how nutrition affects an organism at the molecular level thanks to advancements in the field of molecular biology.

Prebiotics such as yeast cell products regulated the expression of genes related to oxidative phosphorylation and other cellular stress responses in the jejunum tissue. Furthermore, a nutrigenomics data analysis showed that feeding chicken an algae-based diet can reduce its stress profile. Like vitamin E, some minerals—like zinc—can actively regulate the expression of the transporter gene in the intestine of broiler chickens, and there is a strong correlation between the pattern of expression and the zinc content of the feed. Using gene chips that carry animal genetic code, researchers can track the effects of particular dietary supplements and how they alter the body's gene interactions. Its main objective is to alter gene activity, thereby increasing the activation of genes that are beneficial and decreasing the activity of genes that are detrimental.

### 5.4.2. Intestinal Microbiome of Poultry Birds

Sequencing of the 16S rRNA gene, first using Sanger sequencing technology, and more recently NGS technology, has enabled a comprehensive characterization of the poultry gut microbiota, whose sequence information provides, greatly expanded our knowledge of the current bacterial diversity in the intestinal tract of poultry birds. Chicken and turkey caecums have a complex microbiome composed almost entirely of bacteria. Early culture based studies showed low abundance of *Lactobacillus* (>104/g colony forming units, CFU) and *Clostridia* (102–104/g). CFU in the small intestine) and anaerobic bacteria occur at high frequencies 4(1010–1011/g microscopic counts) in the cecum of chickens.

Among the bacteria found were facultative anaerobes, streptococci, and anaerobic gram-negative cocci. The primary genera identified by culture from the caecum microbes are proteostreptococcus, propionibacterium, eubacterium, brachodeletes, and clostridium. Intestinal development and nutrient digestibility can be significantly impacted by insoluble fiber when it makes up 3–5% of the diet. Poultry gizzard development is influenced by dietary fiber. A healthy gizzard is necessary for a healthy intestine. Indigestible food poly-, oligo-, and disaccharides can be hydrolyzed by a variety of intestinal bacteria to produce complex sugars, primarily acetate and propionate, which the bacteria then ferment. It can generate butyrates and other short chain fatty acids (SCFA), such as salts. Additionally, the gut microbiota of poultry provides the host with vitamins, particularly vitamin B. For these animals to reach their full genetic potential, maintaining a healthy digestive system is essential. Through the application of cutting-edge nutrigenome technology, researchers are better understanding the mechanisms and effectiveness of various nutritional management strategies by examining the relationship between the gut immune system and diet.

## 5.4.3. New Research Approaches in Nutrigenomics

Every person has a different expression of the same diet to varying degrees. For instance, dietary cholesterol consumption may be the cause of changes in plasma cholesterol; however, each person's experience with this expression varies. It is feasible to alter the diet and care practices for cattle and poultry such that genotype determines nutrition and that nutrients may affect the expression of a person's genotype. Establishing these fundamental relationships and using them to enhance existing ones as well as the health and performance of all poultry animals at various ages presents the biggest challenge, though. When it comes to non-genetic factors that affect an individual's health and productivity, nutrigenomics can only offer a portion of the solution.

We can now start to comprehend how diet influences gene expression and how this modulation affects animal health and performance thanks to DNA microarray technology. Rapid nutritional strategy evaluation is another benefit of these molecular technologies. Researchers can now ascertain the biological and functional significance of nutrigenomic datasets thanks to developments in bioinformatics. These developments in technology present fresh possibilities for studying poultry, which may enhance the productivity and well-being of these animals. Microbiomics is a relatively new field of study that is expanding our knowledge of the connection between gut health and diet.

The field of microbiomics employs molecular biology techniques, including high-throughput DNA sequencing, to investigate the ecology of microbial communities in the gastrointestinal tract. As the field is still in its infancy, research has only gone so far as to provide a general overview of the various communities that can be found in each section of the gut. The digestive tracts of poultry species are home to a diverse range of microorganisms. The microflora in chickens is vital to their growth and well-being because it boosts their immune system and improves their absorption of nutrients microbes that collaborate closely, e.g., closely to the host and the meal consumed. When the eggs hatch, the complex microbiome that inhabits the poultry's gastrointestinal tract is mostly composed of anaerobic bacteria that interact with external microorganisms. As the host organism develops, this microbiome diversifies more and more, eventually reaching a relatively stable but dynamic state. The host is supported by the gut microbiome, which supplies nutrients from food that would otherwise go wasted and controls the development and operation of the digestive and immune systems. The host gives nutrients and an appropriate environment for the bacteria to settle in and grow in exchange (Anderson, 2005).

# REFERENCES

1. Abd El-Hack, E. M., El-Saadony, M. T., Salem, H. M., El-Tahan, A. M., Soliman, M. M., et al. (2022). Alternatives to antibiotics for organic poultry production: Types, modes of action and impacts on bird's health and production. *Poultry Science*, *101*, Article 101.
2. Abdelhamed, A., El-Dawla, R. E., Karadag, A. S., Agamia, N. F., & Melnik, B. C. (2021). The impact of isotretinoin on the pituitary-ovarian axis: An interpretative review of the literature. *Reproductive Toxicology*, *104*, 85–95.
3. Abdelli, N., Solà-Oriol, D., & Pérez, J. F. (2021). Phytogenic feed additives in poultry: Achievements, prospective, and challenges. *Animals*, *11*(12), Article 3471.
4. Adebo, O. A., Njobeh, P. B., Gbashi, S., Nwinyi, O. C., & Mavumengwana, V. (2017). Review on microbial degradation of aflatoxins. *Critical Reviews in Food Science and Nutrition*, *57*(15), 3208–3217.
5. Ahluwalia, M. K. (2021). Nutrigenetics and nutrigenomics—A personalized approach to nutrition. *Advances in Genetics*, *108*, 277–340.
6. Aiken, C. E., & Ozanne, S. E. (2014). Transgenerational developmental programming. *Human Reproduction Update*, *20*(1), 63–75.
7. Akanbi, O. M. (2019). Animal nutrigenomics: Opportunities and challenges for sustainable animal products in Nigeria. *Emergency Medicine and Trauma Care Journal*, *2*, Article 100007.
8. Alagawany, M., Elnesr, S. S., Farag, M. R., Tiwari, R., Yatoo, M. I., & Karthik, K., et al. (2021). Nutritional significance of amino acids, vitamins, and minerals as nutraceuticals in poultry production and health: A comprehensive review. *Veterinary Quarterly*, *41*(1), 1–29.
9. Alexandre, P. A., Kogelman, L. J., Santana, M. H., Passarelli, D., Pulz, L. H., & Fantinato-Neto, P., et al. (2015). Liver transcriptomic networks reveal main biological processes associated with feed efficiency in beef cattle. *BMC Genomics*, *16*, Article 177.
10. Aloia, L., Di Stefano, B., & Di Croce, L. (2013). Polycomb complexes in stem cells and embryonic development. *Development (Cambridge, England)*, *140*(12), 2525–2534.
11. Andersen, H. J., Oksbjerg, N., Young, J. F., & Therkildsen, M. (2005). Feeding and meat quality—A future approach. *Meat Science*, *70*(3), 543–554.
12. Arif, M., Akteruzzaman, M., Islam, S. S., Das, B. C., Siddique, M. P., & Kabir, S. L. (2021). Dietary supplementation of Bacillus-based probiotics

on the growth performance, gut morphology, intestinal microbiota, and immune response in low biosecurity broiler chickens. *Veterinary and Animal Science*, *14*, Article 100214.

13. Asmelash, B., Mahlet, D., & Brhane, H. (2018). Livestock nutrigenomics applications and prospects. *Journal of Veterinary Science & Technology*, *9*, Article 1–4.
14. Ayuso, M., Fernandez, A., Isabel, B., Rey, A., Benitez, R., et al. (2015). Long-term vitamin A restriction improves meat quality parameters and modifies gene expression in Iberian pigs. *Journal of Animal Science*, *93*(6), 2730–2744.
15. Baharum, S. N., & Azizan, K. A. (2018). Omics applications for systems biology. In *Metabolomics in Systems Biology* (pp. 51–68). Springer.
16. Balkir, P., Kemahlioglu, K., & Yucel, U. (2021). Foodomics: A new approach in food quality and safety. *Trends in Food Science & Technology*, *108*, 49–57.
17. Barros-Rodríguez, A., Rangseekaew, P., Lasudee, K., Pathom-Aree, W., & Manzanera, M. (2021). Impacts of agriculture on the environment and soil microbial biodiversity. *Plants*, *10*(11), Article 2325.
18. Baykalir, Y., Baykalir, B. G., & Simsek, U. G. (2018). Application of some proteome analysis techniques in animal reproduction. In *New Insights into Theriogenology*. IntechOpen.
19. Bellner, L., Nichols, A., Pandey, V., Vanella, L., Gilliam, C., & Gupte, R., et al. (2015). Effect of vitamin B12 and nutrients on adipogenesis-adipogenic markers in 3T3 cells. *The FASEB Journal*, *29*, 996–2.
20. Benítez, R., Núñez, Y., & Óvilo, C. (2017). Nutrigenomics in farm animals. *Journal of Investigative Genomics*, *4*, Article 1.
21. Benitez, R., Nunez, Y., Fernandez, A., Isabel, B., Fernandez, A. I., Rodriguez, C., et al. (2015). Effects of dietary fat saturation on fatty acid composition and gene transcription in different tissues of Iberian pigs. *Meat Science*, *102*, 59–68.
22. Benitez, R., Nunez, Y., Fernandez, A., Isabel, B., Rodriguez, C., et al. (2016). Adipose tissue transcriptional response of lipid metabolism genes in growing Iberian pigs fed oleic acid vs. carbohydrate-enriched diets. *Animal*, *10*(6), 939–946.
23. Berton, M. P., Fonseca, L. F., Gimenez, D. F., et al. (2016). Gene expression profile of intramuscular muscle in Nellore cattle with extreme values of fatty acid. *BMC Genomics*, *17*, Article 972.
24. Bhatti, S. A., Khan, M. Z., Hassan, Z. U., Saleemi, M. K., Saqib, M.,

& Khatoon, A., et al. (2018). Comparative efficacy of Bentonite clay, activated charcoal, and *Trichosporon* mycotoxinivorans in regulating the feed-to-tissue transfer of mycotoxins. *Journal of the Science of Food and Agriculture*, *98*(3), 884–890.

25. Bibikova, M., Barnes, B., Tsan, C., Ho, V., Klotzle, B., Le, J. M., et al. (2011). High-density DNA methylation array with single CpG site resolution. *Genomics*, *98*(4), 288–295.
26. Bishai, J. D., & Palm, N. W. (2021). Small molecule metabolites at the host–microbiota interface. *The Journal of Immunology*, *207*(7), 1725–1733.
27. Blaner, W. S., & Olson, J. A. (1994). Retinol and retinoic acid metabolism. In Sporn, M. B., et al. (Eds.), *The retinoids: Biology, chemistry and medicine* (2nd edn., pp. 229–256). Raven Press.
28. Blumberg, J. M., Tzameli, I., Astapova, I., Lam, F. S., Flier, J. S., & Hollenberg, A. N. (2006). Complex role of the vitamin D receptor and its ligand in adipogenesis in 3T3-L1 cells. *Journal of Biological Chemistry*, *281*(16), 11205–11213.
29. Boddicker, R. L., Koltes, J. E., Fritz-Waters, E. R., Koesterke, L., Weeks, N., Yin, T., et al. (2016). Genome-wide methylation profile following prenatal and postnatal dietary omega-3 fatty acid supplementation in pigs. *Animal Genetics*, *47*(6), 658–671.
30. Bordoni, L., & Gabbianelli, R. (2019). Primers on nutrigenetics and nutri(epi)genomics: Origins and development of precision nutrition. *Biochimie*, *160*, 156–171.
31. Brennan, K. M., Crowdus, C. A., Cantor, A. H., Pescatore, A. J., & Barger, J. L., et al. (2011). Effects of organic and inorganic dietary selenium supplementation on gene expression profiles in oviduct tissue from broiler-breeder hens. *Animal Reproduction Science*, *125*(1 4), 180–188.
32. Bruh, A., & Haïom, B. (2017). Gene transfer in eukaryotic cells: Current applications and implication. *International Research Journal of Biological Sciences*, *6*(1), 1–7.
33. Burton, M. A., & Lillycrop, K. A. (2019). Nutritional modulation of the epigenome and its implication for future health. *Proceedings of the Nutrition Society*, *78*(3), 305–312.
34. Butler, W. R. (1998). Review: Effect of protein nutrition on ovarian and uterine physiology in dairy cattle. *Journal of Dairy Science*, *81*(9), 2533–2539.
35. Castelli, F. A., Rosati, G., Moguet, C., Fuentes, C., Marrugo-Ramírez, J.,

Lefebvre, T., & Junot, C. (2021). Metabolomics for personalized medicine: The input of analytical chemistry from biomarker discovery to point-of-care tests. *Analytical and Bioanalytical Chemistry*, *413*(31), 7657–7682.

36. Ceciliani, F., Eckersall, D., Burchmore, R., & Lecchi, C. (2014). Proteomics in veterinary medicine: Applications and trends in disease pathogenesis and diagnostics. *Veterinary Pathology*, *51*(2), 351–362.

37. Chen, G. (2021). The interactions of insulin and vitamin A signaling systems for the regulation of hepatic glucose and lipid metabolism. *Cells*, *10*(8), Article 2160.

38. Chmurzynska, A., Mlodzik, M. A., Radziejewska, A., Szwengiel, A., & Malinowska, A. M., et al. (2018). Caloric restriction can affect one-carbon metabolism during pregnancy in the rat: A transgenerational model. *Biochimie*, *152*, 181–187.

39. Cordero, P., Gomez-Uriz, A. M., Campion, J., Milagro, F. I., & Martinez, J. A. (2013). Dietary supplementation with methyl donors reduces fatty liver and modifies the fatty acid synthase DNA methylation profile in rats fed an obesogenic diet. *Genes & Nutrition*, *8*(1), 105–113.

40. De Sousa, C. S., Hassan, S. S., Pinto, A. C., Silva, W. M., De Almeida, S. S., et al. (2018). Microbial omics: Applications in biotechnology. In *Omics Technologies and Bio-Engineering* (pp. 3–20). Academic Press.

41. Dervishi, E., Serrano, C., Joy, M., Serrano, M., Rodellar, C., et al. (2010). Effect of the feeding system on the fatty acid composition, expression of the Δ9-desaturase, peroxisome proliferator-activated receptor alpha, gamma, and sterol regulatory element-binding protein 1 genes in the semitendinosus muscle of light lambs of the Rasa Aragonesa breed. *BMC Veterinary Research*, *6*, Article 40.

42. Desoye, G., & Herrera, E. (2021). Adipose tissue development and lipid metabolism in the human fetus: The 2020 perspective focusing on maternal diabetes and obesity. *Progress in Lipid Research*, *81*, Article 101081.

43. Di Girolamo, F., D'Amato, A., Lante, I., Signore, F., Muraca, M., & Putignani, L. (2014). Farm animal serum proteomics and impact on human health. *International Journal of Molecular Sciences*, *15*(9), 15396–15411.

44. Dietrich, B., Neuenschwander, S., Bucher, B., & Wenk, C. (2012). Fusarium mycotoxin-contaminated wheat containing deoxynivalenol alters the gene expression in the liver and the jejunum of broilers. *Animal*, *6*(2), 278–291.

45. Dimofski, P., Meyre, D., Dreumont, N., & Leininger-Muller, B. (2021). Consequences of paternal nutrition on offspring health and disease. *Nutrients*, *13*(8), Article 2818.

46. Ding, R., Zhuang, Z., Qiu, Y., Ruan, D., Wu, J., Ye, J., & Yang, J. (2022). Identify known and novel candidate genes associated with backfat thickness in Duroc pigs by large-scale genome-wide association analysis. *Journal of Animal Science*, *100*(2), Article skac012.
47. Eckersall, D. (2020). Quantitative proteomics in animal and veterinary science: A pipeline for exploiting advanced analytical technology. *Journal of Animal Science*, *98*(Suppl. 4), 55–56.
48. Felisbino, K., Granzotti, J. G., & Bello-Santos, L. (2021). Nutrigenomics in regulating the expression of genes related to type 2 diabetes mellitus. *Frontiers in Physiology*, *12*, Article 1109.
49. Ferguson, J. F., Allayee, H., Gerszten, R. E., Ideraabdullah, F., Kris-Etherton, P. M., Ordovás, J. M., et al. (2016). Nutrigenomics, the microbiome, and gene-environment interactions: New directions in cardiovascular disease research, prevention, and treatment: A scientific statement from the American Heart Association. *Circulation: Cardiovascular Genetics*, *9*(3), 291–313.
50. Fukase, E., & Martin, W. (2020). Economic growth, convergence, and world food demand and supply. *World Development*, *132*, Article 104954.
51. Ghodasara, P., Satake, N., Sadowski, P., Kopp, S., & Mills, P. C. (2022). Investigation of cattle plasma proteome in response to pain and inflammation using next-generation proteomics technique, SWATH-MS. *Molecular Omics*, *18*(2), 133–142.
52. Gilani, S. M. H., Rashid, Z., Galani, S., Ilyas, S., Sahar, S., Al-Ghanim, K., & Al-Mulham, N. (2021). Growth performance, intestinal histomorphology, gut microflora, and ghrelin gene expression analysis of broilers by supplementing natural growth promoters: A nutrigenomics approach. *Saudi Journal of Biological Sciences*, *28*(6), 4383–3447.
53. Goldansaz, S. A., Guo, A. C., Sajed, T., Steele, M. A., Plastow, G. S., & Wishart, D. S. (2017). Livestock metabolomics and the livestock metabolome: A systematic review. *PLOS One*, *12*(5), Article e0177675.
54. Golder, H. M., Thomson, J. M., Denman, S. E., McSweeney, C. S., & Lean, I. J. (2018). Genetic markers are associated with the ruminal microbiome and metabolome in grain and sugar challenged dairy heifers. *Frontiers in Genetics*, *9*, Article 62.
55. Gowda, G. N., Zhang, S., Gu, H., Asiago, V., Shanaiah, N., & Raftery, D. (2008). Metabolomics-based methods for early disease diagnostics. *Expert Review of Molecular Diagnostics*, *8*(5), 617–633.
56. Guo, B., Greenwood, P. L., Café, L. M., Zhou, G., Zhang, W., & Dalrymple, B. P. (2015). Transcriptome analysis of cattle muscle identifies

potential markers for skeletal muscle growth rate and major cell types. *BMC Genomics*, *16*, Article 177.

57. Hamill, R., Aslan, M. O., Mullen, A. M., O'Doherty, J. V., McBryan, J., et al. (2013). Transcriptome analysis of porcine *M. semimembranosus* divergent in intramuscular fat as a consequence of dietary protein restriction. *BMC Genomics*, *14*, Article 453.
58. Han, Y., Gao, S., Muegge, K., Zhang, W., & Zhou, B. (2015). Advanced applications of RNA sequencing and challenges. *Bioinformatics and Biology Insights*, *9*, Article BBI-S28991.
59. Haq, Z., Jain, R. K., Khan, N., Dar, M. Y., Ali, S., & Gupta, M., et al. (2016). Recent advances in the role of chromium and its antioxidant combinations in poultry nutrition: A review. *Veterinary World*, *9*(12), 1392–1401.
60. Hasan, M. S., Feugang, J. M., & Liao, S. F. (2019). A nutrigenomics approach using RNA sequencing technology to study nutrient–gene interactions in agricultural animals. *Current Developments in Nutrition*, *3*(8), Article 82.
61. Hossain, M. (2020). Selected reaction monitoring mass spectrometry (SRM-MS) in proteomics. In *Selected Reaction Monitoring Mass Spectrometry* (pp. 53–88). Springer.
62. Hsu, J. M., & Ding, S. T. (2003). Effect of polyunsaturated fatty acids on the expression of transcription factor adipocyte determination and differentiation-dependent factor 1 and of lipogenic and fatty acid oxidation enzymes in porcine differentiating adipocytes. *British Journal of Nutrition*, *90*(3), 507–513.
63. Ibeagha-Awemu, E. M., & Zhao, X. (2015). Epigenetic marks: Regulators of livestock phenotypes and conceivable sources of missing variation in livestock improvement programs. *Frontiers in Genetics*, *6*, Article 302.
64. Jha, R., & Mishra, P. (2021). Dietary fiber in poultry nutrition and its effects on nutrient utilization, performance, gut health, and on the environment: A review. *Journal of Animal Science and Biotechnology*, *12*(1), Article 51.
65. Jiang, Z. Y., Zhong, W. J., Zheng, C. T., Lin, Y. C., Yang, L., et al. (2010). Conjugated linoleic acid differentially regulates fat deposition in backfat and longissimus muscle of finishing pigs. *Journal of Animal Science*, *88*(5), 1694–1705.
66. Joseph, S. J., Pratt, S. L., Pavan, E., Rekaya, R., & Duckett, S. K. (2010). Omega-6 fat supplementation alters lipogenic gene expression in bovine subcutaneous adipose tissue. *Gene Regulation and Systems Biology*, *4*, 91–101.

67. Joseph, S. J., Robbins, K. R., Pavan, E., Pratt, S. L., Duckett, S. K., et al. (2010). Effect of diet supplementation on the expression of bovine genes associated with fatty acid synthesis and metabolism. *Bioinformation and Biology Insights*, *4*, 19–31.
68. Juniper, D. T., Phipps, R. H., Ramos-Morales, E., & Bertin, G. (2009). Effects of dietary supplementation with selenium-enriched yeast or sodium selenite on selenium tissue distribution and meat quality in lambs. *Journal of Animal Science*, *86*(11), 3100–3109.
69. Kang, S. W., Madkour, M., & Kuenzel, W. J. (2017). Tissue-specific expression of DNA methyltransferases involved in early-life nutritional stress of chicken, *Gallus gallus*. *Frontiers in Genetics*, *8*, Article 204.
70. Kasper, C., Ribeiro, D., Almeida, A. M. D., Larzul, C., & Liaubet, L., et al. (2020). Omics application in animal science: A special emphasis on stress response and damaging behaviour in pigs. *Genes*, *11*(8), Article 920.
71. Katsarou, E. I., Billinis, C., Galamatis, D., Fthenakis, G. C., Tsangaris, G. T., & Katsafadou, A. I. (2021). Applied proteomics in 'One Health'. *Proteomes*, *9*(3), Article 31.
72. Kaya, A., Dogan, S., Vargovic, P., Kutchy, N. A., Ross, P., & Topper, E., et al. (2022). Sperm proteins ODF2 and PAWP as markers of fertility in breeding bulls. *Cell and Tissue Research*, *387*(1), 159–171.
73. Keller, J., Ringseis, R., Priebe, S., Guthke, R., Kluge, H., et al. (2011). Dietary L-carnitine alters gene expression in skeletal muscle of piglets. *Molecular Nutrition & Food Research*, *55*(3), 419–429.
74. Khan, A. A., Ganai, A. M., & ul Haq, Z. (2018). Advances in nutrigenomics and its application in poultry. *International Journal of Current Microbiology and Applied Sciences*, *7*(3), 2866–2872.
75. Khodadadi, M., Sheikhi, N., Nazarpak, H. H., & Brujeni, G. N. (2021). Effects of dietary turmeric (*Curcuma longa*) on innate and acquired immune responses in broiler chickens. *Veterinary and Animal Science*, *14*, Article 100226.
76. Kim, D. K., Lillehoj, H. S., Lee, S. H., Jang, S. I., & Bravo, D. (2010). High-throughput gene expression analysis of intestinal intraepithelial lymphocytes after oral feeding of carvacrol, cinnamaldehyde, or Capsicum oleoresin. *Poultry Science*, *89*(1), 68–81.
77. Kimura, J., Kudo, H., Fukuda, A., Yamada, M., Makita, K., Oka, K., et al. (2022). Decreasing the abundance of tetracycline-resistant *Escherichia coli* in pig feces during nursery using flavophospholipol as a pig feed additive. *Veterinary and Animal Science*, *17*, Article 100241.

78. Ladeira, M. M., Schoonmaker, J. P., Swanson, K. C., Duckett, S. K., Gionbelli, M. P., & Rodrigues, L. M., et al. (2018). Nutrigenomics of marbling and fatty acid profile in ruminant meat. *Animal: An International Journal of Animal Bioscience*, *12*(s2), s282–s294.

79. Laible, G. (2009). Enhancing livestock through genetic engineering—Recent advances and future prospects. *Comparative Immunology, Microbiology and Infectious Diseases*, *32*(2), 123–137.

80. Liang, J., Nabi, M., Zhang, P., Zhang, G., Cai, Y., Wang, Q., & Ding, Y. (2020). Promising biological conversion of lignocellulosic biomass to renewable energy with rumen microorganisms: A comprehensive review. *Renewable and Sustainable Energy Reviews*, *134*, Article 110206.

81. Liao, S. F., Brown, K. R., Stromberg, A. J., Burris, W. R., Boling, J. A., et al. (2011). Dietary supplementation of selenium in inorganic and organic forms differentially and commonly alters blood and liver selenium concentrations and liver gene expression profiles of growing beef heifers. *Biological Trace Element Research*, *140*(2), 151–169.

82. Lopez-Bote, C. (1999). *XV Curso de especialización F.E.D.N.A: Avances en Nutrición y Alimentación animal* (pp. 225–252).

83. Lopez, I. P., Marti, A., Milagro, F. I., & Zulet, Md. M. de L., et al. (2003). DNA microarray analysis of genes differentially expressed in diet-induced (cafeteria) obese rats. *Obesity Research*, *11*(2), 188–194.

84. Lowe, R., Shirley, N., Bleackley, M., Dolan, S., & Shafee, T. (2017). Transcriptomics technologies. *PLoS Computational Biology*, *13*(5), Article e1005457.

85. Lu, M., Taylor, B. V., & Körner, H. (2018). Genomic effects of the vitamin D receptor: Potentially the link between vitamin D, immune cells, and multiple sclerosis. *Frontiers in Immunology*, *9*, Article 477.

86. Lucini, L., Marti-Quijal, F. J., Barba, F. J., Rocchetti, G., Quilez, F., & Cuesta, L., et al. (2020). Nutrigenomics and public health. In *Agri-Food Industry Strategies for Healthy Diets and Sustainability* (pp. 219–233). Academic Press.

87. Lundstrom, K. (2020). *Nutrition and Disease: Prevention and Therapy*. Cambridge Scholars Publishing.

88. Mackin, S. J. (2019). *Novel Insights into the Effects of Manipulating DNA Methyltransferase Levels on the Imprinted and Non-imprinted Regions of the Genome* (Doctoral dissertation, Ulster University).

89. Maestro, B., Dávila, N., Carranza, M. C., & Calle, C. (2003). Identification of a vitamin D response element in the human insulin receptor gene

promoter. *The Journal of Steroid Biochemistry and Molecular Biology*, *84*(2–3), 223–230.

90. Malgwi, I. H., Halas, V., Grünvald, P., Schiavon, S., & Jócsák, I. (2022). Genes related to fat metabolism in pigs and intramuscular fat content of pork: A focus on nutrigenetics and nutrigenomics. *Animals*, *12*(2), Article 150.
91. Manzoni, C., Kia, D. A., Vandrovcova, J., Hardy, J., Wood, N. W., & Lewis, P. A., et al. (2018). Genome, transcriptome, and proteome: The rise of omics data and their integration in biomedical sciences. *Briefings in Bioinformatics*, *19*(2), 286–302.
92. Martyniuk, C. J. (2020). Perspectives on transcriptomics in animal physiology studies. *Comparative Biochemistry and Physiology Part B: Biochemistry and Molecular Biology*, *250*, Article 110867.
93. McPherson, R., & Spiller, G. A. (2020). Effects of dietary fatty acids and cholesterol on cardiovascular disease risk factors in man. In *Handbook of Lipids in Human Nutrition* (pp. 41–50). CRC Press.
94. Mendenhall, E. M., Koche, R. P., Truong, T., et al. (2010). GC-rich sequence elements recruit PRC2 in mammalian ES cells. *PLoS Genetics*, *6*(12), Article e1001244.
95. Miao, Z., Wang, S., Wang, Y., Guo, L., Zhang, J., Liu, Y., et al. (2020). A potential linking between vitamin D and adipose metabolic disorders. *Canadian Journal of Gastroenterology and Hepatology*, *2020*, Article 2020845.
96. Mierziak, J., Kostyn, K., Boba, A., Czemplik, M., Kulma, A., & Wojtasik, W. (2021). Influence of the bioactive diet components on the gene expression regulation. *Nutrients*, *13*(11), Article 3673.
97. Miggiels, P., Wouters, B., van Westen, G. J., Dubbelman, A. C., & Hankemeier, T. (2019). Novel technologies for metabolomics: More for less. *TrAC Trends in Analytical Chemistry*, *120*, Article 115323.
98. Molotoks, A., Smith, P., & Dawson, T. P. (2021). Impacts of land use, population, and climate change on global food security. *Food and Energy Security*, *10*(1), Article e261.
99. Mondal, S., & Panda, D. (2020). Nutrigenomics: An interface of gene-diet-disease interaction. In *Mineral Deficiencies—Electrolyte Disturbances, Genes, Diet and Disease Interface* (pp. 1–18). IntechOpen.
100. Mosca, P., Leheup, B., & Dreumont, N. (2019). Nutrigenomics and RNA methylation: Role of micronutrients. *Biochimie*, *164*, 53–59.
101. Muhammad, I., Wang, H., Sun, X., Wang, X., Han, M., & Lu, Z., et al.

(2018). Dual role of dietary curcumin through attenuating AFB1-induced oxidative stress and liver injury via modulating liver phase-I and phase-II enzymes involved in AFB1 bioactivation and detoxification. *Frontiers in Pharmacology*, *9*, Article 554.

102. Muhanguzi, D., Ndekezi, C., Nkamwesiga, J., Kalayou, S., Ochwo, S., & Vuyani, M., et al. (2022). Anti-tick vaccines: Current advances and future prospects. *Vaccine Design*, *2411*, 253–267.

103. Muller, M., & Kersten, S. (2003). Nutrigenomics: Goals and strategies. *Nature Reviews Genetics*, *4*(4), 315–322.

104. Mullins, Y., Keogh, K., Blackshields, G., Kenny, D. A., Kelly, A. K., & Waters, S. M. (2021). Transcriptome-assisted label-free proteomics of hepatic tissue in response to both dietary restriction and compensatory growth in cattle. *Journal of Proteomics*, *232*, Article 104046.

105. Murdoch, B. M., Murdoch, G. K., Greenwood, S., & McKay, S. (2016). Nutritional influence on epigenetic marks and effect on livestock production. *Frontiers in Genetics*, *7*, Article 182.

106. Muthubharathi, B. C., Gowripriya, T., & Balamurugan, K. (2021). Metabolomics: Small molecules that matter more. *Molecular Omics*, *17*(2), 210–229.

107. Nahed, A., El-Hack, Abd M. E., Albaqami, N. M., Khafaga, A. F., Taha, A. E., Swelum, A. A., et al. (2022). Phytochemical control of poultry coccidiosis: A review. *Poultry Science*, *101*(1), Article 101.

108. Nicastro, H. L., Trujillo, E. B., & Milner, J. A. (2012). Nutrigenomics and cancer prevention. *Current Nutrition Reports*, *1*(1), 37–43.

109. Nicholson, J. K., Holmes, E., Kinross, J. M., Darzi, A. W., Takats, Z., & Lindon, J. C. (2012). Metabolic phenotyping in clinical and surgical environments. *Nature*, *491*(7424), 384–392.

110. Núñez, Y., Fernandez, A., Benitez, R., Arija, I., & Viveros, A., et al. (2016). Dietary supplementation with vitamin E or grape pomace influences antioxidant and lipid metabolism candidate gene expression in broilers. *35th International Society for Animal Genetics Conference* (Utah, USA).

111. Osorio, J. S., Vailati-Riboni, M., Palladino, A., Luo, J., & Loor, J. J. (2017). Application of nutrigenomics in small ruminants: Lactation, growth, and beyond. *Small Ruminant Research*, *154*, 29–44.

112. Osz, J., McEwen, A. G., Bourguet, M., Przybilla, F., Peluso-Iltis, C., & Poussin-Courmontagne, P., et al. (2020). Structural basis for DNA recognition and allosteric control of the retinoic acid receptors RAR–RXR. *Nucleic Acids Research*, *48*(17), 9969–9985.

113. Ovilo, C., Benitez, R., Fernandez, A., Isabel, B., Nunez, Y., et al. (2014). Dietary energy source largely affects tissue fatty acid composition but has minor influence on gene transcription in Iberian pigs. *Journal of Animal Science*, *92*(3), 939–954.

114. Pal, A. (2022). Nutrigenomics. In *Protocols in Advanced Genomics and Allied Techniques* (pp. 559–569). Springer.

115. Pan, Y. Z., Wu, S. G., Dai, H. C., Zhang, H. J., Yue, H. Y., & Qi, G. H. (2013). Solexa sequencing of microRNAs on chromium metabolism in broiler chicks. *Journal of Nutrigenetics and Nutrigenomics*, *6*(3), 137–153.

116. Pan, Y. Z., Wu, S. G., Dai, H. C., Zhang, H. J., Yue, H. Y., & Qi, G. H. (2013). Solexa sequencing of microRNAs on chromium metabolism in broiler chicks. *Journal of Nutrigenetics and Nutrigenomics*, *6*(3), 137–153.

117. Pathak, V., Flatt, P. R., & Irwin, N. (2018). Cholecystokinin (CCK) and related adjunct peptide therapies for the treatment of obesity and type 2 diabetes. *Peptides*, *100*, 229–235.

118. Pavlovic, Z., Miletic, I., Jokic, Z., & Sobajic, S. (2009). The effect of dietary selenium source and level on hen production and egg selenium concentration. *Biological Trace Element Research*, *131*(3), 263–270.

119. Peddinti, D., Memili, E., & Burgess, S. C. (2011). Proteomics in animal reproduction and breeding. In *Methods in Animal Proteomics* (pp. 369–392). Springer.

120. Peng, D. Q., Smith, S. B., & Lee, H. G. (2021). Vitamin A regulates intramuscular adipose tissue and muscle development: Promoting high-quality beef production. *Journal of Animal Science and Biotechnology*, *12*(1), Article 34.

121. Phillips, C. M. (2013). Nutrigenetics and metabolic disease: Current status and implications for personalised nutrition. *Nutrients*, *5*(1), 32–57.

122. Pirgozliev, V., Mansbridge, S. C., Rose, S. P., Lillehoj, H. S., & Bravo, D. (2019). Immune modulation, growth performance, and nutrient retention in broiler chickens fed a blend of phytogenic feed additives. *Poultry Science*, *98*(9), 3443–3449.

123. Popovic, D., Bjelobrk, M., Tesic, M., Seman, S., Jayasinghe, S., & Hills, A. P., et al. (2022). Defining the importance of stress reduction in managing cardiovascular disease—the role of exercise. *Progress in Cardiovascular Diseases*, *72*, 75–82.

124. Qi, W., Chan, H., Teng, L., Li, L., Chuai, S., & Zhang, R., et al. (2012).

Selective inhibition of Ezh2 by a small molecule inhibitor blocks tumor cell proliferation. *Proceedings of the National Academy of Sciences, 109*(52), 21360–21365.

125. Rahman, A., Chakraborty, S., & Kabir, Y. (2020). Harnessing personalized nutrigenomics for cancer prevention and treatment through diet-gene interaction. In *Functional Foods in Cancer Prevention and Therapy* (pp. 387–403). Academic Press.
126. Rajaei-Sharifabadi, H., Ellestad, L., Porter, T., Donoghue, A., Bottje, W. G., & Dridi, S. (2017). *Morinda citrifolia* (Noni) modulates the hypothalamic expression of stress-and metabolic-related genes in broilers exposed to acute heat stress. *Frontiers in Genetics*, *8*, Article 192.
127. Renema, R. A. (2003). Effects of dietary selenium source on egg production, fertility, hatchability, and shell quality of broiler breeders. *Poultry Science*, *82*(Suppl 1), 51.
128. Sevane, N., Bialade, F., Velasco, S., Rebole, A., Rodriguez, M. L., et al. (2014). Dietary inulin supplementation significantly modifies the liver transcriptomic profile of broiler chickens. *PLOS One*, *9*, Article e98942.
129. Woods, V. B., & Fearon, A. M. (2009). Dietary sources of unsaturated fatty acids for animals and their transfer into meat, milk and eggs: A review. *Livestock Science*, *126*(1–3), 1–20.
130. Xiao, R., Power, R. F., Mallonee, D., Crowdus, C., Brennan, K. M., et al. (2011). A comparative transcriptomic study of vitamin E and an algae-based antioxidant as antioxidative agents: Investigation of replacing vitamin E with the algae-based antioxidant in broiler diets. *Poultry Science*, *90*(1), 136–146.
131. Zhang, H. B., Wang, Z. S., Peng, Q. H., Tan, C., & Zou, H. W. (2014). Effects of different levels of protein supplementary diet on gene expressions related to intramuscular deposition in early-weaned yaks. *Animal Science Journal*, *85*(4), 411–419.

Chapter 6

# CANINE NUTRIGENOMICS

## Contents

# INTRODUCTION

The relationship between nutrition, the genome, and an organism's general health is investigated in the developing field of nutrigenomics. Nutrigenomics studies how particular nutrients interact with an animal's genetic makeup to affect health, susceptibility to disease, and longevity in companion animals like dogs and cats. Because it enables the creation of individualized nutrition plans that can maximize these animals' health and well-being, this field is especially significant. Similar to people, companion animals have distinct genetic profiles that affect how their bodies react to various nutrients. The field of nutrigenomics investigates these genetic variations to ascertain their impact on nutrient metabolism and the likelihood of contracting specific diseases. For instance, certain dog breeds may have a genetic predisposition to obesity, diabetes, or heart disease. Knowing these factors can help with developing diets that reduce these risks.

Nutrigenomics is used in companion animals for purposes other than illness prevention; it is essential for maintaining general health, enhancing coat quality, and boosting energy and cognitive performance. Veterinarians and pet owners can make sure that animals receive the best nutrition possible by customizing diets based on genetic information. This could potentially extend the animals' lives and improve their quality of life. Nutritional therapies that are more accurate and successful are becoming possible thanks to nutrigenomics. It is anticipated that as this field of study develops, more advanced diagnostic instruments will be created, enabling even more customized food recommendations. This is a major departure from the conventional one-size-fits-all approach to pet nutrition in favor of a more individualized, scientifically grounded approach that takes into account the particular genetic requirements of each companion animal.

## 6.1. NUTRITIONAL GENOMICS: IMPLICATIONS FOR COMPANION ANIMALS

For biologists, this is an exciting moment as the "omics" era continues to change and influence how science is perceived and carried out. Other mammals are scheduled to have their genomes sequenced as the sequencing of the human genome draws to an end. Dogs are currently on the high priority list for sequencing, along with pigs and cows, and cats might soon join them. Genetic maps can be used to identify significant physical features of a genome until sequence data are available. While gene location can be determined through genome mapping, function is not well-understood. Functional genomics, which encompasses the worldwide examination of RNA and protein expression, protein localization, and protein-protein interactions, will consequently become a significant field of research. Up until now, the study of human and veterinary medicine has been the main application of the dog and cat genome maps. These potent tools can also be used in the nutritional genomics and proteomics fields, improving our knowledge of metabolism and maximizing the nutritional and health status of companion animals. Nutritional research has started to use genomics, but problems that are particularly pertinent to companion animals have not yet been resolved. Proteomics and genomics research will be essential for determining nutrient requirements, treating and preventing disease, testing functional ingredients, and other fields. Proteomics and nutritional genomics will undoubtedly be important components of pet food development in the future.

### 6.1.1. Genome Sequencing and Mapping

There is currently very little knowledge about gene function as the Human Genome Project moves closer to its objective of sequencing the entire human genome. The majority of the 30,000–40,000 protein-encoding genes thought to be present in the human genome have not received much research attention.

Even less is known about the placement and function of genes in the canine and feline genomes. It is estimated that whole genome sequencing of a mammalian genome containing approximately 2.7–3.2 billion bases (similar to that of humans, dogs, and cats) at 5× coverage costs around 50 million U.S. dollars. Therefore, only a few mammalian species have been selected for sequencing thus far, with humans, mice, and rats either completed or in progress. Over the last year, the National Human Genome Research Institute (NHGRI) added cats to the medium-priority list for potential genome sequencing, while dogs, cows, and pigs were placed on the high-priority list of model organisms. The sequencing process does not always begin with high priority, as the NHGRI allocates resources to each organism as capacity becomes available. Consequently, it may take several years before the genomes of dogs and cats are fully sequenced.

Various genome-mapping strategies (genetic, physical, and cytological maps) have been used to reveal the physical characteristics of a given genome because whole genome sequencing has not been feasible until recently (Figure 6.1). "Genetic maps," in which the distance between markers is expressed as units of recombination, depict the relative order of genetic markers in linkage groups. The standard unit of genetic distance is the centiMorgan (cM), which is equivalent to a recombination frequency of approximately 0.01. One cM is roughly equivalent to 1000 kb in humans, dogs, and cats. Polymorphic markers and multigenerational reference populations are necessary for genetic mapping. The process of creating genetic maps is commonly known as radiation hybrid

(RH) mapping. Chromosomes break apart when fibroblast cells are exposed to radiation. A panel of hamster fibroblast cell cultures is supplemented with these fragments by irradiating fibroblast cells from an organism of interest. The loci present in each line are then determined by species-specific PCR amplification, and the frequency at which markers cosegregate serves as a gauge for the markers' physical separation from one another. The ability to map nonpolymorphic markers and achieve gene order is the power of RH mapping. Higher-resolution mapping is made possible by RH mapping, which is difficult, if not impossible, for large populations.

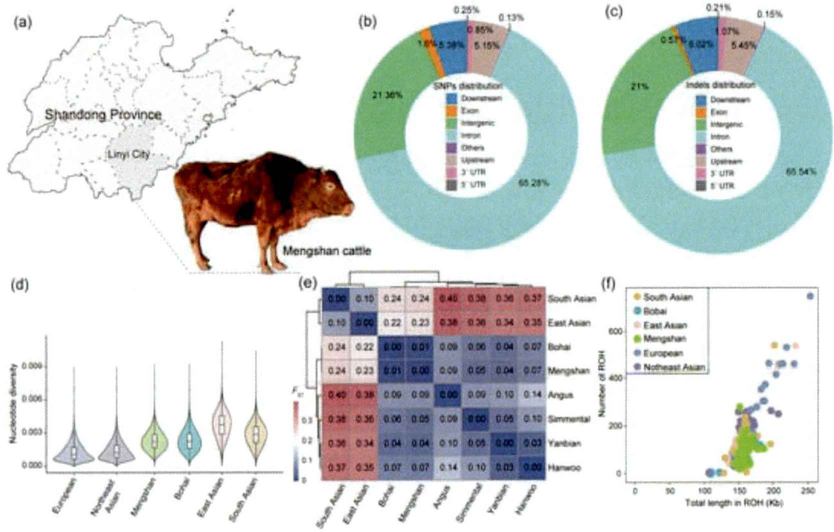


**Figure 6.1.** *Mapping techniques used to describe physical characteristics of a genome.*

***Source:*** *Cheng, H.; Lyu, Y.; Liu, Z.; Li, C.; Qu, K.; Li, S.; Ahmed, Z.; Ma, W.; Qi, X.; Chen, N.; et al. (2024). A whole-genome scan revealed genomic features and selection footprints of mengshan cattle. Genes, 15, 1113. https://doi.org/10.3390/genes15091113.*

"Physical maps" put together contigs, or consecutive regions of chromosomal DNA, where the marker spacing is expressed in kilobases. By aligning randomly separated clones based on shared chromosome fragment profiles, or by a method known as "chromosome walking," physical maps can be created. Bacterial artificial chromosomes (BAC) and yeast artificial chromosomes are frequently used in these procedures. Compared to genetic maps, physical maps allow for finer-scale linkage mapping because they offer a scaffold upon which anonymous polymorphic markers can be positioned. "Cytological maps" show the banding patterns on stained chromosomes that can be seen under a microscope. These maps can be used to distinguish between different chromosomes and assist in the alignment of genetic and physical maps. Lastly, gene-rich species maps of humans, mice, or rats can be used to create maps of species about which little is known through the use of comparative mapping techniques. On the gene-poor species map, syntenic groups—conserved segments of two or more organisms—can serve as anchoring landmarks. A method known as "chromosome painting" can be used to identify synteny regions between two genomes by using fluorescent dyes. Common genomic terms are listed with definitions in Table 6.1.

**Table 6.1.** An Example of Analyses

| SNP categories | Count | Percent |
|---|---|---|
| 3 prime UTR variant | 1234998 | 0.8485% |
| 5 prime UTR premature start codon gain variant | 54610 | 0.0375% |
| 5 prime UTR variant | 314888 | 0.2164% |
| downstream gene variant | 7822149 | 5.3745% |
| initiator codon variant | 84 | 0.0001% |
| intergenic region | 31044794 | 21.3303% |
| intragenic variant | 639 | 0.0004% |
| intron variant | 95034845 | 65.2968% |
| missense variant | 603598 | 0.4147% |
| non coding transcript exon variant | 531970 | 0.3655% |
| splice acceptor variant | 2522 | 0.0017% |
| splice donor variant | 3224 | 0.0022% |
| splice region variant | 195197 | 0.1341% |
| start lost | 824 | 0.0006% |
| stop gained | 4553 | 0.0031% |
| stop lost | 486 | 0.0003% |
| stop retained variant | 542 | 0.0004% |
| synonymous variant | 1214832 | 0.8347% |
| upstream gene variant | 7478213 | 5.1381% |
| **InDel categories** | **Count** | **Percent** |
| 3 prime UTR variant | 187194 | 1.0645% |
| 5 prime UTR truncation | 2 | 0.0000% |
| 5 prime UTR variant | 36451 | 0.2073% |
| bidirectional gene fusion | 79 | 0.0004% |
| conservative inframe deletion | 3444 | 0.0196% |
| conservative inframe insertion | 2772 | 0.0158% |
| disruptive inframe deletion | 6685 | 0.0380% |
| disruptive inframe insertion | 3128 | 0.0178% |
| downstream gene variant | 1056934 | 6.0102% |
| exon loss variant | 2 | 0.0000% |
| frameshift variant | 28777 | 0.1636% |
| gene fusion | 26 | 0.0001% |
| intergenic region | 3687048 | 20.9662% |
| intragenic variant | 79 | 0.0004% |
| intron variant | 11531115 | 65.5710% |
| non coding transcript exon variant | 55755 | 0.3170% |
| non coding transcript variant | 584 | 0.0033% |
| splice acceptor variant | 1744 | 0.0099% |
| splice donor variant | 1653 | 0.0094% |
| splice region variant | 24965 | 0.1420% |
| start lost | 146 | 0.0008% |
| stop gained | 488 | 0.0028% |
| stop lost | 178 | 0.0010% |
| upstream gene variant | 956449 | 5.4388% |

***Source:*** *Cheng, H.; Lyu, Y.; Liu, Z.; Li, C.; Qu, K.; Li, S.; Ahmed, Z.; Ma, W.; Qi, X.; Chen, N.; et al. (2024). A Whole-genome scan revealed genomic features and* *selection footprints of mengshan cattle. Genes, 15, 1113. https://doi.org/10.3390/genes15091113.*

### 6.1.2. Current use of Genome Maps

It is imperative that scientists persist in creating biological methodologies and bioinformatics applications that can unveil gene function, given the ongoing generation of data on genome structure by sequencing and mapping initiatives. "The development and application of global experimental approaches to assess gene function by making use of the information and reagents provided by genome sequencing and mapping" is one definition of functional genomics. Important research topics will include the global analysis of RNA and protein expression, protein localization, interactions between proteins, and chemical inhibition of pathways.

Dog and cat genome maps have traditionally been used primarily for research in human and veterinary medicine. The only animals more closely watched for medical issues than humans are dogs and cats. It is possible to identify the genetic cause of monogenic diseases using molecular biological methods. The disease can then be prevented or treated with the use of this information. In 1989, a canine disease-causing gene was cloned and characterized. At least 20 canine disease genes have since been identified and cloned. Mutation-based tests can be created to test for diagnosis and carrier identification once DNA sequence data for a disease are known. The results of tests can be used to remove carriers from the population that reproduces in order to reduce or eradicate the occurrence of disease. It is far more difficult to identify the genetic loci causing complex genetic diseases. But in the years to come, this might be achievable thanks to the application of single nucleotide polymorphism haplotype maps for linkage disequilibrium research, which has already been carried out on humans.

Cats and dogs make great animal models for hereditary and nutritional disorders in humans. Given that many of the most prevalent diseases affecting purebred dogs also pose serious health risks to humans, the dog may be a very helpful model. Animal models for conditions such as arthritis, cancers of different kinds, deafness, heart disease, blindness, epilepsy, and chronic metabolic diseases (e.g., diabetes, hypercholesterolemia, obesity), and a number of other human illnesses. Nutrition is a major factor in the development and/or treatment of many of these disease states, even though there is a genetic component.

There are noteworthy similarities between humans, dogs, and cats in terms of evolving lifestyles and the rise in obesity and related illnesses over the past 50 years. In the United States, up to 40% of dogs seen by veterinarians are overweight, a substantial increase from only a few decades prior (25). Dogs with diabetes mellitus had a higher incidence of the disease in 1993—from approximately 1.9 cases per 1000 in 1970 to 6.2 cases in 1993. Research on diabetes, pancreatic function, and glucose metabolism has greatly benefited from the use of dogs. Dogs were actually the first animals to develop diabetes in experiments, and Mering and Minkowski used dogs to study the pancreas' function. Dog obesity shares many of the same causes as human obesity, including high-quality animal-based foods consumed in excess and insufficient daily exercise. Dogs experience many of the same detrimental health effects from obesity that are seen in humans. Similar to humans, dogs that gain weight also experience increases in heart rate, blood pressure, cardiac output, plasma volume, and fasting insulin concentration. Canines with diabetes also have many of the same complications as people with the disease, such as hypertension, hypercholesterolemia, atherosclerosis, and retinopathy. Actually, diabetes is a common model for ocular manifestations in dogs because, like in humans, it results in cataracts, which are the primary cause of blindness in dogs. Lastly, the clinical manifestations of diabetes in dogs are comparable to those in humans, with polydipsia and polyuria being the most prevalent symptoms in recently diagnosed cases.

In addition, cats are often used as human models for a wide range of illnesses, such as leukemia, epilepsy, arthritis, Cushing disease, and various metabolic disorders (e.g., diabetes, hypertriglyceridemia, obesity, and glycogen storage disease). Because cats can contract their own strain of the disease, using them as models for AIDS research is crucial. Cats can contract feline immunodeficiency virus, but it is not as deadly as HIV in humans. Thus, this area of research may be very helpful for HIV treatment to be successful. When cats grow used to a sedentary lifestyle indoors, the prevalence of obesity and diabetes in them continues to rise, reaching up to 40% of the population. Cats provide yet another naturally occurring model for studying diabetes and obesity. Many of the detrimental effects of obesity on health that have been seen in dogs and humans also apply to cats, despite some differences. For instance, feline obesity causes hypertriglyceridemia, glucose intolerance, and decreased insulin sensitivity. Comparative studies to find and examine loci influencing disease across species may be carried out as the canine and feline genome maps advance in sophistication.

Dogs and cats are also useful models for human genetic diseases because about half of the genetic diseases found in them are homologous to genetic defects in humans. Companion animal models of human genetics are crucial for determining human disease genes, assessing the conservation of species-to-species synteny, examining the function of proteins, and helping with disease treatment. As animal role models for people, cats and dogs have a number of benefits. First, disease heterogeneity can be prevented because of dog and cat breeding practices. Dogs and cats often have diseases caused by the mutation of a single gene, as opposed to many genetically complex human diseases. Different breeds of dogs and cats with the same phenotype may have distinct disease-causing mutations in addition to monogenic diseases. In this instance, identifying the etiology and therapeutic options for the disease's human form may benefit from an analysis of all genotypes. Second, larger dog and cat families yield higher statistical power than is possible when examining hereditary diseases in humans. Additionally, matching cats and dogs can mate to create more enlightening families for genetic mapping. Third, despite belonging to different evolutionary clades—dogs and cats to clade IV and rodents to clade III—dogs and cats are more closely related to humans in terms of genome structure. In comparison to the mouse genome (which requires 160 translocation steps), cats (13 translocation steps) and dogs (45 translocation steps) require fewer chromosomal rearrangements to match the human genome. Finally, excellent veterinary medicine is used to monitor and treat dogs and cats, yielding valuable clinical data on specific disease states.

### 6.1.3. Methodology Used to Study Functional Genomics

Numerous molecular biological techniques are employed to investigate nutritional genomics and proteomics, in addition to Northern-blot analysis and PCR techniques. Thousands of genes can have their expression simultaneously monitored by microarray technology, which can be used for gene expression profiling. Microarrays are primarily used to detect sequence variations (e.g., single nucleotide polymorphisms, gene mutations) and the assessment of a group of mRNA molecule expression level (abundance). Microarrays are potent substitutes for traditional, classical methods that have historically restricted research to assessing a small number of genes at once. The greatest resolution possible for proteome analysis is achieved through the use of two-dimensional PAGE (2D-PAGE). With this method, proteins are separated by size (molecular mass) using SDS-PAGE in the second dimension, and by charge (isoelectric point) using isoelectric focusing in the first. Proteome analysis also uses MS. Matrix-assisted laser desorption/ionization time-of-flight MS to isolate and identify relevant proteins from the gel. Following ionization and desolvation, a sample is laser-ablated and examined using a mass spectrophotometer to determine its time-of-flight, which can be translated into the molecular mass of the sample's proteins. Computer software can then be used to analyze the mass spectrophotometer data and determine which proteins are present in the sample. These methods can provide scientists vital information that traditional methods just cannot, even though they are expensive and occasionally still problematic.

## 6.2. NUTRITIONAL GENOMICS

Research in human and veterinary medicine has been the main motivation behind the genetic mapping of cats and dogs. Genetic maps, however, will also play a major role in the fields of nutritional genomics and proteomics, improving our knowledge of metabolic pathways and promoting optimal nutritional and health conditions. Some food ingredients (e.g., zinc, fatty acids, vitamins A and D, and others) have the direct ability to affect gene expression, but others (e.g., dietary fiber) may have an indirect impact by altering the way hormones signal, responding to mechanical stimuli, or producing metabolites that are made by the gut microbiota. Through the application of sophisticated molecular biological techniques, researchers are now able to quantify the impact of a single nutrient on a cell or tissue's gene expression profile.

Scientists have started using genomics in the field of nutrition in recent years. A wide range of subjects, such as energy restriction, vitamin and mineral deficiencies, glucose metabolism, and illnesses impacting nutritional status, have been studied using changes in gene expression. However, the field of companion animal nutrition has not yet felt the impact of this one. Many of the dietary needs of dogs and cats are unknown, despite some progress over the past 20 years. Determining the ideal and harmful concentrations of nutrients will be crucial, in addition to understanding the minimal concentrations needed to prevent deficiencies. Companion animal nutrition needs to pay attention to how different life stages and genotypes are affected by genes. Sheep provide evidence of the significance of maternal nutrition during pregnancy for the development and expression of offspring genes. Pig nutrient excretion has been shown to differ by breed, indicating genotype-based variations in metabolism. Analyzing the nutritional needs of dogs engaged in various physical activities (e.g., sprint racing, dog sled racing, or hunting/herding) would also be worthwhile research projects.

Dogs and cats are living longer than ever thanks to the significant advancements in companion animal nutrition and veterinary care in recent years. Currently, intestinal parasites are not bothering people, and viral infections many companion animals are living long enough to experience obesity and many of the same illnesses that afflict their human owners (canine distemper, for example). Diabetes, cancer, and heart disease are examples of highly complex diseases that involve multiple genes and are influenced by environmental factors as well as lifestyle choices, in contrast to diseases that are caused by mutations at a single gene locus. Arthritis, renal disease, and a number of illnesses linked to atypical immune function (e.g., atopic illness, and ulcerative colitis) are also becoming more common in dogs and cats. Biomarkers that can forecast the occurrence and prognosis of diseases are desperately needed. Researchers looking into how energy restriction affects aging have already started their investigation. Dietary interventions may be able to prevent or delay the development of a given disease once the genes responsible for its onset are identified.

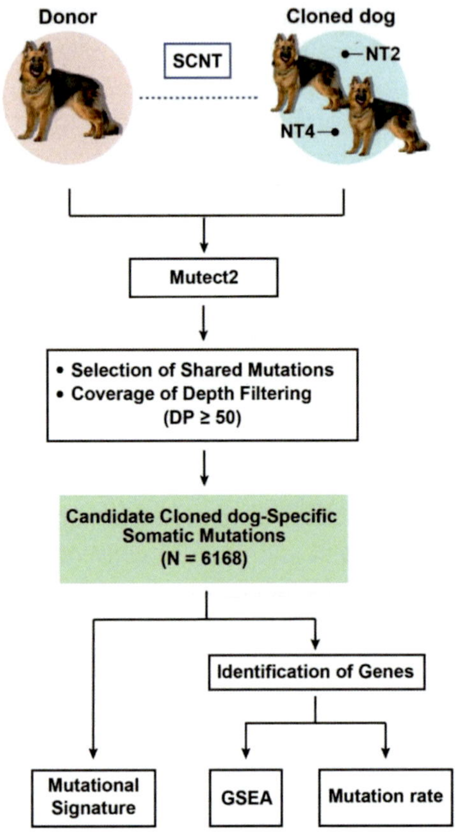


**Figure 6.2.** *Somatic mutation filtering flow chart.*

***Source:*** *Woo, S. W.; Kim, M.; Kang, D.; Choe, Y. H.; Oh, S. J.; You, A. S.; Lee, S. L.; Kim, J. (2024). Genome-wide characterization of somatic mutation patterns in cloned dogs reveals implications for neuronal function, tumorigenesis, and Aging. Genes, 15, 801. https://doi.org/10.3390/genes15060801.*

The development of genotype-specific drug therapy, which is anticipated to have higher efficacy and lower toxicity than existing drugs, may be an alternative strategy if dietary intervention is not the solution. Numerous drug-metabolizing enzyme polymorphisms in humans have been identified. Companion animals most likely share similar polymorphisms, so genotype may soon affect the kind and dosage of medication veterinarians prescribe. Many pharmaceutical companies have high hopes and are heavily investing in this field of research, despite the fact that scientists have differing opinions about the idea of personalized drug therapy. We are currently running a long-term study in our lab to assess how diet affects gene expression in weanling and elderly dogs. Figure 6.2 presents our experimental design.

A diet consisting mostly of premium animal-based components is contrasted with one that consists mostly of plant-based components. Throughout the experiment, samples of blood and liver biopsies are taken, and at the conclusion, tissue samples are taken for RNA isolation. Oligonucleotide microarrays are being used to analyze ribonucleic acid samples in order to produce gene expression profiles. An oligonucleotide microarray comprising 384 genes has been designed, with a primary focus on immune function and metabolic pathways. Diabetes is very common in dogs and positively correlated with age, just like it is in people. This experiment aims to detect disease-predictive biomarkers, which is necessary. Thus, the microarray was expanded to include genes linked to glucose metabolism and homologous to human genes that have been shown to contribute to the onset of diabetes. Apart from nutritional genomics and proteomics, scientists can also identify markers indicating nutritional and/or health status through metabolomics, which is the study of metabolic profiles in a cell, tissue, or organism. Metabolite quantification and cataloging are already being done in the field of metabolomics to learn more about particular biological pathways. Research on cats and dogs will also be necessary to identify species-specific profiles, even though a large portion of the metabolomic research on humans and rodents will probably be applicable to companion animals.

Blood samples were used in our experiment to analyze serum chemistry profiles and complete blood counts. Researchers have found serological indices linked to aging, but they still need to find biomarkers that indicate the presence of disease. It will probably be necessary to measure multiple biomarkers (using both gene expression profiles and serological indices) in order to predict complex disease states. The possibility of a disease developing can then be predicted using statistical analysis of these biomarkers. We plan to start identifying biomarkers that can be used for this purpose with the data generated from this experiment. In order to measure the concentrations of fermentative end products, fecal microbe populations, and total tract nutrient digestibility, feces and digesta were also collected. Significant relationships between alterations in diet and ensuing metabolic changes and gene expression will be found through statistical analyses. Aging has been demonstrated to affect metabolism as well as the expression of genes linked to biosynthesis and the stress response. As a result, more oligonucleotide microarrays are being developed to investigate the genes belonging to these functional groups.

The advent of "functional" ingredients in food for humans and pets has led to the discovery of a new, crucial field of study that may benefit most from gene expression profiling. According to a recent survey, 41% of pet owners had thought about or tried alternative therapies, such as homeopathy (4%), herbal remedies (7%), and nutritional supplements (29%). Despite the fact that the National Cancer Institute's Developmental Therapeutics Program has screened over 70,000 compounds—including botanicals—using microarray technology in *in vitro* experiments, these should be investigated *in vivo* models. Most functional ingredients' mechanisms of action, ideal inclusion levels, and toxicological and/or harmful effects are not fully understood. Pet foods are already incorporating a number of useful ingredients, such as those meant to enhance joint health (e.g., green-lipped mussels, glucosamine, and chondroitin sulfate), shield the body from damage caused by free radicals (e.g., selenium, vitamin E, and β-carotene), and enhance skin (e.g., n-3 fatty acids) and intestinal well-being (e.g., probiotics, oligosaccharides). Many functional ingredients may have health benefits, but more research along the lines of Gohil and Packer's is required to identify mechanisms of action and safe, yet effective, consumption levels. If disregarded, the risk of toxicosis from herbal supplements has been documented by Ooms et al. may see a rise in the number of companion animals.

## 6.3. DIETS FORMULATED FOR GENOTYPE—THE FUTURE OF PET FOODS

In order to maximize nutritional and health status, genomic technologies will be used in the pet food business in the future. This will be achieved by figuring out the minimal requirements for nutrients and the ways in which "functional" ingredients work to treat and prevent illnesses. Interactions between genes and nutrients will make things even more complicated. When creating diet plans for a particular genotype, these interactions must be identified and taken into account. In humans, dogs, and cats, a number of polymorphisms directly related to nutritional and immunological status have already been discovered (Table 6.2). As the field of genomics develops, this list will keep getting longer.

Large-breed dog diets are currently available on the market, but this is only a hint of what the pet food industry may hold in store. Similar to tailored medication therapy, genomics will enable genotype-based diet formulation, which will be far more accurate than some of the dog diets currently marketed for large or elderly animals. While diet plays a major role in the development of skeletal diseases like hip dysplasia, osteochondrosis, and hypertrophic osteodystrophy, these conditions are also highly heritable. For this reason, a large-breed diet's advantages for one dog might not apply to another. As genetic polymorphisms influencing nutritional status and disease, along with the biomarkers associated with them, are discovered, diets can be developed not only to prevent skeletal abnormalities but also to address more complicated conditions like diabetes, cancer, and heart disease.

**Table 6.2.** Macronutrient Content in Dry Extruded Puppy Food Analyzed and Labeled for Growth, Compared to the Recommended Essential Nutrient Concentrations for Growth Maintenance

| Item | DEPF_1 | DEPF_2 | DEPF_3 | DEPF_4 | DEPF_5 | Recommendations * | |
|---|---|---|---|---|---|---|---|
| | | | | | | Early Growth | Late Growth |
| Dry matter (g/100 g fresh food) | 95.4 [b] | 95.8 [b] | 92.7 [a] | 95.4 [b] | 96.6 [b] | NR ** | NR |
| Nutrient (unit/100 g DM) | | | | | | | |
| Crude protein (g) | 30.1 [a] | 31.7 [ab] | 35.3 [c] | 32.8 [bc] | 30.8 [ab] | 25.0 | 20.0 |
| Crude fat (g) | 11.7 [a] | 19.2 [c] | 14.5 [b] | 19.5 [c] | 19.1 [c] | 8.5 | 8.5 |
| Crude fiber (g) | 6.1 [a] | 6.2 [a] | 4.5 [a] | 7.8 [a] | 4.7 [a] | NR | NR |
| Crude ash (g) | 7.7 [c] | 6.4 [a] | 7.2 [b] | 7.6 [c] | 6.4 [a] | NR | NR |
| NFE (g) | 39.8 [c] | 32.3 [ab] | 31.3 [ab] | 27.7 [a] | 35.7 [bc] | NR | NR |
| ME (kcal) | 355.9 [a] | 397.6 [b] | 373.6 [ab] | 379.0 [ab] | 407.8 [b] | NR | NR |

* Minimum recommended level (MRL) for growth of puppies up to 14 weeks of age (early growth) and for puppies over 14 weeks of age (late growth) [7], ** means with at least one some letter in the superscript ([a], [b], [c]) not differ statistically at $p = 0.05$ (for all columns separately), *** NR–No Recommendation; NFE–Nitrogen–Free Extract; ME–Metabolic Energy.

*Source: Jacuńska W, Biel W, Witkowicz R, Maciejewska-Markiewicz D, Piątkowska E. (2023). Comparison of key nutrient content of commercial puppy foods with canine dietary requirements. Applied Sciences, 13(21):11791. https://doi.org/10.3390/app132111791.*

## 6.4. FUNCTIONAL NUTRITION

So far, we've learned that creating optimum health at the cellular level involves two key nutritional steps:

- **Step 1:** feeding a diet high in foods with functional properties, which tell the epigenome to start expressing genes healthily.

- **Step 2:** Limiting or avoiding foods that cause aberrant gene expression and send negative signals to the epigenome (Fekete and Brown, 2007).

Functional foods, as you'll recall, are nutritional ingredients that switch on a gene's expression to fight disease and switch off the expression to promote disease.

### 6.4.1. Functional Carbohydrates

You might be surprised that we're even talking about this food category given how popular "low-carb" or "no-carb" diets are for both humans and dogs. While it's true that dogs don't need to consume a certain amount of carbohydrates, remember that our goal is to design an optimal diet for your dog that will positively influence his epigenome.

Coincidentally, if the right kind of carbohydrates are consumed, they are packed with functional nutrients that are necessary to achieve this goal. Vitamins, minerals, and phytonutrients—also known as phytochemicals—that support health at the cellular level are abundant in functional carbohydrates.

Chemicals found naturally in plants are called phytonutrients (the Greek word "phyto" means "plant"). While phytonutrients have been demonstrated to have significant health benefits for those who consume them, they are not as necessary for survival as proteins, fats, vitamins, and minerals are. Among phytonutrients, carotenoids—which include beta- and alpha-carotene—are arguably the most well-known. Carotenoids are the pigments in red, orange, and yellow that give vegetables their beautiful color, including sweet potatoes, carrots, and pumpkins.

Anthocyanin pigments, which give berries and other dark-colored fruits and vegetables their blue, purple, and red hues, are a type of phytonutrient known as flavonoids. Strong antioxidant qualities found in phytonutrients help shield cells from oxidative damage brought on by free radicals. Studies have also shown that phytonutrients prevent cancer and heart disease and inhibit tumor growth. Phytonutrients may work via a variety of mechanisms, including:

- Serving as antioxidants.
- Enhancing immune response.
- Enhancing cell-to-cell communication.
- Altering estrogen metabolism.
- Converting to vitamin A (retinol).
- Causing cancer cells to die (apoptosis).

- Repairing DNA damage caused by exposure to environmental toxins.
- Detoxifying carcinogens through the activation of the cytocrome
- P450 and Phase II enzyme systems (USDA, 2005).

Researchers demonstrated that exosome-like nanoparticles (EPDENs) made from edible plants, such as carrot and ginger, altered the mice cells' cellular pathways to produce anti-inflammatory and antioxidant effects. This new discovery is one of the most recent in a series of findings about the relationship between gene expression, the epigenome, and nutritional components that will undoubtedly continue to surface. It's yet another justification for making sure your dog consumes enough fruits and vegetables.

We recommend incorporating the following functional carbohydrates into your dog's diet:

- Cruciferous vegetables, such as broccoli, cauliflower, Brussels sprouts, cabbage and Bok choy.
- Whole, fresh fruits like watermelon, bananas, berries, cantaloupe, and apples. It should be noted that although fruits include the simple sugars fructose and glucose, fruits have functional qualities when consumed in their whole, unadulterated state. Fruit has a high fiber content that helps to buffer the sugar. Fruit sugar is absorbed into the bloodstream gradually, preventing abrupt spikes in blood sugar levels because the fiber in the fruit takes longer for the digestive system to break down. Fresh fruit contains fiber, which supports both weight loss and optimal GI functioning. Keep in mind that fruit juice is a sugary disaster if it doesn't contain fiber. Moreover, never feed raisins or grapes to your dog as they are poisonous to dogs.
- Gluten-free grains, such as millet, quinoa, sorghum and gluten-free oats.
- Green leafy vegetables, such as kale and collard greens (see sidebar).
- Legume crops, including lima beans, kidney beans, pinto beans, black-eyed peas, and garbanzo beans (chickpeas).

## 6.4.2. Functional Proteins

Your dog's diet needs protein because he cannot survive without it, making it possibly the most vital nutritious item. Protein gives skin, hair, nails, bones, joints, tendons, ligaments, cartilage, and muscle fibers their structural integrity and is essential for the growth and repair of muscles and tissues (Case et al., 2011).

Here are a few important functions of protein:

- Antibodies made of proteins help protect the body from foreign "attackers" such as viruses and bacteria, keeping the immune system strong.
- Collagen (a protein) forms most of the body's connective tissue.
- Messenger proteins, including some types of hormones, regulate various systems in the body (e.g., insulin, which is essential to controlling blood glucose levels).
- Protein enzymes carry out almost every one of the thousands of chemical reactions that take place in the cells.
- Proteins help regulate muscle action.
- Protein provides essential amino acids.
- Transportation proteins in the blood carry essential vitamins and nutrients throughout the body (e.g., hemoglobin, which carries oxygen to the tissues).

It makes sense that dogs need to eat a lot of protein in order to thrive, considering all of its numerous vital functions. This is particularly true because the body requires a steady supply of protein to replenish it because it is constantly being turned over.

### 6.4.3. High Quality Protein

However, the quality of the protein is more significant than its quantity. About protein quality, a lot has been written in the past. You can actually find out in-depth information about protein and the kinds of foods you should eat for optimal muscle growth by picking up almost any bodybuilding magazine. All of that is crucial, but now let's take a fresh look at protein quality. From now on, your dog's protein will also need to be functional in addition to being of the highest caliber in the conventional sense.

### 6.4.4. Amino Acids: The Building Blocks of Protein

Amino acids, the smaller building blocks of proteins, are arranged in lengthy chains. Amino acids can number in the hundreds or even thousands within a single protein molecule.

Dogs use 22 different amino acids, which combine to form different proteins in different sequences. Twelve of these amino acids are referred to as "non-essential" because a dog's body is capable of producing them in sufficient quantities to meet their needs, negating the need for them in the diet. The remaining ten are referred to as "essential amino acids" since dogs' bodies are unable to produce enough of them to stay healthy. The diet needs to provide the necessary amino acids (Case et al., 2011).

The ten essential amino acids for dogs:

- Arginine
- Histidine
- Isoleucine
- Leucine
- Lycine
- Methionine
- Phenylalanine
- Threonine
- Tryptophan
- Valine

Dietary protein is not only the body's main source of the ten essential amino acids, but also nitrogen. In order for the body to synthesis non-essential amino acids and other molecules required for life, like different hormones and neurotransmitters, nitrogen is required.

Protein synthesis within the body can only proceed if all essential amino acids are available in sufficient quantities. If even a single essential amino acid is inadequate, the entire protein synthesis process comes to a standstill, irrespective of the abundance of every other amino acid. Therefore, the primary dietary need isn't for protein as such; it's for essential amino acids. As the building blocks of protein, amino acids are the fundamental building blocks of life. For a protein source to be considered high-quality, it must provide the appropriate balance of essential amino acids. Moreover, it should possess a high biological value, which signifies that the body can easily absorb and utilize the amino acids and nitrogen the protein delivers.

Based on these criteria, the highest quality proteins for dogs come from animal sources. The winners (in alphabetical order) are:

- Dairy
- Eggs
- Fish
- Muscle meats
- Organ meats

However, if the term "high quality" is redefined to include functional sources, it acquires new significance. By using the concepts of nutrigenomics, we can define a high-quality, functional protein as follows:

- Bioavailable (easily digested and assimilated).
- Free of contaminants such as chemicals, hormones and antibiotics.
- Does not promote a food intolerance/sensitivity.
- Does not contain compounds that send unhealthy messages to the epigenome, triggering unhealthy gene expression.
- Unadulterated (e.g., non-GMO) and unprocessed or minimally processed.

Based on this definition, we encourage you to rotate among the following functional, high quality protein sources:

- Dairy products made from goat or sheep sources, including milk, cheese and yogurt (preferably raw and organic).
- Eggs (preferably free-range and organic).
- Mercury-free fish include sardines, pollack, catfish, and wild-caught Alaskan salmon (avoid farm-raised salmon). Steer clear of fish that are high in mercury, such as swordfish, shark, tilefish, king mackerel, and tuna (especially albacore or white tuna). Generally speaking, we advise against giving your dog shellfish like shrimp, crab, lobster, oysters, and clams because some dogs may develop severe allergic reactions similar to those that occur in humans.
- Meat from muscles and organs from unusual animals, such as venison, bison or buffalo, duck, goat, pig, rabbit, and turkey (ideally raised on grass and without the use of hormones or antibiotics). Comparatively speaking, these sources are less likely than more common animal proteins like beef, chicken, and lamb to induce food intolerances or sensitivities.

Your dog's needs for protein vary depending on his age, activity level, health conditions, and the protein's quality (Kealy, 1999). Keep in mind that your dog will require more protein in order to assimilate the amino acids, the less bioavailable the source is. Therefore, compared to a home-prepared diet that uses fresh, highly bioavailable animal protein sources, a commercial diet that is packed with low-quality, grain-based proteins will need to have a higher percentage of protein if your dog is eating it.

### 6.4.5. Functional Fats and Oils

Diets low in carbohydrates and/or no carbohydrates are popular, and so are diets low in fat. But don't deprive your dog of this vital source of nutrients unless they have a specific medical condition that calls for fat restriction, like pancreatitis. In addition to providing dogs with the most concentrated and easily digested source of energy (dietary fat has more than twice the calories per gram than protein or carbohydrates), it also plays a crucial role in the absorption of fatsoluble vitamins, supports a healthy nervous system, and provides important essential fatty acids like omega-3 fatty acids. Dogs can also eat a larger percentage of fat in their diets because they burn fat for energy much more efficiently than humans do.

Fats are broken down into two main categories:

- Saturated fats, which are mostly found in animal foods (e.g., meat, dairy). Saturated fats are solid at room temperature.
- Unsaturated fats, which are mostly found in plant oils.
- Unsaturated fats are liquid at room temperature. There are two types of unsaturated fat:
  - Monounsaturated fat, e.g., olive oil, peanut oil.
  - Polyunsaturated fat, e.g., sunflower oil, sesame oil known as polyunsaturated fatty acids, or PUFAs). Omega-6 and omega-3 fatty acids are also types of PUFAs.

We recommend incorporating a variety of functional fats into your dog's diet. Here are some of our favorites:

- Chicken fat or lamb fat (as long as your dog does not have a food intolerance/sensitivity to chicken or lamb).
- Fatty fish that are low in mercury and rich in omega-3 fatty acids, such as those listed above.
- Oils, such as fish oil (for omega-3s), borage oil, coconut oil, hemp oil, olive oil, primrose oil, pumpkin seed oil and sunflower oil.

We implore you to never give your dog hydrogenated oils, also known as trans fats, or to consume them yourself. Chronic illnesses like diabetes, obesity, and heart disease have been linked to hydrogenated oils, which are produced when liquid oil turns into a solid. The two most popular hydrogenated oils are vegetable shortening and margarine.

## 6.5. BUILDING THE CANINE NUTRIGENOMICS BASAL DIET

### 6.5.1. The Basal Diet: The Foundation of Optimal Health

We're going to start with what we refer to as the "basal" diet to create your dog's ideal diet. The basal diet provides a healthy foundation upon which you can add other useful ingredients based on your dog's specific needs, including any existing health conditions. Consider the basal diet to be similar to a building's foundation. A weak foundation will collapse even when a rickety old shed is placed on it, but a strong foundation can support a lofty, beautiful skyscraper.

Once more, we're going to challenge your preconceived notions about what constitutes a balanced, healthy dog diet. This is especially true if you've been adhering to the widespread recommendation among veterinarians and mass-market commercial pet food manufacturers to feed your dog only one "balanced" brand of food for the entirety of its adult life. Even though the pet food companies that sell these products might profit from this advice, consider the scenario in which you were compelled to eat the same meal every day of your life. Of course, you would grow bored very quickly. People also question why their dogs have suddenly developed a "finicky" appetite.

### *6.5.1.1. Building a Balanced Diet*

It's not just that dogs must be disappointed and dejected when they eagerly approach their bowl and, gasp, there's that same pile of food again; from a nutritional standpoint, following a single "balanced" diet day in and day out is just plain wrong. It is predicated on the idea that your dog is a static entity whose physiological requirements don't change over time, only depending on life stage (i.e., puppy, grown-up, and elderly). Common sense ought to tell us that this line of reasoning is completely erroneous, yet producers of foods for "all life stages" even assert they have that covered (Hofve, 2013; Straus, 2007).

The pediatrician might place a bag of "kiddie food" in front of you and tell you, "This bag of dry pellets contains all the vitamins, minerals, and nutrients that little Billy will need for his entire life." Imagine this happening to you and your toddler. Furthermore, Billy's health could be seriously jeopardized if you give him any "table foods," such as fresh carrots, broccoli, or apples, as this will upset the diet's delicate balance. Ideally, you would dash out the door to find a new physician right away (Hofve, 2013; Straus, 2007).

Therefore, why do we accept this way of thinking when it comes to our dogs? Why do people avoid giving their dogs a healthy variety of fresh, wholesome foods or live in fear of deviating from one "complete and balanced" (and boring) diet? A dog's physiology is constantly responding to both internal and external factors, such as fatigue, stress, and hormonal fluctuations, just like a person's is. It is absurd to believe that a dog's nutritional needs can be fulfilled by a single diet, not only annually but also on a daily basis.

Many "experts" act as though our understanding of dog food nutrition is at its peak. These professionals chastise and scare concerned dog owners into thinking that deviating from the recommendations that are currently accepted will guarantee their beloved pets' downfall and that they will be solely responsible.

In addition, commercial foods frequently use artificial vitamins, grains, by-products, and the tiniest quantity of low-quality proteins to achieve balance, among other unpleasant ingredients. Of course not, is that the kind of nutrition you would want to give your child? Let's now demand better for our children who have four legs!

Feeding your dog a variety of nutrient-dense whole foods that promote optimal gene expression and nutritional balance over time is the best way to meet his ever-changing nutritional needs.

### *6.5.1.2. The Three Keys*

The ideal canine diet contains three key elements, which we call the Three Keys:

- **Variety:** You must eat foods from all the required food groups—red meat, poultry (including fat), fish, eggs, and organ meats—in order to achieve nutritional balance, e.g., dairy products (sheep and goat), fruits, vegetables, liver from cattle, chicken, lamb, pig, or bison (Straus, 2013).
- **Nutrient-Dense:** Choose whole, fresh foods that are rich in essential fatty acids, antioxidants, phytonutrients, and premium amino acids.
- **Whole Foods:** This means real food, including fresh meats, fish, eggs, dairy, fruits and vegetables—not rendered by-products and synthetic chemicals!

Unfortunately, nutritional analyses of the majority of home-prepared dog diets show that they are often very low in certain nutrients, which is also detrimental to promoting optimal health. In creating homemade dog diets, many people neglect to consider the fact that dogs do have specific nutritional needs that are different from our own. Dogs, for instance, need far more calcium per pound of body weight than do humans. However, we think that (apart from calcium deficiencies) most nutritional deficiencies result from not adhering to one of the three dietary keys mentioned above.

For instance, you may believe that alternating between a few different protein sources—like lamb and pork—enoughens the nutritional diversity. But in actuality, you're leaving out other crucial sources that provide distinct amino acid profiles and vital nutrients, like fish and organ meats. For instance, fish is a significant source of the vital omega-3 fatty acids EPA and DHA, while liver offers protein along with fat, vitamin A, copper, iron, niacin, phosphorus, zinc, and B vitamins (VetInfo, 2012).

To make sure you don't overeat from any one food group, rotate foods from the essential categories mentioned above. In our liver example, giving a small amount a few times a week is fine, but giving too much can lead to an excess of vitamin A, which can cause deformity of the bones, bone spurs on the dog's legs or spine that make him limp, upset stomach, stiffness, weakness, or loss of weight. An intolerance or sensitivity to a particular ingredient is also more likely to develop in your dog if he consumes excessive amounts of a particular food, particularly one high in protein.

So, yes, feeding a properly balanced diet is important for your dog's health, but it doesn't have to come from a bag or can or from a board-certified veterinary nutritionist's "nutrition prescription pad." Instead, it should have a lot of high-GI carbohydrates like white rice, a small amount of one type of lean meat (like chicken, which is problematic for many dogs), a non-functional vegetable oil like canola or corn to provide calories and fat, and an all-purpose synthetic multi-vitamin/mineral mix to balance the diet (Straus, 2013). One cannot achieve optimal gene expression with such a diet.

### ***6.5.1.3. Nutritional Supplementation for Home-Prepared Diets***

The following nutrients need to be supplemented because they are frequently lacking in diets prepared at home (but not in "complete and balanced" commercial foods). Keep in mind that your dog needs calcium supplements to meet his needs, regardless of the quality of his homemade diet. The only exception would be if he consumes raw meaty bones, as opposed to just chewing on them.

#### *6.5.1.3.1. Calcium*

The most common mineral in animals is calcium. The skeleton and teeth contain 99% of the calcium in the body, with the remaining 1 percent found in the blood and soft tissues. Dogs require a lot of calcium, and because meat, vegetables, and grains are all inadequate for a dog's needs, most home-prepared diets fall short in this area. Consequently, supplementing with calcium is required. Adult animals with calcium deficiency develop osteomalacia, which makes their bones brittle and prone to breaking. Research also demonstrates that changes in gene expression that impact the brain can be caused by calcium entering the cytoplasm of the cell. Dairy products (especially goat and sheep's milk), bananas, dark leafy greens, cooked beans, sea vegetables (wakami, kombu, and nori), and bone are all excellent sources of calcium (Case et al., 2011; McDonald et al., 2011) (

**Dose:** 1,000 mg calcium per 1,000 kcal/ME (metabolizable energy) daily (NRC, 2006). We recommend calcium citrate, which is easier to digest than calcium carbonate.

Making your own calcium supplement with ground eggshell (1/2 teaspoon = 1,000 mg calcium) is arguably the simplest and most economical way to give your dog extra calcium in their diet. All you have to do is rinse the shells, let them dry completely, and then use a clean coffee grinder to grind them into a fine powder. If maintained dry, they'll last forever. It is advised that you store them in an airtight, clean, and dry bottle, like an empty supplement bottle.

We also suggest the company Animal Essentials' whole food calcium supplement, which is made from dehydrated seaweed meal. For purchasing information, see the Resources section and our example recipe for an adult maintenance diet for one day. Supplements containing powdered bone meal are also available, but they may contain heavy metals like lead and aluminum. Because of these factors, we do not advise giving dogs bone meal as a calcium source.

Note: Extra supplementation is not required because commercial pet foods already contain calcium, and it may even be harmful because calcium binds to a lot of other minerals and can result in a deficiency. However, in commercial raw diets that do not contain raw meaty bones, calcium supplements are required. You won't need to add extra calcium if your dog consumes at least 20% raw meaty bones.

### *6.5.1.3.2. Linoleic Acid*

Since it is required to produce other Omega 6 fatty acids, linoleic acid is the most significant Omega 6 fatty acid for dogs. Since it makes the skin permeable to water, linoleic acid is also particularly crucial for the health of a dog's skin and coat. It is mostly present in plant oils such as evening primrose, hempseed, olive, safflower, sunflower, and walnut oil as well as poultry fat (Dose: 2,800 mg per 1,000 kcal/ME daily).

### *6.5.1.3.3. Omega-3 Essential Fatty Acids*

In contrast to saturated fats like butter, which are solid at room temperature, omega-3 fatty acids are a form of polyunsaturated fatty acid (PUFA), a dietary fat that is liquid at room temperature. There are three types of PUFAs:

- Omega-3 fatty acids, which can originate from both plant-based and marine sources, as follows:
  - Alpha-linolenic acid (ALA) is a plant-based omega-3 that can

be found in many vegetable oils, nuts like walnuts, and seeds like chia and flaxseeds.

  - Eicosapentaenoic acid (EPA) and Docosahexaenoic acid (DHA) are present in fish oil, various marine organisms, and certain plant-based alternatives. These two compounds, EPA and DHA, are the favored type of omega-3 fatty acids for canine nutrition.
  - Omega-6 fatty acids, which include linolenic acid (LA) and arachidonic acid (AA). Omega-6s are found in seeds and nuts and the oils that are made from them.
- Omega-9 fatty acids, which we will not discuss because they are not required in the diet.

Omega-3 and omega-6 fatty acids are vital fats that our bodies cannot produce on their own, necessitating their acquisition through diet. Conversely, omega-9 fatty acids can be produced within the body and are therefore not categorized as essential. Despite their shared necessity, omega-3 and omega-6 fatty acids play distinct roles in maintaining good health. Omega-3 fatty acids control metabolic processes and gene expression at a cellular level, thereby decreasing inflammation. In contrast, an excessive intake of omega-6 fatty acids, particularly arachidonic acid (AA), can trigger inflammatory responses (Deckelbaum et al., 2006; Laflamme, 2004; Waldron, 2004).

The problem is that the Standard American Diet (SAD)—or, more appropriately in the case of our canine companions, the Standard American Dog Diet (SADD)—contains a significant imbalance in the ratio of omega-6 fatty acids to omega-3 fatty acids, with both people and dogs consuming far too much omega-6s and far too little omega-3s.

EPA and DHA from fish oil and other marine sources are far better sources of omega-3s for dogs than plant-based ALA. This is so for two main reasons:

- Studies clearly show that EPA and DHA are more effective at modulating cellular metabolic functions and gene expression than plant-based ALA (Deckelbaum et al., 2006).
- Although dietary ALA can be metabolized into EPA and DHA in the body, dogs do not efficiently make this conversion. Therefore, they should consume EPA and DHA directly to gain maximum benefit.

We recommend incorporating foods rich in EPA and DHA into your dog's diet. Here are our favorites:

- Anchovies
- Fish oil

- Halibut
- Herring
- Mackerel
- Salmon (select only wild-caught)
- Sardines

High-quality EPA/DHA supplements derived from fish oil or other marine sources can also be beneficial for dogs. Just keep in mind that your dog will require fewer supplements the more omega-3s it gets from fresh food sources (Dose: 100 to 150 mg of EPA and DHA combined per 10 pounds of body weight daily).

The significance of acquiring a human-grade, pure fish oil supplement from a reliable supplier cannot be overstated. Sadly, a large portion of fish in our food supply is tainted with heavy metals and other industrial pollutants; you definitely don't want to add these toxins, which can alter your dog's epigenome, to his diet! Products made from smaller fish, like anchovies and sardines, are better because they don't build up hazardous levels of toxins like their larger, carnivorous relatives (e.g., tuna as well as swordfish).

Although fish oil has many advantageous anti-inflammatory qualities, it can also raise oxidation levels and eventually lower vitamin E levels in the body. When taking fish oil supplements, vitamin E, a potent antioxidant, can help offset these effects. Continue reading for more details on vitamin E. Additionally, avoid giving your dog fish liver oil supplements, like cod liver oil, as these can cause dangerously high vitamin A and D levels.

### *6.5.1.3.4. Selenium*

One of the many vital functions of the trace mineral selenium is to protect the body from oxidative damage and to strengthen the immune system. Additionally, selenium may make vitamin E more effective. The soils of many nations, including the US, are low in selenium, and crops grown in these soils will have comparatively low levels of the mineral. Selenium plays a key role in thyroid health maintenance, and a recent study found a connection between hypothyroidism and selenium deficiency. When selenium is deficient, blood levels of thyroid hormones rise, but not tissue levels, making selenium related hypothyroidism difficult to diagnose. This implies that a dog lacking in selenium may show clinical hypothyroidism symptoms, but his blood thyroid levels will show normal. But keep in mind that selenium is the most toxic mineral and needs to be taken in very small amounts. Eggs, meat, and fish are all excellent sources of selenium (Dose: 87.5 μg (mcg) per 1,000 kcal/ME daily) (NRC, 2006).

### 6.5.1.3.5. Vitamin B-6 (Pyridoxine)

Red blood cell production and function, protein metabolism, a healthy nervous system, and strong cognitive and immunological systems all depend on vitamin B-6. Additionally, it has been demonstrated that vitamin B-6 can either up- or down-regulate the expression of specific genes, such as those that regulate the synthesis of albumin, a liver-produced protein found in blood plasma, and the production of glucocorticoid (steroid) hormones. Albumin (found in egg whites) is a good source of many important molecules that are transported through the blood, including hormones, fatty acids, calcium, medications, and other molecules of importance. Symptoms of a vitamin B-6 deficiency include fatigue, anemia, skin conditions, rheumatoid arthritis, kidney stones, and kidney damage. Unlike fat-soluble vitamins, which are stored in the tissues, excess amounts of B vitamins are excreted from the body through the urine. This is because all B vitamins are water-soluble. The majority of the vitamin B-6 found in raw foods is destroyed during cooking and processing. Whole grains, legumes (eat in moderation due to lectins), green leafy vegetables (e.g., spinach, kale, etc.) are good sources of vitamin B-6 e.g., kale, broccoli, Brussels sprouts, chard, collard greens, bananas, chicken, and fish (Dose: 375 μg (mcg) per 1,000 kcal/ME daily).

### 6.5.1.3.6. Vitamin D

One well-known function of vitamin D, a fat-soluble vitamin, is to keep the body's levels of calcium and phosphorus adequate. However, vitamin D plays a much wider role. Calcitriol, the byproduct of vitamin D synthesis, is essentially a steroidal hormone that attaches itself to thousands of different genes and affects them. From the brain to the bones, almost every kind of cell has calcitriol-responsive receptors. The hormone calcitriol strengthens the body's defenses against infections and long-term inflammation. It also helps the body's organs heal damaged cells, including those caused by cancerous cells.

Since dogs are unable to convert UV rays from the sun into vitamin D, they must obtain their vitamin D from supplements or food sources. While most people get plenty of vitamin D from sun exposure (assuming they don't wear sunscreen every day), it's not as simple for dogs. Vitamin D is not found naturally in many foods; most are fortified. The best sources are the flesh of fatty fish, such as salmon, tuna, and swordfish. However, due to their high mercury content, we strongly advise against feeding tuna and swordfish and only suggest wild caught salmon (as opposed to farm-raised, which is the equivalent of fish raised in factory farms). Sardines and egg yolks are also good options because they contain small amounts of mercury.

Vitamin D comes in two different forms: D2 (ergocalciferol) and D3 (cholecalciferol). Because dogs use cholecalciferol more effectively than humans do, vitamin D3 is more crucial to their diets than vitamin D2. Make sure the product you buy contains natural vitamin D3, or cholecalciferol, rather than synthetic vitamin D2.

As vitamin D is fat-soluble, excess levels are stored in the liver and can cause toxicity, which is potentially fatal, so it's equally important to avoid taking too much of it. Dog toxicity symptoms can include constipation, dark, tarry feces, seizures, weakness, depression, vomiting (sometimes with blood), loss of appetite, increased thirst and urination, and excessive drooling. In addition to causing hypercalcemia, or elevated blood calcium levels, excessive vitamin D intake can also lead to a number of other conditions, from bladder stones to, in extreme circumstances, coma. Please speak with your veterinarian about your dog's appropriate vitamin D levels (Dose: 136 IUs per 1,000 kcal/ME daily. This is equivalent to 3.4 μg (mcg) of cholecalciferol per 1,000 kcal/ME daily).

### *6.5.1.3.7. Vitamin E*

As an antioxidant vitamin that is fat-soluble, vitamin E is best known for its ability to neutralize dangerous free radicals, which are unstable and chemically incomplete atoms that can oxidize and damage cells if left unchecked. Vitamin E slows down the aging process and helps prevent diseases of the circulatory system, such as arteriosclerosis (mineral deposits in artery walls), by acting as an antioxidant. Additionally, it has been demonstrated that vitamin E has a major effect on T-cell expression, which is linked to aging and enhances T-cell function in the elderly.

In addition, vitamin E increases blood oxygenation, strengthens the immune system, enhances the performance of internal organs, stops hormones from oxidizing, lowers inflammation, and aids in the fight against infection. Additionally, it has been used to treat or manage immune-mediated diseases and skin conditions in dogs. Vitamin E deficiency in animals can cause brown bowel syndrome, a disorder in which the tissue degenerates and the intestines ulcerate and bleed. Eggs, most fish, cold-pressed vegetable oils, nuts, seeds, green leafy vegetables, and meats like chicken breast, beef, duck, and turkey legs are good sources of vitamin E (Dose: 7.5 mg of alpha-tocopherol per 1,000 kcal/ME daily).

### *6.5.1.3.8. Zinc*

Zinc is a trace mineral that is essential to the immune system's overall performance. It is also involved in over 300 enzymatic and metabolic processes, such as thyroid hormone synthesis and cell division. It is also essential for maintaining skin health. Zinc responsive dermatosis, a common condition associated with zinc deficiency, is particularly common in breeds such as Samoyeds, Huskies, and Malamutes, which have a genetic predisposition to poor zinc absorption. Zinc responsive dermatosis is characterized by scaly, crusty skin around the legs, head, and face (especially on the nose and encircling the eyes, ears, chin, and mouth); thick, crusty foot pads; and hair loss. Reproduction may also be impacted by zinc deficiency. Liver and red meat are great sources of zinc. Egg yolks are a good source as well (Dose: 15 mg per 1,000 kcal/ME daily).

To make up for any nutritional deficiencies, you can give your dog a high-quality, whole-food human or canine multivitamin. However, since multivitamins are deficient in calcium, you should always supplement your dog's diet with calcium from a separate source. When buying supplements, exercise caution because many brands include known allergens like wheat or corn that can aggravate dietary sensitivities or intolerances. Additionally, avoid products that contain artificial flavors, colors, fillers, sugars, or other ingredients that aren't necessary because they will negate the functional effect of the ingredients. Additionally, some trustworthy producers create premium freeze-dried premixes. These are composed of a blend of nutrients and vegetables that, when rehydrated with your own source of animal protein and fat (i.e., oil) to produce a meal that is nutrient-balanced. If you want to make a healthy, fresh meal but don't have time to buy and cook your own vegetables, this is a great solution.

## 6.5.2. The Basal Diet and High-Performance Dogs

High-performance dogs in terms of nutrition include canine athletes, nursing moms, and pregnant bitches. A bitch's body is subjected to greater nutritional demands during pregnancy and lactation than it is at any other point in her life. Whether the pups are getting these nutrients from their mother's milk when nursing or through the placental wall during pregnancy, a pregnant or nursing dog must provide all the nutrients for both herself and her litter. Thus, yes, nursing and pregnant women require more nutrition than someone who plays frisbee or runs with agility.

### *6.5.2.1. Canine Athletes*

A diet rich in nutrients is necessary for canine athletes to sustain the extra strain that their active bodies endure. To begin with, since exercise speeds up metabolism, the active dog needs more calories to support his lifestyle and reach his maximum potential (Gillette, 1999). However, make sure to honestly evaluate your dog's activity level before you begin feeding him like a "jock."

Is he an agility, farm, herding, or search-and-rescue dog, an intense jogger, or a hiker? Although your dog may relish the occasional beach playdate or strenuous weekend hike, that does not make him an athlete. Overestimating your dog's energy requirements can lead to overfeeding him, which will make him overweight and unwell.

Identifying the activities your dog engages in—those requiring sustained high-intensity energy bursts or endurance sports requiring longer-range stamina—is the next step after confirming that your dog is, in fact, a "athlete." This is crucial because your dog's body uses three different systems to metabolize energy, and the kind of activity he engages in will dictate which system he uses. This will therefore establish whether his calories (i.e., "fuel") ought to be mostly derived from fats or carbs (Gillette, 1999).

The three energy systems are:

- **Adenosine Triphosphate–Creatine Phosphate (ATP-CP) System** (also called the phosphagen system or one enzyme system): This system uses adenosine triphosphate (ATP) and creatine phosphate (CP), which are stored in the muscles, to supply just enough energy for a brief, high-intensity activity such as a single vertical jump. It fuels the body for only about five to 20 seconds (Gillette, 1999; Heffernan, 2012).
- **Glycolytic Energy System:** This system takes over when the ATP-CP system stops and supplies energy for up to about two minutes. It uses the process of glycolysis to convert glucose from carbohydrates into ATP (energy) and is important for strength and power activities (Gillette, 1999; Heffernan, 2012).
- **Oxidative Energy System**: This system is the most complex and takes over where glycolysis leaves off, providing the sustained energy required for endurance activities. The oxidative system is an aerobic system, meaning that it uses oxygen to convert substrates into energy. The oxidative system provides the long, slow energy burn dogs need to delay fatigue, increase stamina and fuel endurance activities (Gillette, 1999; Heffernan, 2102; Hill, 1998).

Your dog is using the oxidative energy system when he engages in endurance sports, so a large portion of fat should be in his basal diet. Generally speaking, dogs are better than humans at burning fat for energy, both during rest and during physical activity (Hill, 1998). In fact, studies reveal that dogs fed high-fat diets have higher energy expenditure than dogs fed a normal-fat diet, even in the absence of additional conditioning. Dogs who are sedentary and who consume a high-fat diet can burn even more energy when they are conditioned.

Naturally, in order to prevent weight gain, we advise giving your dog access to age- and health-appropriate exercise opportunities. Additionally, exercising with your dog—even if it's just a stroll around the neighborhood—deepens your bond with him and provides him with excellent mental and social stimulation.

Therefore, even though it's common for human athletes to "carb load" before a marathon run, dogs cannot benefit from this approach. Dogs can stay cooler by burning fat because they produce less heat in their bodies during this process, which is particularly beneficial in warmer climates. Dogs can consume a lower portion of a high-fat diet and still get the energy they require because fats have more than twice as many calories (kcal) per gram of food as either carbohydrates or protein (8.5 kcal per gram compared to 3.5 kcal for carbs and protein). The ultimate endurance dogs, sled dogs, do best on a diet high in fat (>50% of the energy). To ascertain your dog's ideal fat requirements, you must ascertain where he sits on the "endurance scale."

Engaging in endurance exercises also raises a dog's requirement for premium animal protein, which supplies the amino acids required to rebuild and repair muscle tissue and guard against training-related anemia, or "sports anemia." Hematocrit, or packed red blood cells, decreased in sled dogs fed a diet with 28% protein, but not in those fed a diet with 32% or more protein.

Your dog's nutritional requirements will change if he engages in activities that demand shorter energy bursts because he will be using both the glycolytic and ATP-CP systems. Dogs in this category don't move around for long enough for their bodies to use the oxidative energy system, which burns longer. These athletes need a basal diet that prioritizes protein and a high percentage of carbohydrates, as these nutrients fuel glycolysis. But keep in mind that if your dog engages in these brief spurts of activity repeatedly, his body will eventually start to use the oxidative system, necessitating the addition of more fat to prevent fatigue in subsequent competitions.

Note: Don't expect immediate results when you change your dog's diet; it can take between four to six weeks for his body to adapt, especially when switching to a higher-fat diet.

An additional aspect to take into account is the impact of external factors like temperature, humidity, and wind direction on a dog's energy expenditure during physical activity. Make sure to take these factors into consideration and modify your dog's calorie intake as necessary. For example, if your dog exercises outside in the winter, you'll need to give him more energy to make up for the extra calories he'll burn to stay warm.

### *6.5.2.2. Added Nutrients and Supplements for Athletic Dogs*

The following nutrients will help support your canine athlete's performance and stamina:

- Antioxidants, such as carotenoids, Co-Q10, berries and pomegranate to combat oxidative stress produced during workouts.
- Bioactive nutrients, such as purslane, king of bitters, flame of the forest, milk thistle (Silybum marianum), purple coneflower (Echinacea angustifolia), curcumin, Siberian gingseng, valerian root and chamomile.
- Lutein (2–2.5 mg per pound) to protect and stabilize the cell membrane, regulating the immune response.
- Omega-3s to combat inflammation.
- Niacin for red blood cell production and to help with carbohydrate metabolism.
- Taurine (250 to 1000 mg twice daily) to assist immune function and protect the heart.
- Thiamin to help minimize the effects of stress. Vitamin A for ligament and tendon health.
- Vitamin B12 to help with protein synthesis and formation of red blood cells and hemoglobin.
- Vitamin C to facilitate lactic acid depletion caused by muscle exertion (also antioxidant) (dose: additional 250 to 1000 mg of Ester C daily, based on the dog's weight and degree of exertion).
- Vitamin D to maintain a proper calcium: phosphorous balance.
- Vitamin E to support endurance and agility (also antioxidant) (additional 100 to 600 IU daily, depending upon the dog's weight and degree of exertion).
- Vitamin K to maintain proper blood conditions (not imitation vitamin K3!).

### *6.5.2.3. Water: The Ultimate Nutrient*

Dogs don't sweat through their skin like humans do, so dehydration—which is caused by an excessive loss of water in the body—is a dangerous condition that needs to be treated right away. Active dogs are particularly vulnerable, especially those that perform in hot climates or in the summer. Water consumption will help your dog stay cool on the inside and avoid overheating. To encourage drinking, try adding some broth to your dog's water or freezing it in ice cube trays.

### 6.5.2.4. Pregnant Bitches

In dogs, the gestational period (pregnancy) lasts 56–66 days, with an average of 63 days. Pregnant bitches should continue to feed their regular high-quality maintenance diet in the same amount during the first four to five weeks of pregnancy, as puppies in the womb do not grow much during this time. In actuality, feeding a pregnant dog too much during the first four to five weeks of her pregnancy can cause her to gain unwanted weight and cause problems giving birth. Around the third week of pregnancy, a pregnant dog may also experience an appetite loss; this is common and usually doesn't signal a problem. But after a few days, if she still won't eat, speak with your veterinarian. Puppies enter a critical growth period in the womb after week five and continue to grow until parturition, or delivery; during this time, they will need extra nutrition to ensure healthy prenatal development. Throughout the final three to four weeks of her pregnancy, gradually increase the amount of food the pregnant dog consumes so that, by the time she gives birth, she has gained 15 to 25% of her normal body weight. She will require approximately 25 to 50% more food than her maintenance diet to reach this weight gain, depending on her size and the size of her litter (Case et al., 2011).

Generally speaking, extra calcium supplementation is not necessary—in fact, it may even be harmful—as long as your pregnant dog is fed a diet appropriately balanced for gestation.

Studies on dairy cattle suggest that eating a high-calcium diet during pregnancy may raise the incidence of eclampsia, a condition that manifests during lactation. This condition is also referred to as "milk fever," post-partum hypocalcemia, puerperal tetany, or parturient paresis. However, research on dogs has been limited in this regard. When the body is unable to sufficiently release calcium stored in the bones to replenish the calcium used for milk production, milk fever develops. Low blood calcium levels can cause convulsive seizures, tetany (muscle spasms and cramps), weakness, loss of appetite, and heart failure (Case et al., 2011). Most often, during peak lactation, which occurs two to three weeks after whelping, small breed bitches with large litters are impacted. A cleft palate may result from excessive vitamin A supplementation, which should be avoided (Blasa, Booles, and Burger, 1989; Segal, 2007; Merck Veterinary Manual, 2011). It's unclear whether folic acid deficiency and the development of cleft palates in smaller breeds are related.

Because the physiological stress of pregnancy and lactation lowers the status of these nutrients, omega-3 fatty acid (EPA and DHA) supplements derived from fish oil and other marine sources are especially important for both types of dogs. The number of puppies in the litter decreases with increasing number

(Case et al., 2011). During the final three to four weeks of pregnancy, it's a good idea to feed your pregnant dog a lactation diet. This will ensure that your dog is both nutritionally prepared and able to produce an abundance of high-quality milk for her puppies, while also providing the extra energy, fats, carbohydrates, proteins, and nutrients she needs to maintain a healthy pregnancy.

### *6.5.2.5. Lactating Dams*

The physical demands of lactation are immense for the dam, or mother, as her diet must provide her with energy and nutrients to meet her own needs and enable her to produce enough milk in both quantity and quality to meet the needs of her developing puppies. In actuality, a dog's nutritional needs during lactation are greater than they have ever been during any other period of her adult life, and in some cases, they even surpass those of growing puppies. A lactating dog's milk production is determined by the number of suckling puppies; therefore, the more puppies a dam is nursing, the higher her nutritional needs (Facetti and Delaney, 2012; Hand et al., 2000; Hand et al., 2010; McNamara, 2006).

Around eight weeks is when puppies usually wean themselves naturally from their moms. An insufficient diet given to a nursing dog can cause weight loss, anemia, and/or diarrhea, among other health issues. A diet that only provides a minimal amount of nutrition could also lead her to overeat in an attempt to satisfy her hunger, overwhelming her GI system and causing diarrhea. For a nursing bitch, diarrhea poses a risk because it can dehydrate her, deprive her of vital fluids, and hinder her milk production. It's also possible to get mastitis, an infection of the mammary glands, or agalactia, which is an absence of milk.

### *6.5.2.6. Energy*

A lactating dam's energy requirement is the most significant factor affecting her metabolic and nutritional needs; no other stage of growth or production needs as much energy. A nursing dog may need up to four times as many calories as a dog of the same size in the maintenance stage. Not surprising, given that she yields as much milk as a dairy cow—averaging over 8% of her body weight in milk each day! — but it is crucial that the food's energy density be adequate when creating a lactation diet. The dog won't be able to eat enough to meet her energy needs if the food doesn't have enough calories by volume. This can lead to a number of issues, such as undesired weight loss and the inability to produce enough milk to feed her litter. A lactating dog needs more energy the more puppies it is nursing, and its energy needs are highest during peak lactation, which occurs in weeks three and four (Case et al., 2011; Facetti and Delaney, 2012; Hand et al., 2000; Hand et al., 2010; Schenck, 2010).

### 6.5.2.7. Water

A plentiful supply of water is necessary for lactating dams to produce enough milk to meet the needs of their young, as milk is 78% water. Make sure your nursing mother always has access to fresh, clean water so she can drink as much as she wants (Case et al., 2011; Hand et al., 2000; Facetti and Delaney, 2012; Hand et al., 2010).

### 6.5.2.8. Protein

Since the protein content of a dam's milk is more than twice that of cow's milk and even higher than that of goat's milk, protein is especially important during lactation. Given how rapidly puppies grow—much faster than calves—this makes reasonable. It is even more necessary for nursing mothers to consume protein than energy. Compared to a dog of the same size, lactating bitches usually need almost twice as many essential amino acids for maintenance—more specifically, nearly three times as much leucine and valine. A significant amount of the lactating dam's diet's total energy must therefore come from high-quality protein (Facetti and Delaney, 2012; Hand et al., 2000; Hand et al., 2010; NRC/NAS, 2006).

### 6.5.2.9. Fat

For nursing mothers to produce enough milk, they need an enormous amount of calories, which can only be obtained through a high-fat diet. Keep in mind that fat has the highest calorie density of any food component, providing over twice as many calories per serving as proteins and carbs. A nursing dog may need up to five times the amount of fat for maintenance compared to a dog of the same size (NRC/NAS, 2006). Omega-3 supplements from fish oil or other marine sources not only offer significant health benefits to nursing mothers, but they can also raise the fatty acid content of the mother's milk, allowing EPA and DHA to be transferred to the nursing puppies.

### 6.5.2.10. Carbohydrates

Dogs don't really need carbohydrates in their diet, but they do need glucose. Dogs can produce glucose in their liver and kidneys as long as they eat enough protein. However, contradictory findings from studies on bitches during gestation and lactation have sparked discussion about whether or not dietary carbohydrates are required during these more stressful periods in the canine nutrition community. For instance, lactating bitches do need more glucose in order to produce lactose, the sugar found in breast milk. We suggest using caution and giving your nursing dog 10 to 20% of its daily calories from carbohydrates (Schenck, 2010).

## REFERENCES

1. Allred, C. D., Allred, K. F., Ju, Y. H., Virant, S. M., & Helferich, W. G. (2001). Soy diets containing varying amounts of genistein stimulate growth of estrogen-dependent (MCF-7) tumors in a dose-dependent manner. *Cancer Research*, *61*(13), 5045–5050.
2. Appleton, D. J., Rand, J. S., & Sunvold, G. S. (2000). Feline obesity: Pathogenesis and implications for the risk of diabetes. In G. A. Reinhart & D. P. Carey (Eds.), *Recent Advances in Canine and Feline Nutrition* (Vol. III, pp. 81–90). Orange Frazer Press.
3. Axelsson, E., Ratnakumar, A., Arendt, M. J., Maqbool, K., Webster, M. T., Perloski, M., Liberg, O., Arnemo, J. M., Hedhammar, A., & Lindblad-Toh, K. (2013). The genomic signature of dog domestication reveals adaptation to a starch-rich diet. *Nature*, *495*, 360–364.
4. Barrie, J., Watson, T. D. G., Stear, M. J., & Nash, A. S. (1993). Plasma cholesterol and lipoprotein concentrations in the dog: The effects of age, breed, gender, and endocrine disease. *Journal of Small Animal Practice*, *34*, 507–512.
5. Bauer, J. E. (2001). Evaluation of nutraceuticals, dietary supplements, and functional food ingredients for companion animals. *Journal of the American Veterinary Medical Association*, *218*, 1755–1760.
6. Berdanier, C. D. (1999). Diabetes mellitus: A genetic disease. *Nutrition Today*, *34*, 89–98.
7. Beynen, A. C., & Legerstee, E. (2010). Influence of dietary beta-1,3/1,6-glucans on clinical signs of canine osteoarthritis in a double-blind, placebo-controlled trial. *American Journal of Animal and Veterinary Science*, *5*(2), 97–101.
8. Blanchard, R. K., Moore, J. B., Green, C. L., & Cousins, R. J. (2001). Modulation of intestinal gene expression by dietary zinc status: Effectiveness of cDNA arrays for expression profiling of a single nutrient deficiency. *Proceedings of the National Academy of Sciences of the United States of America*, *98*, 13507–13513.
9. Boermans, H. J., & Leung, M. K. (2007). Mycotoxins and the pet food industry: Toxicological evidence and risk assessment. *International Journal of Food Microbiology*, *119*, 95–102.
10. Breen, M., Bullerdiek, J., & Langford, C. F. (1999). The DAPI banded karyotype of the domestic dog (*Canis familiaris*) generated using chromosome-specific paint probes. *Chromosome Research*, *7*, 401–406.
11. Breen, M., Jouquand, S., Renier, C., Mellersh, C. S., Hitte, C., Holmes, N.

G., Cheron, A., Suter, N., Vignaux, F., Bristow, A. E., Priat, C., McCann, E., Andre, C., Boundy, S., Gitsham, P., Thomas, R., Bridge, W. L., Spriggs, H. F., Ryder, E. J., Curson, A., Sampson, J., Ostrander, E. A., Binns, M. M., & Galibert, F. (2001). Chromosome-specific single-locus FISH probes allow anchorage of an 1800-marker integrated radiation-hybrid/linkage map of the domestic dog genome to all chromosomes. *Genome Research*, *11*, 1784–1795.

12. Breen, M., Reimann, N., Bosma, A. A., Landon, D., Zijlstra, S., Bartnitzke, S., Switonski, M., Long, S. E., de Haan, N. A., Binns, M. M., Bullerdiek, J., & Langford, C. F. (1998). Standardization of the chromosome nos. 22–38 of the dog (*Canis familiaris*) with the use of chromosome painting probes. *Proceedings of the 13th European Colloquium on Cytogenetics of Domestic Animals*, Budapest, Hungary.
13. Brown, L. P. (2001). Pet nutraceuticals: Hype or wave of the future? *Nutraceuticals World*, January/February 2001, 34–41.
14. Case, L. P., Daristotle, L., Hayek, M., & Rausch, M. F. (2011). *Canine and Feline Nutrition: A Resource for Companion Animal Professionals* (3rd ed.). Mosby Elsevier.
15. Cerundolo, R., Court, M. H., Hao, Q., & Michel, K. E. (2004). Identification and concentration of soy phytoestrogens in commercial dog foods. *American Journal of Veterinary Research*, *65*(5), 592–596.
16. Cerundolo, R., Michel, K. E., Court, M. H., Shrestha, B., Refsal, K. R., Oliver, J. W., & Shofer, F. S. (2009). Effects of dietary soy isoflavones on health, steroidogenesis, and thyroid gland function in dogs. *American Journal of Veterinary Research*, *70*(3), 353–360.
17. Cousins, R. J. (1999). Nutritional regulation of gene expression. *American Journal of Medicine*, *106*, 20S–23S.
18. Cox, D. R., Burmeister, M., Price, E. R., Kim, S., & Myers, R. M. (1990). Radiation hybrid mapping: A somatic cell genetic method for constructing high-resolution maps of mammalian chromosomes. *Science*, *250*, 245–250.
19. Crocker, A. W., & Robison, O. W. (2002). Genetic and nutritional effects on swine excreta. *Journal of Animal Science*, *80*, 2809–2816.
20. Da Silva, P., Aitken, R. P., Rhind, S. M., Racey, P. A., & Wallace, J. M. (2002). Impact of maternal nutrition during pregnancy on pituitary gonadotrophin gene expression and ovarian development in growth-restricted and normally grown late gestation sheep fetuses. *Reproduction*, *123*, 769–777.
21. Daniel, H. (2002). Genomics and proteomics: Importance for the future of nutrition research. *British Journal of Nutrition*, *87*, S305–S311.

22. Daniel, H. (2002). Genomics and proteomics: Importance for the future of nutrition research. *British Journal of Nutrition*, *87*(S305-S311).
23. Eizirik, E., Murphy, W. J., & O'Brien, S. J. (2001). Molecular dating and biogeography of the early placental mammal radiation. *Journal of Heredity*, *92*, 212–219.
24. Elliot, R., & Ong, T. J. (2002). Science, medicine, and the future: Nutritional genomics. *British Medical Journal*, *324*, 1438–1442.
25. Elliott, R., & Ong, T. J. (2002). Science, medicine, and the future: Nutritional genomics. *British Medical Journal*, *324*, 1438–1442.
26. Evans, J. P., Brinkhous, K. M., Brayer, G. D., Reisner, H. M., & High, K. A. (1989). Canine hemophilia B resulting from a point mutation with unusual consequences. *Proceedings of the National Academy of Sciences of the United States of America*, *86*, 10095–10099.
27. Fahey, G. C., Jr. (2003). Research needs in companion animal nutrition. In J. Kvamme & T. Phillips (Eds.), *Petfood Technology* (in press). Watt Publishing Co.
28. Fekete, S. G., & Brown, D. L. (2007). Veterinary aspects and perspectives of nutrigenomics: A critical review. *Acta Veterinaria Hungarica*, *55*(2), 229–239.
29. Fort, P., Moses, N., Fasano, M., Goldberg, T., & Lifshitz, F. (1990). Breast and soy-formula feedings in early infancy and the prevalence of autoimmune thyroid disease in children. *Journal of the American College of Nutrition*, *9*(2), 164–167.
30. Galibert, F., Wilton, A. N., & Chuat, J.-C. (2001). The canine model in medical genetics. In A. Ruvinsky & J. Sampson (Eds.), *The Genetics of the Dog* (pp. 505–520). CAB International.
31. Gannon, M. C., & Nuttall, F. Q. (1997). Effect of feeding, fasting, and diabetes on liver glycogen synthase activity, protein, and mRNA in rats. *Diabetologia*, *40*, 758–763.
32. German, J. B., Roberts, M. A., & Watkins, S. M. (2003). Genomics and metabolomics as markers for the interaction of diet and health: Lessons from lipids. *Journal of Nutrition*, *133*, 2078S–2083S.
33. German, J. B., Roberts, M. A., Fay, L., & Watkins, S. M. (2002). Metabolomics and individual metabolic assessment: The next great challenge for nutrition. *Journal of Nutrition*, *132*, 2486–2487.
34. German, J. B., Roberts, M. A., Fay, L., & Watkins, S. M. (2002). Metabolomics and individual metabolic assessment: The next great challenge for nutrition. *Journal of Nutrition*, *132*, 2486–2487.

35. Gibson, G., & Muse, S. V. (2002). *A Primer of Genome Science*. Sinauer Associates.
36. Gohil, K., & Packer, L. (2002). Bioflavonoid-rich botanical extracts show antioxidant and gene regulatory activity. *Annals of the New York Academy of Sciences*, *957*, 70–77.
37. Green, E. D. (2001). Strategies for the systematic sequencing of complex genomes. *Nature Reviews Genetics*, *2*, 573–583.
38. Guptill, L., Glickman, L., & Glickman, N. (1999). Is canine diabetes on the increase? In *Recent Advances in Clinical Management of Diabetes Mellitus*, Proceedings from a Symposium at North American Veterinary Conference (Orlando, FL), pp. 24–27.
39. Guyon, R., Lorentzen, T. D., Hitte, C., Kim, L., Cadieu, E., Parker, H. G., Quignon, P., Lowe, J. K., Renier, C., Gelfenbeyn, B., Vignaux, F., DeFrance, H. B., Gloux, S., Mahairas, G. G., Andre, C., Galibert, F., & Ostrander, E. A. (2003). A 1-Mb resolution radiation hybrid map of the canine genome. *Proceedings of the National Academy of Sciences of the United States of America*, *100*, 5296–5301.
40. Hawken, R. J., Murtaugh, J., Flickinger, G. H., Yerle, M., Robic, A., Milan, D., Gellin, J., Beattie, C. W., Schook, L. B., & Alexander, L. J. (1999). A first-generation porcine whole-genome radiation hybrid map. *Mammalian Genome*, *10*, 824–830.
41. Hieter, P., & Boguski, M. (1997). Functional genomics: It's all how you read it. *Science*, *278*, 601–602.
42. Hill, R. C. (1998). The nutritional requirements of exercising dogs. *The Journal of Nutrition*, *128*(12), 2686S–2690S.
43. Howell, W. M., Turner, S. J., Hourihane, J. O., Dean, T. P., & Warner, J. O. (1998). HLA class II DRB1, DQB1, and DPB1 genotypic associations with peanut allergy: Evidence from a family-based and case-control study. *Clinical and Experimental Allergy*, *28*, 156–162.
44. Hu, W. T., Murray, J. A., Greenaway, M. C., Parisi, J. E., & Josephs, K. A. (2006). Cognitive impairment and celiac disease. *Archives of Neurology*, *63*, 1440–1446.
45. International Human Genome Sequencing Consortium. (2001). Initial sequencing and analysis of the human genome. *Nature*, *409*, 860–921.
46. Jamadar-Shroff, V., Papich, M. G., & Suter, S. E. (2009). Soy-derived isoflavones inhibit the growth of canine lymphoid cell lines. *Clinical Cancer Research*, *15*(4), 1269–1276.
47. Kealy, R. D. (1999). Protein in life stage nutrition: Factors influencing

lean body mass in aging dogs. *Compendium on Continuing Education for the Practicing Veterinarian*, *21*, 34–37.

48. Kerns, N. (2012). Canned controversy: Consumers are demanding BPA-free cans, but industry isn't ready with an alternative. *Whole Dog Journal*, *15*(5), 4–5.
49. Lee, C.-K., Klopp, R. G., Weindruch, R., & Prolla, T. A. (1999). Gene expression profile of aging and its retardation by caloric restriction. *Science*, *285*, 1390–1393.
50. Li, T.-K., Yin, S.-J., Crabb, D. W., O'Connor, S., & Ramchandani, V. A. (2001). Genetic and environmental influences on alcohol metabolism in humans. *Alcoholism: Clinical and Experimental Research*, *25*, 136–144.
51. Lingaas, F., Sorenson, A., Juneja, R. K., Johannson, S., Fredholm, M., Wintero, A. K., Sampson, J., Mellersh, C., Curzon, A., Holmes, N. G., Binns, M. M., Dickens, H. F., Ryder, E. J., Gerlach, J., Baumle, E., & Dolf, G. (1997). Towards construction of a canine linkage map: Establishment of 16 linkage groups. *Mammalian Genome*, *8*, 218–221.
52. Lu, C., Barr, D. B., Pearson, M. A., & Waller, L. A. (2008). Dietary intake and its contribution to longitudinal organophosphorus pesticide exposure in urban/suburban children. *Environmental Health Perspectives*, *116*(4), 537–542.
53. Mäki, K., Liinamo, A.-E., & Ojala, M. (2000). Estimates of genetic parameters for hip and elbow dysplasia in Finnish Rottweilers. *Journal of Animal Science*, *78*, 1141–1148.
54. Mellersh, C. S., Hitte, C., Richman, M., Vignaux, F., Priat, C., Jouquand, S., Werner, P., Andre, C., DeRose, S., Patterson, D. F., Ostrander, E. A., & Galibert, F. (2000). An integrated linkage-radiation hybrid map of the canine genome. *Mammalian Genome*, *11*, 120–130.
55. Mellersh, C. S., Langston, A. A., Acland, G. M., Fleming, M. A., Ray, K., Weigand, N. A., Francisco, L. V., Gibbs, M., Aguirre, G. D., & Ostrander, E. A. (1997). A linkage map of the canine genome. *Genomics*, *46*, 326–336.
56. Mering, J. V., & Minkowski, O. (1889). Diabetes mellitus nach pancreas extirpation. *Archives of Experimental Pathology and Pharmacology*, *26*, 371.
57. Moyers, S., & Bailey, L. B. (2001). Fetal malformations and folate metabolism: Review of recent evidence. *Nutrition Reviews*, *59*, 215–235.
58. Nakamura, E. M., Lane, A., Roth, G. S., & Ingram, D. K. (1998). A strategy for identifying biomarkers of aging: Further evaluation of hematology and

blood chemistry data from a calorie restriction study in Rhesus monkeys. *Experimental Gerontology*, *33*, 421–443.

59. Nebert, D. W. (1997). Polymorphisms in drug-metabolizing enzymes: What is their clinical relevance and why do they exist? *American Journal of Human Genetics*, *60*, 265–271.
60. Neff, M. W., Broman, K. W., Mellersh, C. S., Ray, K., Acland, G. M., Aguirre, G. D., Ziegle, J. S., Ostrander, E. A., & Rine, J. (1999). A second-generation genetic linkage map of the domestic dog, *Canis familiaris*. *Genetics*, *151*, 803–820.
61. Nur, T., Peijnenburg, A. A. C. M., Noteborn, H. P. J. M., Baykus, H., & Reifen, R. (2002). DNA microarray technology reveals similar gene expression patterns in rats with vitamin A deficiency and chemically induced colitis. *Journal of Nutrition*, *132*, 2131–2136.
62. O'Brien, S. J., Menotti-Raymond, M., Murphy, W. J., & Yuhki, N. (2002). The feline genome project. *Annual Review of Genetics*, *36*, 657–686.
63. O'Brien, S. J., Menotti-Raymond, M., Murphy, W. J., Nash, W. G., Wienberg, J., Stanyon, R., Copeland, N. G., Jenkins, N. A., Womack, J. E., & Graves, J. A. M. (1999). The promise of comparative genomics in mammals. *Science*, *286*, 458–462, 479–481.
64. Ooms, T. G., Khan, S. A., & Means, C. (2001). Suspected caffeine and ephedrine toxicosis resulting from ingestion of an herbal supplement containing guarana and ma huang in dogs: 47 cases (1997–1999). *Journal of the American Veterinary Medical Association*, *218*, 225–229.
65. Ostrander, E. A. (2012). Both ends of the leash: The human links to good dogs with bad genes. *New England Journal of Medicine*, *367*, 636–646.
66. Ostrander, E. A., & Kruglyak, L. (2000). Unleashing the canine genome. *Genome Research*, *10*, 1271–1274.
67. Patterson, D. F. (2000). Companion animal medicine in the age of medical genetics. *Journal of Veterinary Internal Medicine*, *14*, 1–9.
68. Plotnick, A. N., & Greco, D. S. (1995). Diagnosis of diabetes mellitus in dogs and cats. *Veterinary Clinics of North America: Small Animal Practice*, *25*, 563–570.
69. Priat, C., Hitte, C., Vignaux, F., Renier, C., Jiang, Z., Jouquand, S., Cheron, A., Andre, C., & Galibert, F. (1998). A whole-genome radiation hybrid map of the dog genome. *Genomics*, *54*, 361–378.
70. Ramadan, M. F., & Al-Ghamdi, A. (2012). Bioactive compounds and health-promoting properties of royal jelly: A review. *Journal of Functional Foods*, *4*(1), 39–52.

71. Ricci, R., Granato, A., Vascellari, M., Boscarato, M., Palagiano, C., Andrighetto, I., & Mutinelli, F. (2013). Identification of undeclared sources of animal origin in canine dry foods used in dietary elimination trials. *Journal of Animal Physiology and Animal Nutrition*, *97*, 32–38.
72. Rocchini, A. P., Marker, P., & Cervenka, T. (1997). Time course of insulin resistance associated with feeding dogs a high-fat diet. *American Journal of Physiology*, *272*, E147–E154.
73. Rocchini, A. P., Moorehead, C., Wentz, E., & Deremer, S. (1987). Obesity-induced hypertension in the dog. *Hypertension*, *9*, III-64–III-68.
74. Roth, G. S., Lane, M. A., Ingram, D. K., Mattison, J. A., Elahi, D., Tobin, J. D., Muller, D., & Metter, E. J. (2002). Biomarkers of caloric restriction may predict longevity in humans. *Science*, *297*, 811.
75. Saarinen, K., Jantunen, J., & Haahtela, T. (2011). Birch pollen honey for birch pollen allergy—A randomized controlled pilot study. *International Archives of Allergy and Immunology*, *155*, 160–166.
76. Scearce, L. M., Brestelli, J. E., McWeeney, S. K., Lee, C. S., Mazzarelli, J., Pinney, D. F., Pizarro, A., Stoeckert, C. J., Jr., Clifton, S. W., Permutt, M. A., Brown, J., Melton, D. A., & Kaestner, K. H. (2002). Functional genomics of the endocrine pancreas. The pancreas clone set and PancChip, new resources for diabetes research. *Diabetes*, *51*, 1997–2004.
77. Sforcin, J. M., & Bankova, V. (2011). Propolis: Is there a potential for the development of new drugs? *Journal of Ethnopharmacology*, *133*(2), 253–260.
78. Sloth, C. (1992). Practical management of obesity in dogs and cats. *Journal of Small Animal Practice*, *33*, 178–182.
79. Sottiaux, J. (1999). Atherosclerosis in a dog with diabetes mellitus. *Journal of Small Animal Practice*, *40*, 581–584.
80. Stewart, T. L., & Ralston, S. H. (2000). Role of genetic factors in the pathogenesis of osteoporosis. *Journal of Endocrinology*, *166*, 235–245.
81. Stride, A., & Hattersley, A. T. (2002). Different genes, different diabetes: Lessons from maturity-onset diabetes of the young. *Annals of Medicine*, *34*, 207–216.
82. Struble, A. L., Feldman, E. C., Nelson, R. W., & Kass, P. H. (1998). Systemic hypertension and proteinuria in dogs with diabetes mellitus. *Journal of the American Veterinary Medical Association*, *213*, 822–825.
83. Sunvold, G. D., & Bouchard, G. F. (1998). Assessment of obesity and associated metabolic disorders. In G. A. Reinhart & D. P. Carey (Eds.), *Recent Advances in Canine and Feline Nutrition* (Vol. II, pp. 135–148). Orange Frazer Press.

84. Swanson, K. S., Kuzmuk, K. N., Schook, L. B., & Fahey, G. C., Jr. (2003). Effects of diet and age on metabolic characteristics and gene expression profile in the dog. Part 2: Gene expression profiling. *Journal of Animal Science*, *81*(Suppl. 1), 259.
85. Swanson, K. S., Kuzmuk, K. N., Schook, L. B., & Fahey, G. C., Jr. (2003). Effects of diet and age on metabolic characteristics and gene expression profile in the dog. Part 1: Metabolic characteristics. *Journal of Animal Science*, *81*(Suppl. 1), 259.
86. Swanson, K. S., Schook, L. B., & Fahey, G. C. (2003). Nutritional genomics: Implications for companion animals. *Journal of Nutrition*, *133*(10), 3033–3040.
87. Switonski, M., Reimann, N., Bosma, A. A., Long, S., Bartnizke, S., Pienkowska, A., Moreno-Milan, M. M., & Fischer, P. (1996). Report on the progress of standardization of the G-banded canine (*Canis familiaris*) karyotype. *Chromosome Research*, *4*, 306–309.
88. Ueland, P. M., Hustad, S., Schneede, J., Refsum, H., & Vollset, S. E. (2001). Biological and clinical implications of the MTHFR C677T polymorphism. *Trends in Pharmacological Sciences*, *22*, 195–201.
89. Uyeda, K., Yamashita, H., & Kawaguchi, T. (2002). Carbohydrate responsive element-binding protein (ChREBP): A key regulator of glucose metabolism and fat storage. *Biochemical Pharmacology*, *72*, 1–6.
90. Werner, P., Mellersh, C. S., Raducha, M. G., DeRose, S., Acland, G. M., Prociuk, U., Wiegand, N., Aguirre, G. D., Henthorn, P. S., Patterson, D. F., & Ostrander, E. A. (1999). Anchoring of canine linkage groups with chromosome-specific markers. *Mammalian Genome*, *10*, 814–823.
91. Willett, B. J., Flynn, J. N., & Hosie, M. J. (1997). FIV infection of the domestic cat: An animal model for AIDS. *Immunology Today*, *18*, 182–189.
92. Wilson, A. G., di Giovine, F. S., & Duff, G. W. (1995). Genetics of tumor necrosis factor-alpha in autoimmune, infectious, and neoplastic diseases. *Journal of Inflammation*, *45*, 1–12.
93. Wyman, M., Sato, S., Akagi, Y., Terubayashi, H., Datiles, M., & Kador, P. F. (1988). The dog as a model for ocular manifestations of high concentrations of blood sugars. *Journal of the American Veterinary Medical Association*, *193*, 1153–1156.
94. Zhu, X., Yan, D., Cooper, R. S., Luke, A., Ikeda, M. A., Chang, Y.-P. C., Weder, A., & Chakravarti, A. (2003). Linkage disequilibrium and haplotype diversity in the genes of the renin-angiotensin system: Findings from the family blood pressure program. *Genome Research*, *13*, 173–181.

*Chapter* **7**

# NUTRIGENOMICS APPROACHES IN CANINE DISEASE, TESTING AND PREVENTION

## Contents

# INTRODUCTION

Nutrigenomics, the study of how diet influences gene expression, is emerging as a transformative field in animal health, particularly in the prevention and management of diseases. By understanding the interactions between nutrients and an animal's genome, nutrigenomics offers the potential to tailor diets that not only meet basic nutritional needs but also promote health and mitigate disease risks at the genetic level.

In the context of animal disease, nutrigenomics provides insights into how specific nutrients can modulate the expression of genes involved in disease pathways. For example, certain dietary components may influence genes that regulate inflammation, immune response, or metabolic processes, which are critical factors in the development and progression of diseases such as obesity, diabetes, cancer, and inflammatory conditions. By identifying these nutrient-gene interactions, veterinarians and animal nutritionists can develop customized dietary interventions aimed at preventing or managing specific health conditions in animals.

Moreover, nutrigenomics has the potential to improve animal health on a population level by identifying genetic markers associated with disease susceptibility. This allows for the development of precision nutrition strategies that can be applied across herds or flocks, reducing the incidence of disease and enhancing overall productivity and welfare. For instance, livestock producers could use nutrigenomic insights to optimize feed formulations that reduce the risk of common diseases, such as mastitis in dairy cows or respiratory diseases in poultry.

As research in nutrigenomics advances, the field is expected to play an increasingly important role in animal health management. The ability to integrate genetic information with nutritional science offers a powerful tool for disease prevention, potentially reducing the reliance on pharmaceuticals and improving the longevity and quality of life for animals. Ultimately, nutrigenomics represents a shift towards more personalized and proactive approaches to animal health, with significant implications for veterinary medicine, animal husbandry, and food production systems.

## 7.1. FOOD INTOLERANCE/SENSITIVITY TESTING

### 7.1.1. Food Allergies versus Food Intolerances/Sensitivities

The terms "food allergy" and "food intolerance/sensitivity" are sadly often used interchangeably by non-veterinary professionals, despite the fact that they are very different and have very different health implications for your dog.

Companion animals will continue to suffer and not receive the proper diagnosis unless reliable veterinary professionals clear up this confusion regarding terms.

Actually, true food allergies are extremely uncommon (Kraft and Rothbert, 1967; Challacombe, 1987). Based on professional experience, WJD has estimated that food sensitivities outnumber food allergies by a minimum of 10 to 15 times. Therefore, your dog is most likely experiencing an intolerance or sensitivity to one or more ingredients in his diet rather than an acute food allergy if he scratches constantly or has persistent gastrointestinal issues.

According to (research, food allergies are indicative of a more severe immunological reaction (Jeffers, Shanley and Meyer 1991; Buchanan and Frick, 2002; Day, 2005). For example, a peanut allergy can result in anaphylactic shock. Individuals who are allergic to peanuts experience an immediate closure of their airways, making breathing difficult for them. This answer comes back almost immediately. An immediate, sometimes fatal, immunological and physiological response is elicited by the antigen (in this case, peanuts). Allergy reactions that are less severe but nonetheless dangerous include rashes, hives, and swollen eyes. Type I hypersensitivity reactions are what these are all known as. According to Foster et al. (2003), they manifest as antibodies to immunoglobulins E (IgE) and G (IgG), two of the body's immune proteins, in the blood. Refusing to eat the food that triggers your allergy is the only way to prevent its effects.

On the other hand, food sensitivities and intolerances are usually long-term conditions that develop gradually, sometimes even after months or years of exposure to the offending ingredient or ingredients, and they don't always involve an immune reaction. Types II and III hypersensitivity reactions, which manifest as antibodies to immunoglobulins A (IgA) and M (IgM) in saliva or feces, are the cause of food intolerances and sensitivities (Rinkinen et al., 2003; Lee and Wong, 2009; Miller et al., 2010; Reeves and Robinson, 2013).

Dogs' physical and mental health can be negatively impacted by food intolerances and sensitivities, even though they are usually not life-threatening. Common signs include:

- GI tract issues similar to IBD.
- Chronic itching.
- Chronic burping and gas rumblings (borborigmi).
- Chronic skin, ear and foot infections (especially with the presence of yeast) (Dodds, 2014).

Finding the source of the problem and eliminating the offending ingredient or ingredients from your dog's diet is the only way to "cure" a food intolerance or sensitivity. Since food sensitivities and intolerances usually develop gradually, it can be challenging to link your dog's symptoms to a specific food based only on observation. This is particularly true since, before any obvious skin or gastrointestinal symptoms manifest, your dog may have been eating the offending food for months or even years. Over time, a dog is more likely to develop an intolerance or sensitivity to a particular food if he eats it more frequently and in greater quantities. For this reason, when we hear people argue that, "My dog's itchy skin (or whatever other problem he may have) can't be caused by chicken because he's been eating it for years," they fail to recognize that the possibility that a food may be the cause of the issue arises from consuming it consistently over an extended period of time.

In order to examine the most widely used types of food intolerance/sensitivity testing, let's start with the most antiquated and laborious and progress to the most recent and precise.

## 7.1.2. Food Elimination Trials: The Jigsaw Puzzle of Food Intolerance/Sensitivity Testing

Your veterinarian may advise you to carry out your own food elimination trial if your dog exhibits symptoms of a food intolerance or sensitivity. This entails giving your dog a diet that alternates between one novel protein and one source of carbohydrates, usually for eight to twelve weeks, and watching to see if he reacts in any way. However, keep in mind that food elimination trials present a number of difficulties. As we've already mentioned, mass-market, commercially prepared foods may contain ingredients not disclosed on the label. This is particularly true in this case (Ricci et al., 2013).

Think about kibble. Multiple listed or unlisted protein and carbohydrate sources can readily be found in one bag of kibble (Ricci et al., 2013). For example, the product may contain beef, fish, etc., even though chicken may be the main source of protein. The same is true for carbohydrates, which can be made up of a single tiny nugget of corn, wheat, soy, or other components that could react. Even though wet foods typically have fewer ingredients, they may also include multiple protein and carbohydrate sources. We advise against attempting to ascertain whether your dog is sensitive to dry or wet commercial dog food because of this. We, along with other experts, advise you to stick to your own home-prepared diet, where you control every ingredient, when conducting a food elimination trial because there are just too many variables in place and you risk driving yourself insane in the process (Ricci et al., 2013).

Veterinarians typically advise starting your food elimination trial on a diet consisting solely of potatoes and lamb. However, what if potatoes and/or lamb are the issue? Feeding him these items could backfire rather than help. Assume his symptoms don't go away after 72 hours on the lamb and potato diet. Until the symptoms do go away, you can then continue and try a different mix of one unique protein and one carbohydrate. Even so, you're not quite done because it's still unclear which ingredient—the carb or the protein—is really the source of the issue. This means that each week, you have to add one protein and one carbohydrate back into your dog's diet. You should be able to identify the culprit food when the symptoms of itchy skin return, right? Not always. It may take up to five weeks for food intolerances or sensitivities to manifest clinically. Assume, therefore, that after your dog's symptoms go away, you give him beef again in week five and chicken again in week six. However, by the end of week six, your dog starts to scratch once more. It's hard to tell if he's responding to the chicken or the beef. That might even be a different protein from a few weeks ago. You should be confused. Ultimately, there are simply too many variables in food elimination trials. In addition, they can be highly costly due to the amount of food that is wasted, veterinary visits, and medications needed to manage your dog's symptoms while you try to identify the issue.

### 7.1.3. Skin Patch and Prick Testing

It's hardly surprising that "skin patch and prick testing," which has long been regarded as the gold standard for determining the cause of allergies and food intolerances/sensitivities in humans, has also gained popularity in the canine population. However, there are drawbacks, including:

- Testing must be performed at the veterinarian's office.
- Testing takes approximately two hours to perform.
- You must withhold food and water for four to five hours in advance.
- Your dog must be sedated.
- The area being tested must be shaved.
- The procedure is invasive.
- Antihistamines are required after the procedure.
- Steroid treatment might be required.
- It's expensive (approximately $500 for 20 antigens).

The main issues with these techniques, however, are that they only test for acute or immediate reactions (i.e., food intolerances or sensitivities) rather than long-term ones. e. g., food allergies)—and even in those cases, their accuracy in determining the reaction's actual cause is questionable. If long-term food-related health problems are not identified and treated, our cherished dogs will continue to suffer. This is the responsibility of veterinarians and dog parents.

### 7.1.4. NutriScan: The Most Accurate and Predictive Test to Identify Canine Food Intolerances/Sensitivities

The new gold standard for determining the cause of food sensitivities and intolerances in dogs is NutriScan, which is only available through author WJD's Hemolife testing laboratory. Not only is NutriScan the most scientifically accurate method, but it's also the most economical, convenient, and least intrusive for you and your dog. You can promptly eliminate the ingredient(s) from your dog's diet thanks to NutriScan, which can identify the reactive food(s) in 10 days (Dodds, 2014).

IgA and IgM antibodies in a dog's saliva are measured by NutriScan. NutriScan detects changes in the dog's gene expression when exposed to the reactive food by looking for high levels of IgA and IgM antibodies. This allows the test to pinpoint the exact ingredient or ingredients causing the issue. Because food allergies are often mediated by different antibodies (IgE and IgG) than food intolerances/sensitivities, NutriScan can also distinguish between a food allergy and a sensitivity to a food (Dodds, 2014).

### 7.1.5. Why Is NutriScan Revolutionary?

Let's examine more closely how canine food intolerance testing has been transformed by NutriScan.

#### *7.1.5.1. NutriScan Can Help Dogs That Appear Healthy as Well as Dogs That are Suspected to Suffer from Food Reactivity*

A saliva test can detect a food sensitivity or intolerance in your dog even before they exhibit any obvious symptoms, like skin irritation or digestive problems. This is conceivable because salivary antibodies to food ingredients manifest in patients before the standard clinical or bowel biopsy diagnosis of leaky gut syndrome or inflammatory bowel disease (Lee and Wong, 2009; Vojdani, 2009; Miller et al., 2010). This implies that NutriScan has the ability to anticipate a food reaction before it materializes.

#### *7.1.5.2. NutriScan Can Now Identify Intolerances/Sensitivities to 24 Different Foods*

Currently, NutriScan comes with two distinct panels, each of which tests twelve food antigens.

**Panel 1:** beef, chicken, corn, cow's milk, duck, lamb, pork, soy, turkey, venison, wheat and white fish.

**Panel 2:** barley, hen's egg, lentils, millet, oatmeal, peanut (peanut butter), potato, quinoa, rabbit, rice, salmon and sweet potato.

Both panels can be ordered simultaneously or in reverse order. Currently, cat NutriScan testing with a patent is also accessible).

The panels contain 18 more antigens in addition to the so-called six primary food antigens, which include the glutens found in the pet foods and treats that are most frequently fed. One of these antigens is probably the cause of a food sensitivity or intolerance in a dog (or cat).

### *7.1.5.3. NutriScan Is Simple to Perform Right in Your Own Home*

The test can be performed in duplicate because it only requires two milliliters of saliva to be extracted from your dog's mouth (wet about one-third of the dental cotton rope's length with saliva). It only takes a few simple steps to obtain the saliva: (Note: Before collecting the saliva, your dog must have fasted for at least three hours).

1. Lift your dog's upper lip.
2. Place the cotton rope in your dog's mouth at the corner of the mouth.
3. Leave the free end of the rope hanging out of the mouth, but hold on to it.
4. If you need to promote salivation, wave a treat over your dog's nose (but do not feed it).
5. Keep the rope in your dog's mouth for two minutes.

Simply cut off the wet portion of the cotton rope after gathering the saliva, insert it into the shipping tube, seal it, and mail it along with the completed test request form. You can mail the sample to Hemolife using FedEx, UPS, the Postal Service, or another shipping company. It keeps for up to 30 days and doesn't require refrigeration. Hemolife will provide you with the results (as well as your veterinarian, upon request) in ten days.

### *7.1.5.4. NutriScan Identifies the Specific Foods to Avoid Feeding Your Dog*

Just keep your dog away from any foods, treats, vitamins, grooming products, or drug vehicles (things like pill pockets used to administer drugs) that contain any of the food extracts listed in the panel if your dog tests positive for any of them. As of right now, Nutriscan stands for the most technologically sophisticated diagnostic stage of determining a dog's functional nutrition. The dog's altered gene expression in response to particular food extracts is indicated by the presence (shown by an intermediate, medium, or strong reaction) or absence (shown by a negative or weak reaction) of salivary antibodies. Therefore, NutriScan is dependent on factors that can change gene expression and nutritional influences (Swanson, Schook and Fahey, 2003; Fekete and Brown, 2007).

### 7.1.6. Sample NutriScan Diagnostic Report

Hemolife generated a sample NutriScan report based on Panel 1 (12 foods) and Panel 2 (12 foods) that are shown on the following page. More robust reaction columns are displayed in different grayscale tones. As you can see, the dog's medium-level food intolerance reactions were strongest when it came to corn and venison. A number of foods, including chicken, milk, soy, turkey, wheat, whitefish, and oatmeal, also showed intermediate reactions; it was suggested to stay away from these foods. The dog didn't exhibit any discernible intolerance to the other foods that were tested.

### 7.1.7. NutriScan Interpretation

Treats or food containing ingredient(s) exhibiting strong, medium, or intermediate reactions should be avoided by pets. Recheck salivary food sensitivity or intolerance levels every 12 to 18 months, or more frequently if intolerances are found and the diet is modified appropriately. Preferably, this should be done every six months.

#### *7.1.7.1. Degree of Reactivity*

- <10 U/mL indicates a normal food antigen tolerance level = negative result
- 10–11.4 U/mL indicates a weak reaction; clinical significance unclear
- 11.5–11.9 U/mL indicates a borderline reaction
- 12–12.9 U/mL indicates an intermediate reaction
- 13–14.9 U/mL indicates a medium reaction
- >/= 15 U/mL indicates a strong reaction

#### *7.1.7.2. Differences Between Antibodies to IgA and IgM*

The secretory immunity to bodily secretions (tears, saliva, feces, urogenital tract) is measured by antibodies to IgA. They serve as the first line of defense or a mechanical barrier to help keep the bowel from being invaded by chemicals, foreign substances, infectious agents, and some foods that it cannot or will not tolerate. The body's primary immune response to an exposure that occurred within the last six months or so is measured by antibodies to IgM (to a specific culinary component).

### 7.1.7.3. Saliva Testing Versus Serum (Blood) Testing

Many people are curious about the distinctions between serum, or blood, testing and salivary food intolerance/sensitivity testing. It's natural to think that a blood test would be more accurate and dependable than a saliva test. After all, our doctor routinely takes our blood during a check-up and sends it to a diagnostic lab to find out if we have any undiagnosed health conditions. Saliva testing is more successful in detecting long-term food reactions than blood testing, which is frequently regarded as the window to our health (Robinson and Reeves, 2012).

As you are now aware, the gut is the body's largest immune organ, and leaky gut syndrome develops when the gut's cellular lining is damaged. Keep in mind that big food particles can enter the bloodstream through a leaky gut and be recognized by the body as antigens. To defend itself against what was once a harmless food ingredient, the body's immune system launches into action and produces protein antibodies. Immunoglobulin A (IgA) and immunoglobulin G (IgG), the produced antibodies, circulate in the bloodstream and have the ability to enter tissues and inflict harm.

IgA and IgG antibody levels are measured by blood tests. Elevated antibody levels signify increased intestinal lining permeability, or a leaky gut, whereas normal levels indicate a healthy gut with a healthy mucosal barrier. As we've previously stated, food sensitivities and intolerances are present in the body long before symptoms manifest, and a weakened gut environment is usually the root cause of these issues. Saliva testing is even more sensitive, so why not use blood tests to look for underlying food intolerances/sensitivities if the blood can measure elevated levels of IgA and IgG antibodies before outward signs appear? Months before IgA and IgG antibodies show up in the blood and several months before lesions form on the GI tract's surface, which allow veterinarians to clinically diagnose inflammatory bowel disease (IBD) or leaky gut syndrome, are detectable IgA and IgM antibodies produced in saliva (and feces). Therefore, the pathological process of gut destruction and permeability can be detected by the saliva testing system far sooner than it can be by blood testing or clinical diagnosis. Additionally, saliva collection is noninvasive, painless, reasonably priced, and patient-friendly. Published studies comparing serum and saliva methods for human food sensitivity testing have demonstrated that the latter yields results that are less clinically relevant and less accurate than the former.

## 7.2. A NUTRIGENOMICS APPROACH TO WEIGHT CONTROL

According to APOP (2013) and Byers et al. (2013), obesity is the most common preventable medical condition that veterinarians treat and the biggest threat to the health of companion dogs and cats. 2011). If your dog is overweight or obese, no matter what else you do for him, he will not experience optimal health.

Regretfully, there is a sharp increase in companion animal obesity. The Association for Pet Obesity Prevention (APOP) found that, according to the veterinarians who treat them, 52.5% of dogs and 58.3% of cats are overweight or obese in its 2012 National Pet Obesity Awareness Day Survey.

This indicates that 43.2 million cats and 36.7 million dogs are more likely to develop weight-related conditions such as diabetes, osteoarthritis, hypertension, and various cancers (APOP, 2013). The 2012 State of Pet Health Report from Banfield Pet Hospital offers shocking new information about the connection between obesity and disease. The study discovered the following after analyzing data from 430,000 cats and more than two million dogs.

- 42% of dogs with diabetes are overweight (and 40% of cats).
- 40% of dogs with arthritis are overweight (and 37% of cats).
- More than 40% of dogs with high blood pressure are overweight.
- 61% of hypothyroid dogs are overweight.

Overweight dogs are at increased risk for many diseases, including:

- Cardiorespiratory diseases, including airway obstruction syndrome (seen in the brachycephalic dog breeds) and laryngeal paralysis.
- Endocrine disorders, including hyperadrenocorticism (Cushing's disease) and hypothyroidism.
- Functional alterations, such as decreased respiratory capacity, exercise intolerance, heat intolerance/stroke and decreased immune functions.
- Metabolic abnormalities, such as hyperlipidemia/dyslipidemia (high or abnormal blood lipid levels).
- Neoplasia (abnormal growth or division of cells as seen in cancer), including transitional cell carcinoma (tumor of the bladder).
- Orthopedic disorders, such as osteoarthritis, anterior cruciate ligament rupture and intervertebral disk disease.
- Urogenital system conditions, including transitional cell carcinoma bladder tumor.

Additionally, research indicates that dogs who are overweight or obese typically live two years shorter than their ideal-weight counterparts (Kealy et al., 2002). Conversely, it has been demonstrated that calorie restriction lengthens life expectancy by raising the quantity of dietary activators of sirtuins—proteins that control metabolism and longevity—that dogs are fed, such as resveratrol and other polyphenols (grape seed extract, green and black tea) (Lawler et al., 2005; Lawler et al., 2007; Frantz, Yamka, and Friesen, 2007; Allard et al., 2009).

Even more depressingly, 76% of dog guardians and 69% of cat guardians in Banfield's separate survey of pet caretakers mistakenly thought their animals were at a healthy weight (Banfield Pet Hospital, 2012; Howard, 2012). To prevent obesity-related illnesses and premature deaths in our companion animals, it is imperative that we identify the problem and subsequently take appropriate action to address it. Recently, researchers have started looking for a genetic connection between obesity and they have discovered one. The first "obesity" gene variants were found in 2007 in what is known as the "fat mass and obesity-associated" gene (HSPH, 2013; Sjøgren, 2013). Individuals with one of these gene variants have a 20–30% increased risk of obesity compared to those without it (HSPH, 2013). Additionally, two more recent studies have found a number of gene variants associated with obesity in adults and children (Dotinga, 2013). These results lend credence to the hypothesis that a polygenic (multiple gene) predisposition to obesity is most likely the result of several genes rather than a single gene (Swanson, 2007; HSPH, 2013; Norris, 2013).

Certain breeds of dogs have been identified as genetically predisposed to obesity, including:

- Basset Hounds
- Beagles
- Boxers
- Cairn Terriers
- Cavalier King Charles Spaniels
- Cocker Spaniels
- Dachshunds
- Doberman Pinschers
- Golden Retrievers
- Labrador Retrievers
- Scottish Terriers
- Shetland Sheepdogs
- West Highland White Terriers

However, sighthounds seem to be less vulnerable (Byers et al., 2015), and German Shepherds had the lowest incidence of obesity in the APOP study (APOP, 2013). Even though scientists are uncovering more genetic susceptibilities to obesity in both dogs and humans, the question of how these genes function to influence body weight over time remains unanswered (Swanson, 2007; Yamka et al., 2007). Studies have indicated that while some individuals with specific genetic variants are more likely to become obese, others with the same variants tend to remain thin throughout their lives (Dotinga, 2013). Similarly, every dog breed mentioned above includes both lean and overweight/obese variations.

If many lean individuals carry the same obesity gene variants as those who are overweight, it suggests that other factors are at play. Environmental factors and lifestyle choices may influence an individual's susceptibility to obesity. According to a 2011 peer-reviewed meta-analysis, physical activity reduced the risk of weight gain in individuals carrying a gene that predisposed them to obesity—a study involving over 200,000 people. Researchers found that individuals with the obesity gene who engaged in physical activity had nearly one-third lower risk of becoming overweight or obese compared to those who did not exercise (Kilpeläinen, 2011).

It is obvious that a person is not destined to be obese due to genetic predisposition. As we've been talking about, a lot of the environmental and lifestyle factors that impact how our genes express themselves can be changed, even though we can't change our DNA (Veerman, 2011; Lavebratt et al., 2012).

### 7.2.1. Like Dog Parent, Like Dog

An unsettling parallel has emerged over the past fifty years between the rise in obesity in humans and companion animals. We are convinced that this parallel is related to comparable changes in the environment and way of life, such as:

- Decrease in physical activity.
- Progression away from fresh, wholesome foods to highly processed foods packed with ingredients that send unhealthy messages to the genes (as previously discussed at length).
- Proliferation of readily available, energy-dense convenience foods.

Scientists can now screen for "obesity genes" in humans and many other animals, including dogs, thanks to genomic mapping. However, as you have just discovered, there are several reasons why screening for obesity does not address the issue of obesity.

- Genetic screening for obesity does not take into account environmental and behavioral contributions.

- Individuals who know they are genetically predisposed to obesity might actually feel discouraged to exercise and eat healthy.
- It fails to take into account the larger societal influences at play.
- It has a low predictive power to determine whether an individual will actually become obese.

We are aware that some people are predisposed to being overweight due to "obesity genes," but let's look at the opposite situation. Numerous studies have shown that obesity can impact gene expression, which can lead to disease (Endocrine Today, 2011; Franks and Ling, 2010; Medical College of Georgia, 2013; Demerath et al., 2013; Fiore, 2013).

### 7.2.2. Food Sensitivities: A Hidden Cause of Weight Gain

Food sensitivities and intolerances are major contributors to cellular inflammation, as we have previously discussed. The intestinal lining becomes inflamed throughout this process. Small fissures between the cells are created by this inflammation, which in turn allows foreign invaders like bacteria and partially digested food molecules to pass from the gastrointestinal tract into the bloodstream. This triggers an additional immune system response to eliminate the "enemy." Before you know it, the immune system is at war, and inflammation spreads throughout the entire system (Hyman, 2012; Hyman, 2013).

Eating reactive foods can contribute to obesity because inflammation is a result of food intolerances and sensitivities, and obesity is an inflammatory disease.

If your dog only occasionally consumes the offending food, this might not be a significant concern. However, imagine the inflammation that would build up in his body if he were to consume it daily—for example, in his commercial dog food—and his immune system had to battle the antigen every day! If this process goes undetected, it can persist for months or even years, causing your dog to become overweight and develop other chronic diseases (Hyman, 2012; Hyman, 2013). For this reason, we highly recommend using NutriScan saliva-based food intolerance/sensitivity testing to identify potential antigens specific to your dog. If his body is a "smoldering volcano" of inflammation, it cannot maintain true cellular health, and he will find it much more difficult to lose excess weight.

By now, you are aware that despite his adorable appearance, your dog's excess fat is negatively impacting his long-term cellular health and vitality. Now is the time to take action and help your dog overcome the bulge. According to Byers et al. (2011), he will not only feel better but also live a longer, happier, and healthier life (Byers et al., 2011a; Yamka, Frantz, and Friesen, 2007;

Roudebush, Schoenherr, and Delany, 2008). Given that dog (and cat) parents frequently have a distorted perception of their pet's weight, we suggest adopting a more objective approach and measuring standards that are reliable (unlike our loving eyes). If your dog is small enough, a good old-fashioned scale is one such method. But even the scale misses the crucial aspect of determining body fat composition.

We recommend doing a Body Condition Score (BCS) on your dog at least every four weeks, preferably every two. This is a straightforward method that we prefer. Simply observe your dog from the side and above, then palpate (gently press) his shoulder blades, spinal column, ribs, hips, and the underside of his belly to determine how much fat is on top of them. This is how a BCS is performed. BCS is based on a 5- or 9-point scale, with the middle number (3 out of 5 or 5 out of 9) representing the ideal body condition, which is linked to 15% to 25% body fat. Less than ideal body fat, or "under-condition," is reflected in lower numbers. Greater values correspond to levels of "over-condition," or excess body fat. An obese dog is one with a score of 5 out of 5 or 9 out of 9, which denotes more than 35% body fat (Bartges, 2013).

- **Here's How to Check Your Dog's BCS**

1. With your dog standing on all fours, palpate his shoulder blades, spinal column, ribs, hips and underside of his belly. You should be able to feel them through a thin layer of fat, but they should not protrude. This equals a BCS of 3 on the 5-point scale.
2. View your dog from the top when he's standing on all fours. He should resemble an oblong hourglass—wider at the ribs, narrower at the waist and wider at the hips. If you look down and see two straight lines, your dog is carrying too much fat.

The most obvious signs your dog is overweight are:

- A large body relative to the legs.
- Excess fat around the neck and underside of the belly.
- Round appearance, especially when viewed from above.

Other signs include:

- Decreased activity level.
- Difficulty rising or climbing stairs.
- Excessive panting during activity.

Your veterinarian should also perform a BCS evaluation as part of your dog's wellness exams.

Make sure your dog is examined by a veterinarian to rule out any underlying medical conditions, such as hypothyroidism or Cushing's disease, that could be contributing to his weight gain before starting him on a diet. If you don't do this, your dog may try unsuccessfully to lose weight and you may become very frustrated. Your dog's ideal weight will be determined by your veterinarian once he has given the all-clear, and he will also be a valuable source for ongoing health checks, weigh-ins, and encouragement along the journey.

Finding out how many calories (kcal) your dog needs to consume each day to safely reach its target weight is the next step. Instead of going straight for his ideal weight if he has a lot of weight to lose, you could set an interim goal weight (Ward, 2007). You should also get advice on any necessary supplements from your veterinarian or nutritionist to make sure your dog is getting enough nutrients in its diet.

A monthly weight loss of 3 to 5 percent of body weight is deemed safe for the majority of dogs (Ward, 2007). Feeding your dog 100% of his Resting Energy Requirements (RER) will accomplish this. The daily energy required by your dog for resting and basic metabolic processes like breathing, digestion, and heart function is represented by his RER (metabolism accounts for approximately half of an average adult dog's daily energy requirements).

The following formula is used to calculate RER in animals weighing between 2 to 45 kg (5 to 99 pounds):

1. **Step 1:**

Determine his ideal weight in kilograms (kg):

- Ideal weight in pounds (as determined by your veterinarian or canine nutrition expert) divided by 2.2 gives you weight in kilograms (kg).
- Example: If your dog's ideal weight (remember that this is not his current weight, it is his goal weight) is 20 pounds, divide 20 by 2.2. His ideal weight in kilograms is 9 kg.

2. **Step 2**

Determine his Resting Energy Requirements (RER) based on this ideal weight:

- RER in kcal/day = 30(ideal body weight in kilograms) + 70.
- 30(9) + 70 = 340 kcal/day = RER.

Eating all of their RER calories in a day should help most dogs lose weight. If your dog doesn't react, though, his diet will be changed by your veterinarian or nutritionist (Ward, 2007). Keep in mind that your dog's resting energy requirement (RER) is his daily energy requirement; feeding him at this level will help him lose weight safely. But after he hits his target weight, you should feed him enough to keep him there. His Maintenance Energy Requirement (MER) is the term for this. The factors listed in the table below determine your dog's MER. Just find the matching factor and multiply the RER of your dog—which you already know—by the relevant figure. But keep in mind that no diet is "one size fits all," which has been the main argument of this book! Nor is there a single daily calorie intake. When it comes to their daily MER, dogs are far more unique than humans. Unless you're an elite athlete, chances are you need about the same amount of calories as other people your height and gender. On the other hand, a variety of factors, such as breed, age, health, lifestyle, and even coat thickness, can impact a dog's MER. To make sure that you're feeding your dog companion for optimal weight and health, understanding the numbers is just one more (though still a useful) tool.

### 7.2.3. Not All Calories are Created Equal

Naturally, calorie restriction is necessary for weight loss, but it's also critical to keep in mind that not all calories are created equal. Since inflammation generates obesity, be sure to follow our two-step process to help your dog conquer the battle of the bulge and achieve optimum health:

- **Step 1:** Feed lots of fat-fighting, anti-inflammatory functional foods.
- **Step 2:** Remove pro-inflammatory ingredients such as those containing artificial colors, flavors and preservatives, added antibiotics and hormones, sugar, high levels of pesticides and GMO foods, as well as high-glycemic (high-GI) carbohydrates.

### 7.2.4. Go for "Good" Carbs

A recap of the distinctions between high-GI and healthy carbohydrates. The most important thing to keep in mind is that healthy carbohydrates come from whole, fresh foods like fruits, vegetables, legumes, and whole, gluten-free grains. Conversely, processed foods with a high glycemic index, like bread, pasta, and cereal, as well as many of the refined grains in mass-market commercial pet foods, are the source of unhealthy carbohydrates. High-GI diets are detrimental to weight control and actually encourage the storage of body fat, according to both short- and long-term research in humans and animals (Brand-Miller et al., 2002). In addition to being pro-inflammatory, sugary and starchy foods can seriously hinder your dog's weight loss efforts (Hyman, 2013).

### 7.2.5. Fat-fighting Functional Foods

It's time to introduce fat-fighting functional foods to your dog's diet after you've eliminated as many pro-inflammatory ingredients as you can (Yamka, Frantz and Friesen, 2007; Roudebush, Schoenherr and Delany, 2008). Some of the most significant ones are listed below:

#### *7.2.5.1. High-Quality, Bioavailable Novel Protein*

An essential component of any weight-loss plan is protein. We have discussed in great detail the role that protein plays in the development and repair of tissue, which is crucial to preserving lean body mass while losing weight. Diets that are high in protein, moderate in fat, and low in carbohydrates are ideal for weight loss.

#### *7.2.5.2. Coconut Oil (Virgin, Expeller-Pressed, Preferably Organic)*

Especially in light of its critical role in maintaining brain health. Coconut oil was disapproved of for many years as a bad saturated fat. However, that is not even close to the reality. The health benefits of coconut oil, which is also referred to as medium-chain triglycerides (MCT) and medium-chain saturated fat (MCFA), are numerous and include heart-protective and immune-boosting qualities. Additionally, coconut oil digests differently than other fats, increasing the thermogenic effect—which raises body temperature and speeds up metabolism—that aids in losing weight (Perricone, 2010). Dogs adore the flavor as well.

#### *7.2.5.3. Omega-3 Fatty Acids*

We'll talk a lot about omega-3 fatty acids, another potent canine functional superfood. Because obesity is an inflammatory condition, as we just discussed, omega-3s have a potent anti-inflammatory effect and are therefore important for weight loss. Additionally, omega-3s trigger a particular kind of protein that activates genes linked to fat-burning (Perricone, 2010).

#### *7.2.5.4. L-Carnitine*

Lysine and methionine are converted into the amino acid L-carnitine in the liver and kidneys. L-carnitine has been demonstrated to enhance nitrogen retention when taken as a dietary supplement, increasing lean mass and decreasing fat mass. Research indicates that adding 50–300 ppm of L-carnitine to companion animal weight loss regimens decreases the loss of lean tissue (German, 2006).

#### *7.2.5.5. White Kidney Bean Extract*

An extract made from the white kidney bean is called *Phaseolus vulgaris*, or white kidney bean extract. *Phaseolus vulgaris* has been demonstrated in human studies to function as a starch blocker by obstructing the action of the salivary digestive enzyme alpha-amylase, which breaks down starch. Overweight individuals lost a significant amount of body fat without sacrificing lean body mass when they took a daily *Phaseolus vulgaris* extract and followed a typical diet of 2,000–2,200 calories (Cellano et al., 2007). Although dogs' pancreas produces amylase but not their saliva, *Phaseolus vulgaris* supplements have been shown to have a significant positive clinical impact in patients, according to author WJD. As previously stated, nutrigenomics is still in its infancy as a field of study, and not all of the benefits we observe in our experiences have been thoroughly investigated in canines. However, it functions.

Kidney beans should never be given raw to dogs due to the possibility of toxicity. Only buy white kidney bean extract from a reliable supplier that employs a standardized extract heated to inactivate the beans' trypsin-inhibiting activity (TIA) and hemagglutinating activity (HA), which cause red blood cells to clump together, to ensure safety (Cellano et al., 2007). Dosage: 500 mg extract daily for small dogs, 1000 mg extract for medium to large dogs, and 1500 mg extract for large to giant breed dogs. Dose should be given 10 minutes before each meal.

#### *7.2.4.6. Antiangiogenic Foods*

The process of producing new blood vessels from preexisting ones is known as angiogenesis (Mercola, 2012). For the time being, wc will simply state that fat tissue and cancer cells both depend on angiogenesis. By severing the blood supply to fat cells, antiangiogenic foods (foods that inhibit angiogenesis) can actually cause fat cells to shrink (Li, 2010).

### 7.2.6. Lean Dogs Have "Happier" Cells

Recall that optimal gene expression and vibrant health are signaled by epigenetic tags, which are supported by a lean body. Thus, assist your dog in losing weight, lessen inflammation associated with obesity, eliminate chronic illness, and take pleasure in a body full of happy, healthy cells. If your dog has a medical issue that isn't related to obesity, it will probably be addressed in one of the sections that follows. Continue reading for specifics and the most recent dietary adjustments, tips, and techniques to avoid, treat, or even cure these conditions at the cellular level. Recall that cells are constantly renewing themselves. With a lot of assistance from his epigenome, today's "ill" dog can become tomorrow's vibrantly healthy dog.

## 7.3. A NUTRIGENOMICS APPROACH TO ARTHRITIS

As you will recall, fat causes inflammation, and arthritis is a classic inflammatory disease. Additionally, a heavier dog puts more strain on their stiff, achy joints. In addition, you should take care of any possible dietary sensitivity or intolerances your dog may have. Once more, foods high in antigens cause inflammation.

### 7.3.1. What Is Arthritis?

As many as 20% of adult dogs suffer from arthritis, the most common joint disease in dogs (Laflamme, 2012). Reports of arthritis rose 38% in dogs and 67% in cats between 2007 and 2011. All dogs are susceptible, particularly as they get older; in 2011, 13% of senior dogs had an arthritis diagnosis (Banfield Pet Hospital, 2012; Howard, 2012).

Numerous related conditions involving inflammation of one or more joints are referred to as arthritis. One of the most common causes of lameness in dogs is osteoarthritis (OA), also known as degenerative joint disease (DJD), which is the most commonly diagnosed type of arthritis in dogs. Osteophytes, or bony growths or spurs on the joints, deterioration of the articular cartilage, alterations in the subchondral bone (the bone beneath the articular cartilage), and joint inflammation are the hallmarks of osteoarthritis (OA). These modifications follow changes in the joint's biomechanical and biochemical characteristics (Sanchez and Balligand, 2005; Zhang et al., 2007).

Pain, stiffness, decreased mobility, and—in more severe and persistent cases—debilitation are the symptoms of arthritis. A dog with arthritis may have trouble getting up, jumping, running, or walking. Please take your dog to your veterinarian for a thorough examination if you notice any changes in his behavior or activity level. It's common to mistake the symptoms of arthritis with the natural "slowing down" that comes with aging.

Because large breeds like Labrador Retrievers, German Shepherds, and Mastiffs frequently have inherited or congenital conditions like luxating patellas, hip dysplasia, and elbow dysplasia, they may be more susceptible to developing OA. In general, larger breeds experience more arthritis than smaller and toy breeds because they bear more weight on their joints.

### 7.3.2. OA: It's in the Genes

Though this perspective is evolving, historically, physicians have treated and understood arthritis as a structural disease involving the degeneration of cartilage and its subsequent impact on the joints. We can now ask, "Which came first, the structural changes or the changes in gene expression that led to the structural changes?" because of new methods for measuring gene expression, such as differential display, reverse transcription-polymerase chain reaction (RT-PCR), and DNA microarray analysis (Middleton and Hannah, 2004). Numerous inflammatory cell markers that are activated in dogs with

OA have been discovered by researchers. These indicators are associated with inflammation within the articular chondrocytes, the cells that comprise articular cartilage. A completely new therapeutic strategy, which targets particular genes implicated in the disease, becomes possible with a deeper comprehension of how OA develops at the cellular level (Middleton and Hannah, 2004).

### 7.3.3. The OA-Obesity Link: It's More Than Just Weight

It should come as no surprise that obese dogs have a higher chance of developing arthritis because their weight is disproportionate to their size, putting undue strain on their joints. Indeed, a significant 40% of dogs with arthritis are overweight, and their symptoms are more severe than those of their leaner counterparts (Howard, 2012).

Yet, there is a systemic inflammatory connection between obesity and OA that goes far beyond the wear and tear brought on by gaining excess weight. You now know that pro-inflammatory cytokines are secreted by adipose (fat) tissue. It has now been established by researchers that these cytokines cause articular cartilage inflammation. Excess fat tissue raises the symptoms as well as the risk of osteoarthritis (OA) (Koonce and Braverman, 2013; Laflamme, 2004; Pollack, 2013). We'll discuss a potent functional component in the fight against pro-inflammatory cytokines in a moment. Additionally, obesity increases oxidative stress, which is brought on by an imbalance between pro- and anti-oxidants in the body. Increased oxidative stress has been linked to arthritis. According to research, dogs with OA who lose weight report feeling less pain and lameness as well as having more joint mobility (Laflamme, 2004; Laflamme, 2012). One of the most crucial steps in helping to manage your dog's arthritis is getting him to his ideal weight. A trimmer dog will not only put less strain on his musculoskeletal system from his weight, but he will also experience less oxidation and inflammation throughout his body, which are linked to OA.

### 7.3.4. OA a Cartilage Disease or a Bone Disease

Although osteoarthritis has traditionally been thought of as a disease of the cartilage, new research indicates that it may also be a disease of the bones. As we just discussed, sclerosis (abnormal bone density) and the development of osteophytes and cysts are among the changes in subchondral bone that accompany osteoarthritis (OA). Researchers have found that osteoblasts, which are cells that form bone, appear differently in the vicinity of sclerotic subchondral bone (subchondral bone with abnormal bone density) than they do in the nearby non-sclerotic area. Certain genes linked to sclerotic osteoblasts showed markedly up-regulated expression, whereas the expression of other genes was down-regulated. This indicates that bone cells with arthritis behave and look differently from healthy bone cells. It also implies that particular dietary and pharmaceutical treatments for osteoarthritis may be directed toward arthritic bone cells (Logar et al., 2007; Sanchez et al., 2008).

### 7.3.5. Functional Foods for OA

For the treatment of OA in dogs, the conventional veterinary community usually uses Adequan® injections, commercial prescription diets, and non-steroidal anti-inflammatory drugs (NSAIDs). Some veterinarians also suggest glucosamine/chondroitin (see below). Ironically, grains, gluten, and/or white potatoes are found in most commercial prescription OA diets—all of which have the potential to exacerbate inflammation. Additionally, a proinflammatory ratio of omega-6 to omega-3 fatty acids may be present in these diets. Long-term use of NSAIDs may result in harm to the bone marrow, kidneys, and liver tissues (Dodds and Lassin, 2013). Fortunately, there are plenty of safe, useful nutritional ingredients that can improve bone health and cartilage repair/improvement, effectively reduce inflammation, and detoxify toxins that trigger an inflammatory response (Dodds and Lassin, 2013). Let's examine those that we believe to be among the greatest.

#### *7.3.5.1. Avocado-Soybean Unsaponifiables (ASU)*

The portions of fats that aren't suitable for making soap are called "unsaponifiable lipids." It has been demonstrated that avocado-soybean unsaponifiables (ASUs) have anti-inflammatory, antioxidant, and tissue-building effects on articular chondrocytes. ASUs have been investigated for their potential benefits to treat OA (Laflamme, 2012; Lippiello, 2008).

Anterior cruciate ligament rupture (ASU) can delay the development of subchondral bone lesions and early osteoarthritic cartilage, according to a study involving 16 dogs. The daily dosage of ASU administered to the dogs was 10 mg/kg body weight, which is roughly twice as much as what is advised for people with OA (Boileau et al., 2009; Laflamme, 2012). Powdered elk or deer velvet antler (EVA, DVA).

#### *7.3.5.2. Elk or Deer Velvet Antler (EVA, DVA) Powder*

Nutritional supplements called EVA and DVA have been used for thousands of years in Traditional Chinese Medicine. They are believed to help reduce the symptoms of arthritis by promoting the regeneration of cartilage, enhancing joint fluid, speeding up the healing of damaged tissue and cells, and enhancing circulation. The inner core of the antler during its velvet stage of growth is used to make EVA and DVA. These products are packed with healthy ingredients, such as chondroitin sulfate (see below), which helps reduce inflammation, cartilage degradation, and pain associated with osteoarthritis (OA); and "pilose antler peptide," a peptide known for its anti-inflammatory qualities (Moreau, 2004; Laflamme, 2012).

Collagen, fats, minerals/trace elements (calcium, phosphorus, sulfur, magnesium, potassium, sodium, manganese, zinc, copper, iron, selenium, and cobalt), growth factors, glucosamine, glycosaminoglycans (also known as GAGs; a major component of joint cartilage, joint fluid, and other soft connective tissue), phospholipids, prostaglandins, and hyaluronic acid are also present in EVA and DVA. Furthermore, the velvet of deer and elk contains naturally occurring IGF 1 and IGF 2. IGF 1 affects every cell in the body, including bone, muscle, cartilage, and numerous other organs, in terms of cellular growth. Together with IGF 1, IGF 2 stimulates the growth of cells and the development of organs. EVA powder successfully improved the symptoms of 38 client-owned dogs with OA in a double-blind, placebo-controlled study conducted in 2004 (Moreau, 2004). Although research has shown that EVA and DVA powder are effective in treating OA in dogs, we do not personally use or recommend these products because of ethical concerns regarding their sourcing. Dosage: As directed by the manufacturer, or 14–21 g EVA or DVA powder/kg body weight per day (Laflamme, 2012).

### *7.3.5.3. Glucosamine and Chondroitin Sulfate*

GAGs, or glycosaminoglycans, were briefly discussed earlier (also known as mucopolysaccharides). GAGs are essentially lengthy chains of sugar molecules. Chondroitin sulfate is a GAG, and glucosamine is a glycoprotein that is required for the synthesis of GAGs. GAGs give the body structural support in a variety of places, such as the cartilage, tendons, and ligaments. Additionally, they are present in synovial fluid, which lubricates joints.

Glucosamine and chondroitin sulfate help dogs with OA in a variety of ways, including:

- Work similar to NSAIDs to decrease production of proinflammatory compounds.
- Reduce pain and inflammation.
- Stimulate cells found in cartilage (chondrocytes).
- Serve as the building blocks of new joint cartilage.
- May promote proper joint structure and function by increasing the synthesis of proteoglycans, hyaluronic acid (joint lubricant) and collagen.

Supplements containing glucosamine are usually derived from chitin, which is the tough outer shell of shellfish like crab and shrimp. If your dog has a shellfish allergy, make sure your veterinarian knows about it. Some researchers think there is no risk of shellfish allergy associated with taking glucosamine supplements because shellfish allergies are thought to be caused by the meat of the shellfish. Still, it's advisable to err on the side of caution. Most commonly, the trachea of cows or sharks are used to make chondroitin.

In dogs with OA, chondroitin sulfate and glucosamine together lessen inflammation, delay cartilage degradation, and lessen pain signals. Studies on dogs suggest a therapeutic benefit of combining glucosamine and chondroitin sulfate, despite contradictory research regarding the relative efficacy of the two supplements (Laflamme, 2012).

- **Glucosamine Daily Dosages:** 250 to 500 mg for small dogs, 500 to 1000 mg daily for medium to large dogs and 1000 to 2000 mg daily for large and giant breed dogs.
- **Chondroitin Sulfate Daily Dosages:** 200 to 400 mg for small dogs, 400 to 800 mg for medium to large dogs and 800 to 1600 mg daily for large dogs and giant breed dogs.

### *7.3.5.4. Green Lipped Mussel (GLM) Extract*

The "green lipped" mussel, a commonly cultivated species in New Zealand, gets its name from the green lip that surrounds the edge of its shell. Important nutrients for dogs with OA can be found in green lipped mussel (GLM) powder, such as GAGs (chondroitin sulfate), glutamine, the amino acid, and the omega-3 fatty acids DHA and EPA. g. manganese, copper, zinc, and manganese) as well as the antioxidant vitamins C and E (Rialland et al., 2013). GLM extract is clinically proven to ease the signs of dogs with OA. Here are two examples:

- A 2006, double-blind, placebo-controlled study involving 81 dogs with mild-to-moderate degenerative joint disease found that the dogs significantly benefited from long-term (eight weeks or longer) supplementation with a tablet containing 125 mg GLM extract (The tablet also contained 52.86 mg brewer's yeast, 191.50 mg lactose and 10.64 mg of a tableting aid made of magnesium stearate, acacia and aerosil) (Pollard et al., 2006).
- A 2013 study published in the Canadian Journal of Veterinary Research reported that a diet enriched with GLM extract significantly improved the gait of 23 companion dogs clinically afflicted with OA, compared with a regular control diet. Dogs fed the GLM-enriched diet also absorbed high levels of EPA and DHA omega-3

fatty acids into their blood. The researchers concluded that GLM extract showed a strong ability to benefit dogs with OA, and that it likely works by providing anti-inflammatory effects, translating to better comfort for the dog and probable benefit to the joint structures (Rialland et al., 2013).

### *7.3.5.5. Omega-3 Fatty Acids*

Because inflammatory mediators—substances that cause an inflammatory response—stimulus the synthesis of additional inflammatory mediators, osteoarthritis leads to a self-perpetuating inflammatory cycle (Middleton and Hannah, 2004). By blocking these inflammatory mediators, particularly the COX-2 enzyme, NSAIDs are frequently prescribed to interrupt this cycle (Laflamme, 2012). But NSAIDs can also have more serious side effects, like gastrointestinal bleeding, ulcers, perforations, kidney damage, and liver problems, in addition to less serious side effects like vomiting, loss of appetite, depression, lethargy, and diarrhea (Bren, 2006).

Long-term NSAID use can be safely and effectively avoided by consuming fish and other marine sources of EPA and DHA, which also down-regulate COX-2 expression (Laflamme, 2004; Middleton and Hannah, 2004; Laflamme, 2012).

Benefits of EPA and DHA for arthritis include:

- Decrease lameness
- Improve weight-bearing ability.
- Significantly improve ability to rise, play and walk.
- Reduce the need for NSAIDs.

These advantages are confirmed by a 2012 study that was written up in the Journal of Animal Physiology and Animal Nutrition. A diet rich in omega-3 fatty acids from rainbow trout considerably improved the locomotor disability and day-to-day functioning of thirty companion dogs with OA, according to the trial results. Actually, the improvement was similar to what was seen in dogs with osteoarthritis receiving NSAIDs (Moreau et al., 2012).

### *7.3.5.6. Omega-3 to Omega-6 Fatty Acid Ratio for Dogs with OA*

#### *7.3.5.6.1. SAMe*

The body uses adenosyl-triphosphate (ATP) and the amino acid methionine to produce SAMe. Supplementing with SAMe is sometimes recommended because it is not present in food.

In several areas, including the following, SAMe supplementation proved to be just as beneficial as NSAIDs like ibuprofen and naproxen in clinical trials involving individuals with osteoarthritis (OA) of the knee, hip, or spine.

- Decreasing pain.
- Reducing swelling.
- Reducing morning stiffness.
- Improving range of motion.
- Increasing walking pace.

SAMe may also promote cartilage repair (UMMC, 2011).

#### *7.3.5.6.2. Spirulina*

Spirulina is a class of ancient, microscopic blue-green algae that is named for the spiral shape of its strands. Spirulina grows in warm, fresh water. Chlorophyll gives spirulina its rich green hue, while phycocyanin (C-PC), a substance with potent anti-inflammatory, anti-arthritic, anti-cancer, and hepato (liver)-protective qualities, gives it its blue color (Reddy et al., 2000; Romay et al., 2003; Messonnier, 2006; Wolf, 2009; UMMC, 2011a).

Similar to EPA and DHA, C-PC inhibits the COX-2 enzyme, reducing OA pain and inflammation without having the unfavorable effects of NSAIDs. Even a class of NSAID medications launched by two significant pharmaceutical companies was found to be less effective at inhibiting COX-2 activity than CPC (Reddy et al., 2000). A wide range of minerals, chlorophyll, phytonutrients, carotenoids, gamma linolenic acid, superoxide dismutase (a powerful free radical scavenger), B-complex vitamins, vitamin E, and many other essential nutrients are also abundant in spirulina (Wolf, 2009; UMMC, 2011a; Bensimoun, 2013).

Spirulina has more protein by weight than any other food, ranging from 65 to 71% depending on the variety. About 16 grams of protein can be found in just one ounce of spirulina, and each gram of protein is four times more absorbable than the same amount of protein in beef (Wolf, 2009; Bensimoun, 2013). It's crucial that dogs with OA eat enough protein because, as you may remember, protein gives bones, joints, tendons, ligaments, cartilage, muscle fibers, and collagen their structural integrity.

### *7.3.5.7. Vitamins C, E and Pro-Vitamin A Carotenoids (Antioxidants)*

It has been demonstrated that oxidative stress and reduced antioxidant capacity exacerbate the severity of arthritic lesions (Laflamme, 2012). Thus, we think that experience studies show the usefulness of antioxidants in managing OA, despite the fact that human clinical trials have yielded contradictory results in this regard.

Since a dog's body can produce its own vitamin C (ascorbic acid), unlike a human's, supplementing with the vitamin is particularly contentious when it comes to dogs. Nevertheless, research conducted over the past 20 years has demonstrated that dogs can benefit from additional vitamin C beyond what their bodies can produce. Their increased antioxidant capacity helps them manage the stresses of their contemporary lifestyles, such as poor diets and environmental pollution. However, be aware that large doses of ascorbic acid may cause upset stomach and loose stools, so proceed with caution and gradually increase the dosage (Straus,2007; Sanghi et al., 2009). Mineral ascorbates, which are buffered and less acidic, are another option. Examples of these are sodium and calcium ascorbates. But you also need to consider that dose because the mineral portion of the supplement is also absorbed. For instance, 111 mg of sodium are normally present in 1,000 mg of sodium ascorbate (Linus Pauling Institute, 2013). Giving a high dosage of sodium ascorbate to a dog on a low-sodium diet could cause his sodium intake to skyrocket.

Start your dog's vitamin C regimen with the following dosages: 250–500 mg for small dogs, given twice a day with food; 500–1000 mg for medium–large dogs, given twice a day with food; and 1000–2000 mg for large and giant breed dogs, given twice a day with food.

It may also surprise you to learn that carotenoids (e.g., among the vitamins that act as antioxidants is beta-carotene. Carotenoids are considered antioxidants rather than vitamins because they are transformed by the body into retinol, which is the active form of vitamin A. As you may remember from our previous talks, colorful fruits and vegetables like carrots, pumpkins, squash, and beets are rich sources of beta-carotene and other carotenoids (Evert, 2013). However, keep in mind that retinol is not an antioxidant and that excessive amounts of vitamin A can be hazardous, as we previously discussed.

## 7.4. A NUTRIGENOMIC APPROACH TO CANCER

Although cancer is a word that no dog parent wants to hear from the veterinarian, it does happen to companion animals occasionally. According to Heinze, Gomez, and Freeman (2012) and Tremayne (2010), cancer is the primary cause of death for dogs older than two years old and accounts for up to half of all dog (and feline) fatalities in the United States. When a cherished dog receives a cancer diagnosis, it's normal to experience confusion, fear, and overwhelm. With the knowledge in this chapter, we hope you will be better equipped to make decisions that will improve the outcome and quality of life for your companion.

And while this information offers strong nutritional tools that can be a significant part of an integrated approach to prevention and healing, we in no way intend to discourage you from seeking conventional veterinary cancer treatment. If your veterinarian tries to talk you out of adding functional food therapy to your dog's cancer treatment plan, we strongly advise you to work with a more accepting healthcare provider who is knowledgeable about the most recent research on this subject. Cancer-related genes are among the most prominent examples of the parallel evolution that scientists have recently found between the canine and human genomes, as you may remember from the Introduction. Dogs and humans can both profit from the most recent advancements in cancer research because of this genetic similarity.

### 7.4.1. Epigenetics and Cancer: Evidence too Strong to Ignore

If you believe that a person or animal's inherited DNA somehow predisposes them to cancer, you should reconsider. Only 5–10% of cancer cases, according to researchers, are thought to be genetically predisposed; the remaining 90–95% are thought to be caused by environmental and lifestyle factors, including stress, diet, obesity, exposure to pollutants and other toxins, and lack of physical activity. These elements produce epigenetic alterations that can activate oncogenes, a class of aberrant gene that predisposes cells to become cancerous, or turn off genes that suppress tumors (Broad Institute, 2013).

Numerous studies demonstrate that specific dietary components target epigenetic modifications in cancer cells. These components have the ability to inhibit the growth of numerous cancerous tumor types as well as protect against their development. Remarkably, researchers have found that dietary modifications alone can prevent 30 to 40% of cancer cases (Donaldson, 2004; Divisi et al., 2006; Anand et al., 2008; Ziech et al., 2010; Hardy and Tollefsbol, 2011).

Let's first consider what needs to be cut out of your dog's cancer-protective diet before you start adding items. Foods containing any of the hazardous ingredients previously mentioned should be removed, please. This is significant because the benefits of the functional foods you'll soon be adding to your dog's diet may be offset by these ingredients.

### 7.4.2. Obesity and Cancer: One More Reason to Keep Your Dog Trim

Dogs that are overweight or obese are more susceptible to many chronic, serious diseases, including cancer. In actuality, there is a direct link between cancer and overeating (Donaldson, 2004).

The rates at which cells divide and proliferate, as well as the types of cells they become and undergo, are all properly balanced in normal cells. Cancer may result from obesity's alteration of the metabolism of several hormones necessary to maintain this balance. Furthermore, NK-fB, a transcription factor type that is involved in the development of cancer and other inflammatory diseases like asthma, arthritis, cardiovascular disease, and inflammatory bowel disease, is activated by specific pro-inflammatory cytokines generated by fat cells. It's also been demonstrated that NK-fB is activated by obesity-related hyperglycemia, or elevated blood sugar (Anand et al., 2008; Gilmore, n.d.; Bhatti, Nahleh, and Mal, 2011).

Additionally, recent research has found a connection between insulin resistance and the expression of certain genes in adipose (fat) tissue (Wein, 2012). Type 2 diabetes is linked to insulin resistance and is a significant cancer risk factor (see below).

The following types of cancers have been linked to obesity in people:

- Breast
- Cervical
- Colon
- Endometrial
- Esophageal
- Gastric
- Gall bladder
- Liver
- Multiple myeloma
- Non-Hodgkin's lymphoma
- Ovarian
- Pancreatic
- Rectal
- Renal
- Uterine

The bottom line is that there is a well-established link between obesity and inflammatory diseases, including cancer.

### 7.4.3. Cancer and High-glycemic Foods: A Recipe for Disaster

High-GI foods encourage obesity, which contributes to the development of cancer. High-GI carbohydrates also contribute to hyperinsulinemia, or the overproduction of insulin linked to Type 2 diabetes. Diabetes has been associated in numerous studies with an increased risk of various cancers, including a three-fold increased risk of colon-rectal cancer. Make sure that your dog's cancer-protective diet no longer includes any refined sugars, refined flour products, or other harmful carbohydrates that we previously covered (Nahleh, Bhatti, and Mal, 2011).

### 7.4.4. Get Rid of Gluten

The connection between leaky gut syndrome, inflammation, and gluten has already been discussed in great detail. You may recall that when gluten is consumed, the intestines release a protein known as zonulin, which makes the intestinal lining more permeable, or "leaky," and permits foreign particles to enter the bloodstream from the intestines (Hyman, 2012; Kressler, 2012; Hyman, 2013). When the body detects these particles as potentially harmful foreign invaders, it mobilizes all of its immune-fighting cells, or "troops," to combat them. Simultaneously, these cells inadvertently target the body's own tissue, setting off a systemic firestorm of inflammation linked to numerous illnesses, including cancer. Your dog doesn't need to consume an ingredient known to increase inflammation linked to cancer if he has cancer.

### 7.4.5. Fluoride and Canine Osteosarcoma

The most frequent malignant bone tumor in dogs (and young growing boys) is osteosarcoma, which affects over 8,000 dogs annually in the US and is most common in large dogs with quickly growing bones.

Dogs are almost ten times more likely than people to develop this kind of cancer, with about 900 cases reported annually (EWG). What's even more startling is that your dog may get this deadly bone cancer from his food or water—literally, right under his nose (EWG, 2009; EWG, 2009a; EWG, 2009b).

Eight of the ten major national brands of dry dog food sold for puppies and adults had fluoride levels between one and 2.6 times higher than the maximum permissible amount in drinking water set by the Environmental Protection Agency, according to an independent laboratory test of the products that the Environmental Working Group (EWG) had commissioned. Additionally, the amounts found were greater than those associated with childhood bone cancer in boys (EWG, 2009; EWG, 2009b).

Fluoride accumulates in the bones after consumption. The sources of the fluoride contamination were identified as meat meal, bone meal, or meat byproduct meal, all of which were present in some form in all eight brands. To further raise the level, foods were prepared using fluoridated tap water (EWG,

2009; EWG, 2009b). By eating roughly a cup of dry dog food a day, a 10-pound puppy would take in 0.25 milligrams of fluoride per kilogram of body weight daily, which is five times the amount recommended by the U.S. Agency for Toxic Substances and Disease Registry (ATSDR) as safe. S. Health and Human Services Department (EWG, 2009b).

Fluoride likely contributes to osteosarcoma in two main ways:

- It accumulates in the bones after eating or drinking.
- It stimulates mitosis (cell division), causing rapid proliferation of osteoblasts (cells that form new bone).

Repetitive exposure to high fluoride levels can be regarded as a risk factor for osteosarcoma, since many adult and puppy dogs have a tendency to eat the same food every day.

Large breed puppies who are rapidly forming new bone may be especially vulnerable (EWG, 2009; EWG, 2009a). If your dog has cancer, we strongly suggest giving up all foods that contain animal by-products or bone meal and switching to bottled water that is either low-fluoride or fluoride-free.

### 7.4.6. Environmental Toxins

Detoxifying your dog's environment is a crucial component of any anti-cancer protocol because environmental factors unrelated to food also have a significant impact on the epigenome. Here are some common sources of hazardous substances that may be lurking in and around your home:

- Lawn and garden products, such as chemical fertilizers, pesticides, herbicides, swimming pool products, etc. (Glickman et al., 2004).
- Household products, such as chemical cleaning products, laundry detergent, bleach, dishwashing detergent, air fresheners, carpet cleaners, etc.
- Construction materials, such as drywall, asbestos, insulation, paints, varnishes, adhesives, etc.
- Automotive products, such as antifreeze, break fluid, lubricants, sealants, etc.
- Batteries, especially battery fluids.
- Personal care products, such as antiperspirants, shampoos, sunscreens, hair sprays, perfume, cologne, etc.
- Pet care products, such as insect repellents, cat litter, flea and tick collars and shampoos, etc. (Raghaven et al., 2004).

### 7.4.7. PBDEs (Polybrominated Diphenyl Ethers)

Since the 1970s, flame-retardant chemicals known as PBDEs have been used in electronics, furniture, and apparel. They are particularly common in electronics like cell phones, computers, printers, scanners, copiers, hair dryers, toner cartridges, TVs, TV remote controls, and video equipment, as well as foam products like mattresses, pillows, upholstered goods, and carpet padding (Becker, 2013). PBDEs are known to interfere with thyroid function and other endocrine systems. They can seep into the environment. Additionally, there is evidence connecting them to developmental issues and potentially even cancer in animals. Although foam products manufactured in 2005 and later ought to be devoid of PBDEs, some modern TVs and computer monitors still contain them (EWG, 2007).

Dog and cat serum, food, and household dust have all been linked to PBDE residues (Venier and Hites, 2011; Becker, 2013). Dogs' levels of PBDEs were five to ten times higher than those of cats; this difference can be attributed to the fact that dogs can metabolize these chemicals more quickly than cats and may have an enzyme that aids in their breakdown (Venier and Hites, 2011; Becker, 2013).

The group dedicated to environmental work offers a plethora of information to individuals seeking to minimize or completely remove harmful items from their surroundings, such as PBDEs and common household chemicals. From cosmetics and colognes to baby formula and bath products for kids, EWG will give you the information you need to choose the healthiest products for every part of your home. Even better, EWG's Guide to Healthy Cleaning offers safety ratings for 2,000 household cleaning products so you can choose the brands that will best support the health of your family—two- and four-legged—and yourself.

### 7.4.8. Second-Hand Smoke

If you smoke, you may not consider the risks secondhand smoke poses to your pets. However, studies have shown that dogs who are exposed to high levels of secondhand smoke over time experience significant alterations to their lung tissue. The changes include precancerous and even cancerous lesions, as well as fibrosis (scarring of the lung tissue). Please think about smoking outside if you must (Wilson-Robles, 2013).

### 7.4.9. Topical Chemical Flea and Tick Preventives

Of course, we don't want fleas, ticks, and other insects in our homes, nor do we want our dogs to endure the suffering and possible diseases they may cause.

However, many of the products used to get rid of these pests, such as monthly spot-on flea and tick treatments that are squeezed between the shoulder blades and come in a tube or vial, contain toxic chemicals that can harm a dog's health both immediately and over time, even to the point of death (Raghaven et al., 2004).

In 1989, a study conducted by the Pathobiology Department at Purdue University School of Veterinary Medicine and published in the Journal of Toxicology and Environmental Health revealed that dogs who received one to two topical pesticide applications annually had a 60% higher risk of bladder cancer. Dogs who received more than two applications annually had a 3–5-fold increased risk of bladder cancer. Dogs that were overweight or obese had an even higher risk (Glickman et al., 1989; Glickman et al., 2004; Raghaven and colleagues. 2004).

Organophosphate, carbamate, and pyrethrin/pyrethroid products present especially serious health risks and, in our opinion, shouldn't be used on cancer-stricken dogs or any other dogs.

Most flea and tick problems can be avoided without the use of toxic chemicals that are absorbed through your dog's skin. To get rid of pests, try the following:

- Bathe your dog frequently to control minor infestations.
- Comb your dog with a special fine-tooth flea comb and check him for ticks after outdoor romps.
- Frequently wash your dog's bedding in hot water.
- Keep your grass raked and cut short in areas your dog frequents.
- Use natural flea and tick repellants that contain essential oils rather than toxic chemicals.
- Vacuum and wash your floors regularly.

If you must use a spot-on flea and tick treatment, please make sure your veterinarian is guiding and supervising you; stay away from buying over-the-counter products at pet supply stores or supermarkets, as these frequently contain ingredients that could be extremely dangerous (Raghaven et al., 2004).

### 7.4.10. Hold That Needle: Risks of Vaccination to Immuno-compromised Dogs

According to Tizard and Ni (1998), Twark and Dodds (2000), and Dodds (2001), a vaccine is a biological preparation of altered live or killed pathogens (viruses, bacteria, or parasites) that is injected into the body to promote immunity to a specific disease.

Vaccines have achieved many important benefits for companion animals, including:

- Saved more animals' lives than any other medical advance.
- Significantly reduced canine distemper, hepatitis and parvovirus.
- Significantly reduced feline panleukopenia.
- Eliminated rabies in Europe.

However, those working in the animal healthcare industry now have an obligation to reevaluate and enhance the current vaccination protocols for the patients' health and safety, having spent many years monitoring the results of vaccinosis (reactions and side effects of vaccines). This is particularly true for animals whose immune systems are already weakened, as vaccinations add another stressor that may serve as the "tipping point" between health and illness. Vaccination-related side effects in dogs can manifest immediately or weeks or months later. According to Dodds (2001), vaccinations may even make a dog more susceptible to chronic illnesses that manifest much later in life.

Mild reactions associated with vaccines include:

- Fever
- Malaise
- Urticaria [hives]
- Facial swelling
- Anorexia
- Vomiting
- Stiffness
- Sore joints
- Abdominal tenderness

Severe and fatal adverse reactions include:

- Susceptibility to infections.
- Neurological disorders and encephalitis.
- Aberrant behavior, including unprovoked aggression.
- Collapse with autoagglutinated (clumped) red blood cells and icterus (jaundice); autoimmune hemolytic anemia (AIHA) or the synonym immune-mediated haemolytic anemia (IMHA), when red blood cells are damaged and destroyed; or petechiae (pin-point) and ecchymotic (splotchy) hemorrhages from immune-mediated thrombocytopenia (ITP), when the blood platelets are destroyed. Hepatic enzymes may be markedly elevated, and liver or kidney failure may occur by itself or accompany bone marrow suppression.

### 7.4.11. The Two Types of Vaccines

Vaccines can be broadly classified as either killed or modified live virus (MLV). MLV vaccines, as the name implies, employ a weakened but altered version of the live virus. Injecting the virus into the body causes it to multiply several times over and triggers the immune system to produce antibodies, resulting in an immunological response that shields the body from further illness exposure. Killed vaccines combine an adjuvant—a material added to a vaccine to increase its effectiveness without inducing an immune response—with an inactivated, or "dead," version of the virus. Dogs with weakened immune systems are more vulnerable to the effects of both the MLV and killed vaccines.

Although all dogs are susceptible to vaccine-related side effects, certain breeds are more pre-disposed to vaccinosis. These include:

- Akita
- American Cocker Spaniel
- German Shepherd
- Golden Retriever
- Irish Setter
- Great Dane
- Kerry Blue Terrier
- Dachshunds (all varieties, but especially the long-haired)
- Poodles (all varieties, but especially the Standard Poodle)
- Old English Sheepdog
- Scottish Terrier
- Shetland Sheepdog
- Shih Tzu
- Vizsla
- Weimaraner

Moreover, breeds with dilution in coat color, like fawn (Isabella) or blue Dobermans, the merle coat color, blue Yorkshire Terriers, grey Collies, harlequin Great Danes, and Australian Shepherds, as well as breeds with white or mostly white coats are at higher risk (Dodds, 2001).

We strongly advise against giving your cancer-stricken dog any more shots (assuming he has already finished his puppy series), as this will only exacerbate his already weakened immune system. Your veterinarian can do a titer test, which is a straightforward blood test to assess your dog's immunity, to ensure that he is safe from specific diseases (Tizard and Ni, 1998; Twark and Dodds,

2000). The only vaccination that is legally required is the rabies shot, but your veterinarian may be able to provide you with a medical waiver if the condition is serious. Just be aware that your dog will not be allowed to carry a valid vaccination if he bites someone.

After discussing some significant environmental non-nutritional factors that impact the epigenome, let's move on to discussing functional foods for your dog's cancer-preventive diet. It makes sense to start this discussion with two key findings about the formation, growth, and multiplication of cancer cells: angiogenesis and DNA methylation.

### *7.4.11.1. DNA Methylation: An Important Epigenetic Cancer Marker*

When the DNA methylation process goes awry, the epigenome is altered, which can result in illnesses like cancer. Indeed, aberrantly methylated DNA is frequently present in tumor cells (Davis and Uthus, 2004; Hardy and Tollefsbol, 2011). Oncogenes are abnormal genes that predispose cells to become cancerous. Aberrant DNA methylation can either activate or inactivate these genes. Oncogenes can also suppress tumor growth. According to research, at least half of all the genes involved in tumor suppression are inactivated during the formation of tumors due to epigenetic mechanisms (Hardy and Tollefsbol, 2011).

Fortunately, certain bioactive food components have been shown to support optimal DNA methylation and support cancer prevention (Esteller, 2008; Hardy and Tollefsbol, 2011). Here are some of the most notable:

- **Curcumin:** A polyphenol compound found in the spice turmeric.
- **Folate:** A B vitamin that controls the repair and methylation of DNA. Folate deficiencies lead to DNA hypomethylation (low methylation), which is associated with tumor development. Many types of cancer, including brain, breast, cervix, colorectal, ovary and lung, have been linked to folate deficiencies. The richest sources of folic acid are leafy greens such as bok choy, escarole, kale, watercress, dandelion, mustard and collard greens. Beans and whole grains are also sources (Hardy and Tollefsbol, 2011; Hyman, 2011).
- Isothiocyanates are phytochemicals that exert an anti-cancer effect on the epigenome by inhibiting cancer cell proliferation and inducing apoptosis (cell death). Isothiocyanates are found in cruciferous vegetables such as broccoli, Brussels sprouts, cabbage, cauliflower, collards, kale, radish, turnip and watercress (Hardy and Tollefsbol, 2011).

- SAMe is the sole methyl donor in numerous cellular methylation reactions and is vital to cellular metabolism and function. SAMe is not found in food and must be taken via supplementation (Bottiglieri, 2002).
- Vitamin B-12 (cobalamin) acts as a methyl donor and plays a role in making DNA. Foods rich in vitamin B-12 include fish (opt for low-mercury, wild-caught fish), meat (opt for antibioticfree, novel proteins), milk (opt for sheep or goat), poultry and eggs (opt for organic, pasture-raised) (WebMD, 2012).

### *7.4.11.2. Angiogenesis: Eating to Starve Cancer Cells*

The process by which pre-existing vessels give rise to new blood vessels is known as angiogenesis (Mercola, 2012). The formation of new blood vessels is necessary for several processes, including the formation of the placenta during pregnancy and the promotion of wound healing beneath a scab. The body releases certain proteins in these situations, which promote angiogenesis. However, after these recently created vessels have completed their task (i.e., the birth of a child, or the healing of a wound), the body restores them to their pre-injury state by secreting chemicals known as antiangiogenic substances, which prevent angiogenesis.

Researchers have found that a number of illnesses can arise when the angiogenesis process is out of balance, producing either too few or too many blood vessels. For instance, a cut won't heal properly if your body is unable to produce enough blood vessels to heal the wound. Inadequate blood vessel density can also result in disorders like coronary artery disease and stroke as well as poor circulation. Conversely, an excessive number of blood vessels results in an alternative kind of issue that causes conditions like cancer, rheumatoid arthritis, and obesity. All forms of cancer are associated with angiogenesis, or the expansion of blood vessels excessively (Li, 2010).

Cancer cells need oxygen and nourishment to grow and thrive, just like any other type of living thing. So how do they get their nourishment? It's supplied by the blood vessels that are created during the angiogenesis process. We can literally "starve" tumor cells of the nutrients they require to proliferate and become dangerous by inhibiting angiogenesis. Cancerous tumors could not grow larger than the tip of a ballpoint pen without these new blood vessels (Li, 2010). The issue lies in the fact that cancer cells have the ability to release substances that trigger angiogenesis, giving them the nourishment they require to proliferate. The blood vessels that develop during angiogenesis also give cancer cells a "superhighway" to the bloodstream, which allows them to spread (Li, 2010). The good news is that some medications and food components have antiangiogenic properties, which prevent angiogenesis and starve cancer cells of nourishment. More than 600 dogs have received antiangiogenic therapy for cancers of different kinds, including mast cell tumors, with a 60% response rate overall (Li, 2010). It's interesting to note that in many cases, nutritional ingredients were just as effective at reducing angiogenesis as drugs were, if not more so (Li, 2010).

Here are a few "dog-approved" antiangiogenic foods to add to your canine companions anti-cancer diet:

- Apples
- Artichokes
- Berries (blackberries, blueberries, raspberries)
- Bok choy
- Cherries
- Curcumin (turmeric)
- Ginseng
- Kale
- Lavender
- Maitake mushrooms (see below for more on medicinal mushrooms)
- Olive oil
- Parsley
- Pumpkin
- Sea cucumber

There are other antiangiogenic foods that are not suitable for canines, such as, grapes, nutmeg, and dark chocolate), so we haven't included them here. Nevertheless, you are more than welcome to benefit from their anti-cancer properties. Furthermore, since angiogenesis is a prerequisite for fat tissue, feeding your dog antiangiogenic foods will also help them maintain a healthy weight, which is crucial in the fight against cancer and all other diseases (Li, 2010).

### 7.4.12. Functional Nutrients Offer Powerful Anti-Cancer Protection

The most promising functional foods for your dog's cancer-preventive diet are those listed below. Some of these have been discussed previously and are mentioned repeatedly in the book due to their importance in the prevention, treatment, and even reverse of a wide range of chronic health conditions.

### 7.4.12.1. Berries

One significant antiangiogenic food is berries. They are abundant in anthocyanins, which have been demonstrated to impede the development of tumors and reduce the proliferation of cancer cells at different stages of carcinogenesis (Lila, 2004). Anthocyanins also have anti-inflammatory properties and have been clinically proven to:

- Improve memory by increasing neuronal signaling in the brain.
- Display cardio-protective benefits.
- Improve signs of metabolic syndrome, a group of factors that increase the risk of cardiovascular disease and diabetes (Basu et al., 2010; Krikorian, Shidler, and Nash, 2010).

Favorite dog-friendly berries are:

- **Blueberries:** Blueberries also contain pterostilbene, a phytochemical compound, in addition to anthocyanins. Pterostilbene is a strong antioxidant that has been approved by dogs. It is a derivative of resveratrol, which is present in grapes—which dogs cannot eat—and has been demonstrated in animal studies to have strong anti-cancer effects. It is particularly toxic to breast cancer cells. Researchers found that eating blueberries changed the expression of genes linked to inflammation, cancer, and metastasis in mice given whole blueberry powder, suggesting that blueberries have anti-tumor and anti-metastasis properties. According to Adams et al., (2011) and Pons (2006) 10.6 oz (300 g) of fresh blueberries per day is the human equivalent of the quantity of blueberries fed to the mice.
- **Cranberries:** Many people know that cranberries help prevent urinary tract disease and support healthy urinary tract function, but did you know that they also have anti-cancer properties? Recent research has demonstrated that cranberries can stop the growth of cancer cells in the esophagus, colon, and breast (Howell, 2012; Neto, 2011). Please take note that due to possible allergic reactions, we do not advise strawberries for dogs.

### 7.4.12.2. Cruciferous Vegetables

Cruciferous vegetables have strong anti-cancer and anti-tumor properties. According to the American Institute for Cancer Research, certain compounds in cruciferous vegetables have shown the ability to stop the growth of cancer cells for tumors in the breast, uterine lining (endometrium), lung, colon, liver, and cervix. Broccoli, cauliflower, Brussels sprouts, cabbage, and bok choy are examples of cruciferous vegetables (Magee, 2007). One of the many phytochemicals present in cruciferous vegetables, sulforaphane, has been demonstrated in lab experiments to activate the body's detoxifying enzymes, thereby preventing carcinogens from having an opportunity to cause cell damage. Sulforaphane is present in very high concentrations in broccoli sprouts. Two other cruciferous vegetable compounds, indole 3-carbinol (I3C) and crambene, are also suspected of activating detoxification enzymes; in fact, one sprout contains as much sulforophane as an entire fully grown broccoli plant (Donaldson, 2004; Magee, 2007).

Shawn Messonnier, DVM, discusses the many anti-cancer benefits of I3C, including:

- Protects against hormone-related cancers.
- Blocks receptors on cells that allow estrogen to enter.
- Converts dangerous estrogens into safer estrogens.
- Restores function of the p21 tumor suppressor gene.
- Induces apoptosis of cancer cells.

Supplementation with I3C is very safe for dogs (Messonnier, 2006).

Additionally high in antioxidants, cruciferous vegetables may help prevent cancer by lowering oxidative stress brought on by an excess of free radicals. In a three-week study on humans supported by the National Cancer Institute, it was found that consuming one to two cups of cruciferous vegetables daily decreased oxidative stress by 22%. In comparison, a three-week course of multivitamins combined with fiber resulted in a mere 0.2 percent decrease in oxidative stress (Magee, 2007).

To improve your dog's digestibility without overcooking his cruciferous vegetables, try lightly steaming them. To facilitate easier digestion, break down the cellulose in them by pureeing them first in a food processor or blender if you would rather serve them raw. Buying frozen or already chopped vegetables adds convenience.

Goitrogens in cruciferous vegetables can exacerbate hypothyroidism in dogs, so steer clear of giving them.

### *7.4.12.3. Curcumin*

The deep yellow color of turmeric, an Indian spice, is attributed to a phytochemical called curcumin. Curcumin has been utilized for centuries in Ayurvedic medicine. In recent years, it has been extensively researched in Western medicine and discovered to have numerous advantageous qualities, including strong anti-cancer effects (Hardy and Tollefsbol, 2011).

Cancer is a complicated illness that "outwits" its host's (human or animal) defenses by using a variety of targets and pathways, but cancer treatments frequently target just one of these pathways. Curcumin, on the other hand, can influence a variety of pathways, such as the downregulation of growth factors, enzymes, pro-inflammatory cytokines, transcription factors, and carcinogenesis-promoting factors (Robinson, 2008; Wilken et al., 2011; Lawenda, 2013). It has also been discovered that curcumin modifies the DNA methylation of colorectal cancer cells (Link et al., 2013).

Curcumin's anti-cancer properties include:

- Alters DNA methylation.
- Antiangiogenic properties.
- Anti-inflammatory properties.
- Antioxidant properties.
- Antitumor properties.
- Boosts the immune system.
- Decreases cell growth.
- Decreases metastasis.
- Induces cancer cell apoptosis.
- Inhibits cancer cell division and growth.
- Inhibits enzymes that aid in the spread of cancer.

### 7.4.13. Green Leafy and Yellow-Orange Vegetables

Vegetables with green leaves and a yellow-orange color include healthy phytochemicals called carotenoids (e.g., lutein, zeaxanthin, lycopene, beta-cryptoxanthin, alpha-carotene, and beta-carotene), and flavonoids (e.g., quercetin). Despite not being vital nutrients (nutrients needed to sustain life, like vitamins and minerals), phytochemicals do contain significant bioactive compounds with health-promoting properties, such as antioxidant and anti-cancer properties.

Quercetin, for instance, suppresses the p53 gene mutation, which encourages cell division and cancer. Additionally, it inhibits the expression of the RAS gene, which can become an oncogene that causes cancer when it becomes hyperactive (Armstrong, 2013).

According to the findings of a 2005 study involving 92 Scottish Terriers, eating particular vegetables may help to prevent or delay the onset of bladder cancer, or transitional cell carcinoma (TCC), in this breed. In the study, 83 Scottish Terriers without a recent history of urinary tract disease were compared to the 92 Scottish Terriers older than six years old who had TCC. The owners of the dogs answered questions about their diet and vitamin supplement use in the preceding year (Raghavan, Knapp, Bonney et al., 2005).

While some of the dogs in both groups also ate fresh vegetables, over 95% of the dogs in both groups were fed commercial dry dog food. When the dogs' ages, weights, neuter status, and coat colors were taken into account, the dogs who ate vegetables at least three times a week had an overall 70% lower risk of developing TCC. Carrots were the most often fed vegetable, belonging to the yellow-orange category. Consuming any green leafy vegetable and any yellow-orange vegetable reduced the risk of developing TCC by 90% and 70%, respectively, for individual vegetable types. According to Raghavan, Knapp, Bonney et al. (2005), these results are thought to be related to the pro-vitamin A carotenoids found in vegetables, which also appear to have a protective effect against bladder cancer in humans and chemically-induced bladder tumors in rodents. Carrots, collard greens, dandelion greens, mustard greens, turnip greens, squash, pumpkin, yams, sweet potatoes, and kale are foods high in carotenoids. To reduce the cell wall and boost the carotenoids' bioavailability, steam or grind the veggies (Drake, 2009). Green leafy vegetables provide dogs with significant health benefits, including anti-cancer benefits, due to the presence of bioactive compounds in them. Despite concerns about oxalates, we believe these vegetables are too important to ignore. Moderation and thoughtful vegetable selection are key to feeding your dog.

### *7.4.13.1. Herbs*

Australia's Dr. Barbara Fougère is a renowned holistic veterinarian and expert on bioactive substances used to treat sick animals. As per Dr. Fougère, some herbs have anti-tumor properties and can be used in conjunction with traditional cancer treatments. These treatments include helping the body recover from chemotherapy, radiation, or surgery; acting as a supplement or substitute for conventional treatment (while exercising caution); preventing cancer; and supporting different systems that are impacted by cancer. Herbal remedies that target particular mechanisms and are combined in moderate, nontoxic doses may complement or even enhance conventional therapy, according to Dr. Fougère (Fougère, 2012).

Here are some important herbs with anti-cancer properties:

#### *7.4.13.1.1. Ginger*

An ethanol extract of ginger was tested for its potential therapeutic benefits. The results of the study indicated that the extract caused changes in gene expression that led to the death of breast cancer cells. Specifically, the extract up-regulated the expression of genes involved in the process of apoptosis and down-regulated the expression of genes that prevent it. The researchers found that the ginger extract downregulated the expression of hTERT, a gene involved in the immortalization of most cancer cells, in addition to suppressing the progression of breast cancer cells. These findings led the authors to conclude that ginger may be useful as a supplemental agent in cancer prevention (Ayman, 2012).

### *7.4.13.1.2. Milk Thistle*

The Mediterranean region is home to the stunning purple-flowering herb plant known as milk thistle. Silymarin, the term used to refer to the active component in medicine, is actually a combination of three flavonoids that were isolated from the seeds of the plant: silicristin, silidianin, and silibinin (Ehrlich, 2011). According to recent research, milk thistle may be able to prevent several different types of cancer. Researchers discovered in 2011 that silibinin inhibited the expression of COX-2 and iNOS enzymes involved in tumor growth, as well as the migration of pre-existing lung cancer cells, thereby halting the spread of lung cancer in mice. Comparing silibinin's effects to expensive drugs that are presently undergoing clinical trials for the treatment of lung cancer, the researchers discovered that silibinin was just as effective as the drugs at targeting lung cancer (Colorado State University, 2011; Tyagi et al., 2011).

More recently, it has been demonstrated that silibinin can prevent skin cancer brought on by exposure to UV rays from the sun in two different ways: either by killing or repairing damaged cells. When UVB rays cause damage to cells, silibinin boosts the expression of interleukin-12, a protein that rapidly repairs the damaged cells and gets them back to normal. Silibin quickly destroys the mutated cells in the event of damage brought on by UVA radiation, which makes up around 95% of the solar radiation that reaches Earth. Because the damaged cells are either eliminated or repaired in both situations, skin cancer is avoided (Colorado State University, 2013; Narayanapillai et al., 2013). Herbal remedies are still very promising for both cancer prevention and treatment; however, administering them in a safe and efficient manner is both an art and a science. We highly recommend consulting with a holistic veterinarian who is knowledgeable about the mechanisms and interactions of herbal formulas if you intend to include them in your dog's cancer-protective diet (Wynn and Fougère, 2006; Fougère, 2012).

Some things to consider when working with herbs and supplements:

- **More Is Not Always Better:** The key is adding the correct supplements to the diet, not just piling one on top of another.

- **Interactions May Occur:** Most supplements are safe, however, there is the potential for certain ingredients to negatively interact with certain types of cancers (Fougère, 2012). For example, the amino acid glutamine can potentially worsen a brain tumor and should not be used for dogs with this type of cancer (Messonnier, 2006).
- **Not All Products Are Reliable:** Since supplements are not regulated by the FDA, the quality can vary widely. Many supplements have been found to contain far less of the active ingredients than claimed on the label. The danger of contamination from lead, mercury and other toxins in certain products—most notably herbal products from China—also exists (Tremayne, 2010). Whenever possible, purchase supplements from your holistic veterinarian or opt for products carrying the National Animal Supplement Council (NASC) label, a non-profit trade organization whose members pledge to comply with good manufacturing practices for product safety and quality assurance.

## 7.4.14. Using Raw Foods to Help Fight Cancer

Dr. Sue E. has kindly provided the following information about raw foods and cancer. Armstrong, a well-known holistic veterinarian cancer specialist with international recognition:

Food processing, including cooking, modifies the food's bioavailability of certain essential nutrients as well as its digestibility. Inadequate ingredients categorized under headings like "meat and meat derivatives" may also be present, along with hidden ingredients like artificial coloring and preservatives (discussed previously) (Armstrong, 2013).

A cautious switch to a raw food diet may be best for dogs who are still energetic and have a healthy immune system. But the quality of the ingredients is crucial, as well (e.g., only human-grade meat), as well as giving careful consideration to the nutritional balance of the diet. A raw food diet can be easily thrown off course by eating the wrong vegetables, too much bone, or not enough variety in your nutrients. Some reputable commercial companies offer pre-made, balanced raw food diets that include fully ground bone. These are not only simpler to prepare than diets from scratch, but they also help ensure that, if you're not used to feeding raw, you don't get the nutritional balance incorrect (Armstrong, 2013).

Always switch your dog's diet gradually over a period of 10 to 14 days, giving him more of the new raw diet each day in place of the old one. To support the dog's digestive system during the transition period, administer digestive enzymes and probiotics (Armstrong, 2013).

For a dog that is not emaciated or wasting away, 2–3% of his body weight should be consumed daily (e.g., a 20-pound dog ought to consume about seven ounces of food every day). Note: The weight, not the volume, of the food is seven ounces.

This is just one example of a raw food diet that can be beneficial for cancer patients. Other proteins (e. g., fish, eggs) and other vegetables can be changed to accommodate your dog's unique dietary needs and preferences. Once more, please be advised that the quality of the ingredients is critical; low-quality raw food poses an equal or greater risk of injury than low-quality kibble.

After your dog's digestive tract is used to breaking down ground bone, you can start introducing whole bone (e.g., chicken carcass, turkey neck). If your dog has always eaten kibble, it may take some time for him to learn how to properly crush and tear bone, so make sure you keep an eye on him until you are certain (Armstrong, 2013).

Sample diet for a 65-pound dog (adjust amounts according to your dog's size) (Note: These measurements are by weight, not volume):

- Skinned chicken and ground bone (10.5 oz)
- Lamb, ground bone and liver (6.5 oz)
- Cottage Cheese (1 oz)
- Liver (calf or lamb) (6 oz a week)
- Butternut squash/apples/pears/carrots/spinach/blueberries. Root vegetables and tougher leafy vegetables should be chopped and lightly steamed or liquidized. Vary the vegetables used over a two-week period, trying to achieve a rainbow spectrum of fruits and vegetables to ensure a wide range of vitamins and minerals.
- Sea kelp/spirulina/alfalfa (1/ 4 to 1/ 2 teaspoon)
- Omega-3 essential fatty acids: fish oil/krill oil (5000 mg daily) (NOT to be used if your dog has any tumor with a bleeding tendency, such as hemangiosarcoma)
- Oyster shell calcium (750 mg tablet)
- L- Arginine (700 mg)
- Bio-available full spectrum vitamin/mineral supplement two to three times a week (natural food should provide most of these, but the addition of a high-quality supplement a few times a week can help ensure that there are none lacking).

In order to give your dog a readily utilized energy source, you may need to increase the amount of carbohydrates in your diet if your dog is cachexic. This can be done by adding rice, oats, or pasta. At this point, an animal that has already used up all of its energy should be fed a warm, cooked diet because it requires less energy to digest. The meat and veggies mentioned above can be used to create a cooked version of this diet; however, bone should never be cooked, and more roughage must be added to help with digestion (Armstrong, 2013).

## 7.5. COMMON CANINE HEALTH DISEASE

You now know that certain of the most severe medical conditions can be lessened or even completely cured with the help of functional nutrients. We have the chance to "start fresh" with the environmental signals we send our cells through the epigenome. One of the many wonders of life is that cells are constantly repairing, growing, and renewing. This implies that you can alter your dog's course in life by providing for his cellular needs, even if he is currently dealing with one or more health problems.

Chronic conditions that commonly affect our canine companions. We will discuss:

- Chronic kidney disease (CKD)
- Diabetes mellitus (sugar diabetes)
- Gastrointestinal tract disease (Inflammatory Bowel Disease)
- Heart disease
- Liver disease (hepatopathy)
- Seizures/epilepsy
- Thyroid disease (hypothyroidism)

### 7.5.1. Chronic Kidney Disease (CKD)

The kidneys are bean-shaped organs that carry out several vital tasks, such as eliminating waste and extra water from the blood and excreting it in the form of urine through the bladder. In addition, the kidneys secrete several hormones, including calcitriol, the bioactive form of vitamin D, which is actually a steroidal hormone that influences thousands of different genes, renin, which controls blood pressure, and erythropoietin (EPO), which is essential for the synthesis of red blood cells (NKUDIC, 2012).

The progressive condition known as chronic kidney disease, or CKD, causes long-term kidney damage over a period of time ranging from a few months to several years, ultimately leading to an irreversible decline in kidney function. One of the main causes of death for dogs, chronic kidney disease (CKD) is the most common kidney issue seen in both dogs and cats. While kidney disease is seven times more common in cats than in dogs, from 2007 to 2011, the number of older dogs receiving a kidney disease diagnosis increased by nearly 22%. Given that CKD is progressive and permanent, it is imperative to diagnose it as soon as possible. If untreated, CKD can become a silent killer (Banfield Pet Hospital, 2012; Howard, 2012; Banfield Pet Hospital, 2013; Brown, 2013).

Since dogs don't initially exhibit any clinical symptoms, it can be challenging to diagnose CKD in its early stages. Usually, increased thirst (polydipsia) and increased urination (polyuria) are the initial signs. Other signs include:

- Bad breath
- Inactivity
- Poor coat condition
- Reduced appetite
- Vomiting
- Weight loss

CKD can result from any condition that damages the kidneys over time, including:

- Abnormal protein deposits in the kidneys (amyloidosis).
- Bacterial infections, such as leptospirosis.
- Chronic high blood pressure (hypertension).
- Chronic urinary tract infections.
- Glomerulonephritis, an inflammatory condition causing dysfunction of the glomeruli (networks of capillaries in the kidneys that help filter blood into the urine).
- Kidney stones.
- Toxins, such as antifreeze.
- Tumors.

The breeds most at risk are older dogs and those with genetic predispositions, like Basenji, English Cocker Spaniel, and English Springer Spaniel (Brown, 2013).

#### *7.5.1.1. Obesity and CKD*

Even though obesity causes diabetes and hypertension, two conditions that are major causes of kidney disease, obesity itself is a risk factor for chronic kidney disease (CKD). Although the precise mechanism underlying the relationship between obesity and chronic kidney disease (CKD) is still unknown, theories include inflammation, lipotoxicity (the build-up of excess lipids in non-adipose tissues that causes cell dysfunction or death), hemodynamic (circulatory) effects, and other mechanisms (Schaffer, 2003; Wahba and Mak, 2007).

#### *7.5.1.2. Beware of Kibble*

Since kibble only contains 6–10% moisture, compared to fresh meats that contain 70% or more moisture, feeding your dog those dry kibble nuggets can put stress on their kidneys. When digesting kibble, a dog's body needs to use a lot of water just to get the food through his GI tract. Although canned foods can also contain a significant amount of carbohydrates, dry foods are generally high in these nutrients, which can lead to kidney inflammation. Your veterinarian may suggest a dry "kidney diet" if your dog is suffering from kidney failure. We usually advise against these diets because they can worsen the condition by adding too many carbohydrates and causing dehydration (more on this later) (Case et al., 2011).

#### *7.5.1.3. Dietary Management of CKD*

Despite what some "experts" claim, dogs with moderate to severe kidney failure should not follow a low-protein diet because an excess of protein can produce toxic nitrogen-based byproducts that the damaged kidneys are unable to filter. The majority of dogs already eat low-protein diets. For the reasons mentioned above, low-moisture, high-carb dry diets are detrimental to dogs with kidney problems; in addition, they deprive the dog of essential high-quality protein. Foods appropriate to the species should still be fed to sick dogs; providing less protein than necessary to maintain a healthy body will not lessen the strain on the kidneys (Case et al., 2011).

#### *7.5.1.4. Omega-3 fatty acids and CKD*

Omega-3 fatty acids EPA and DHA are helpful in maintaining kidney health and treating renal issues. Long-term (two years) use of fish oil has been shown in human research to reduce the rate at which high-risk patients with glomerulonephritis, or IgA nephropathy, lose kidney function (see above bulleted list). Additionally, individuals with diabetes-related kidney disease appear to have less protein in their urine when using fish oil (MedlinePlus, 2012).

There is evidence that dogs with kidney disease can directly benefit from omega-3 fatty acids. While supplementing with omega-6 fatty acids accelerated the decline of kidney function in the early stages of the disease, supplementing with omega-3 fatty acids had a protective effect, according to a study done on eighteen adult mixed-breed dogs (Brown et al., 2000).

#### *7.5.1.5. Kidney Cleansing Diet*

The following diet is designed for dogs with kidney dysfunction:

1. **Ingredients**
   - White and/or brown rice OR white potato
   - White fish (or sardines, pork or duck)

   If you're using rice, cook it as directed on the package. Cook the potato, if using, until it's tender (e.g., bake in the skin (or take off the skin, chop, boil, or steam). Peel and mash very lightly. If using canned sardines, drain the fish and bake, broil, or steam it until it's cooked through. Cook the pork or duck thoroughly if using it. In a ratio of two-thirds rice or potato to one-third protein, mix the rice or potato with the protein (duck, pork, or fish). Use parsley or a mixture of Italian herbs to season. When your dog starts to feel better later on, you can progressively add in finely chopped, gently cooked (e.g., steamed) veggies like green beans, carrots, zucchini, yellow squash, spinach, kale, and scrambled eggs, if your dog can eat these items.
2. **Supplements:** Include the supplements that are indicated in the sample diet, such as a premium probiotic. Make sure to take a supplement containing omega-3 fatty acids. Adapt the dosage of supplements based on your dog's weight. Feed your child three to four small meals a day, or at least two if your schedule doesn't permit more. Your dog should receive enough food to meet their daily caloric needs.

### 7.5.2. Diabetes Mellitus (Sugar Diabetes)

When you learn that your dog has diabetes mellitus, you may feel confused and afraid. Providing long-term care may initially seem like a daunting task if you haven't personally dealt with diabetes. Unfortunately, since many owners give up before their dogs even have a chance, euthanasia is the leading cause of death for diabetic dogs. We want to reassure you that taking care of a dog with diabetes is completely manageable and that you shouldn't make any snap decisions that could have unintended consequences. We know of a veterinarian who treats a lot of dogs with diabetes, and she advises her clients to commit to trying the treatment for three months and seeing how it goes instead of even considering how long the treatment will take. Like her clients, we're confident that once you pick up this new skill of caring for your diabetic dog, it will become second nature rather quickly. She claims that her advice has never failed. You'll be relieved that you didn't do anything rash when you wake up one day and discover that giving your dog his insulin injections is just another everyday routine, much like walking or feeding him (Murray, 2012).

### 7.5.2.1. Diabetes in Dogs Increasing

Diabetes is becoming more common in both humans and dogs. Actually, during the last five years, the prevalence of diabetes in dogs has doubled (Banfield Pet Hospital, 2013). Because dogs usually only develop Type 1 diabetes, it has not historically been regarded as a lifestyle disease like it is in people (and cats).

Juvenile diabetes, or type 1 diabetes in humans, is a disease that develops when the pancreatic beta-cells are unable to produce enough insulin, the hormone that regulates blood glucose levels. Glucose is transported from the bloodstream into the cells by insulin, where it is converted into energy. In the absence of insulin, glucose builds up to harmful levels in the blood and urine instead of entering the cells. Diabetes type 1 is not regarded as a condition linked to a particular way of living.

Insulin resistance, another name for type 2 diabetes, is a condition in which the pancreas makes insulin, but the body is unable to properly use it. Because type 2 diabetes develops later in life and is linked to lifestyle factors like diet and obesity, it is also known as adult-onset diabetes.

Herein lies the intriguing aspect. Dogs do not get diabetes when they are young, even though canine diabetes usually stems from an issue producing insulin, which happens in Type 1. Diabetes in dogs develops later in life, much like Type 2 diabetes in people. Diabetes in dogs is therefore an adult-onset condition influenced by diet and lifestyle, despite being classified as Type 1 due to its relationship to insulin production rather than insulin resistance (Becker, 2013).

### 7.5.2.2. Obesity and High-GI Carbs: Leading Contributors to Diabetes

It should come as no surprise to hear that eating high-GI carbohydrates and being obese are the two main lifestyle factors that contribute to diabetes in humans. Here's an intriguing fact about which you probably were unaware. Although it is generally believed that the inability of the pancreatic beta cells to produce insulin is irreversible, a 2011 study in humans demonstrated that insulin resistance and beta cell failure can be reversed in a matter of weeks with just calorie restriction. Diabetes is "a potentially reversible condition and not one that is inevitably progressive," according to the authors' conclusion. According to Lim et al., 2011). Although insulin resistance, a defining feature of diabetes, is known to be brought on by obesity in dogs (Swanson, 2007), a 2012 study involving 35 obese dogs in England revealed that 20% of them also had metabolic syndrome. This is a remarkable discovery because metabolic syndrome affects obese people more often than dogs and can have devastating effects on the liver, pancreas, immune system, heart, and nervous system, among other systems (see below). Insulin resistance and Type 2 diabetes are closely associated with metabolic syndrome in humans (Bugianesi et al., 2005). The study's obese dogs also had higher blood insulin levels, which suggests that the pancreas is under more strain to regulate blood sugar levels. Similar to the previously mentioned human study, these dogs' metabolic problems improved as they shed pounds (University of Liverpool, 2012).

The emergence of metabolic syndrome in dogs is additional evidence that the diseases associated with the lifestyle choices we make as humans are killing off our canine friends, and that a significant nutritional intervention is required to stop and reverse this trend.

According to recent research, pro-inflammatory cytokines produced by adipose tissue may be the cause of insulin resistance, which is actually thought to be an inflammatory disorder (Bugianesi et al., 2005). Obesity is bad for your dog—as it relates to diabetes and every other chronic health condition, regardless of the underlying mechanism.

We won't reiterate here what we've already covered in-depth regarding how to maximize your dog's weight and carbohydrate intake. For all the details, just refer back to previous conversations. But don't lose hope—you can significantly improve your diabetic dog's health just by helping him lose weight and reducing his consumption of bad carbohydrates.

### *7.5.2.3. DNA Methylation: A New Risk Factor for Type 1 Diabetes?*

Although Type 1 diabetes has historically been attributed to genetics rather than lifestyle, more recent studies are identifying the role of epigenetic factors like DNA methylation. A fascinating study examined the genome-wide DNA methylation patterns of identical twins, one of whom had Type 1 diabetes and the other did not. This finding is noteworthy in and of itself because, if Type 1 diabetes were solely genetic, identical twins, who share the same genome at birth, ought to either have the disease or not. The results of the study demonstrated that, in contrast to the non-diabetic twins, the diabetic twins had 132 distinct DNA sites where methylation significantly correlated with the diabetic state. It has been suggested that DNA methylation may be a significant risk factor for Type 1 diabetes since some of these abnormal methylation profiles were found before the twins' official diagnosis (Rakyan et al., 2011). This is just one more justification for feeding your diabetic dog a diet high in foods high in curcumin and folate, which both support optimal DNA methylation.

#### *7.5.2.4. Gene Therapy: A Future Cure for Canine Diabetes?*

Research on diabetes is making exciting strides that may one day allow humans and dogs to completely avoid insulin injections. For over four years, researchers from the Universitat Autonoma de Barcelona in Spain have effectively controlled Type 1 diabetes in dogs through gene therapy, without the hypoglycemia or other risks that come with traditional treatments. The research was released in the Diabetes Journal's May 2013 issue (American Diabetes Association, 2013).

The scientists gave insulin and glucokinase injections to five diabetic laboratory Beagles during a single treatment session. These genes are essential for sustaining appropriate blood glucose levels. The scientists used a vector, or vehicle for DNA transmission, made up of a non-pathogenic virus to "infect" cells in the dogs' skeletal muscle with the two genes. The genes could be injected just once and continue to function normally for years to maintain normal blood glucose levels because skeletal muscle cells do not divide (American Diabetes Association, 2013). After fine-tuning the procedure in companion dogs, the researchers plan to move on to testing on human patients (American Diabetes Association, 2013).

### 7.5.3. Gastrointestinal Tract Disease (Inflammatory Bowel Disease)

As the name suggests, GI tract inflammation is a feature of Inflammatory Bowel Disease (IBD), which is another manifestation of the harm resulting from a leaky gut. You may remember that certain substances can cross the intestinal wall and enter the bloodstream when the mucosal barrier of the gut is compromised (Woodruff Health Sciences Center, 2012). As a result of the body attacking the foreign invaders with white blood cells, intestinal inflammation results, which inevitably leads to IBD (Robinson and Reeves, 2013). Leaky gut syndrome and the IgA/IgM antibody response that results from food intolerances/sensitivities are closely related. If you think your dog may have IBD, we suggest using NutriScan to find out which particular foods are offending so you can cut them out of his diet. If your dog is still eating antigenic foods that worsen his GI inflammation and make the disease worse, it will be very difficult to treat his IBD.

#### *7.5.3.1. Dietary Modifications for Dogs with IBD*

The ideal diet for dogs with IBD is:

- **Extremely Digestible:** to promote bowel rest, lessen the quantity of antigenic dietary protein in the lower small intestine, lessen the quantity of potentially hazardous bacteria in the intestine, and lessen osmotic diarrhea, which is a condition that happens when the GI tract is unable to adequately break down and absorb the food, leading to the component moving through the digestive system too quickly. This leads to loose and watery stools because the dog is unable to properly absorb the nutrients from the food or drink and the waste products do not have time to combine.
- **Gluten-Free:** to prevent the GI inflammatory response and leaky gut caused by eating gluten.
- **Hypoallergenic:** to reduce the potential for food sensitivities that result in GI inflammation.
- **Lactose-Free:** to reduce the inflammatory response to lactose.
- **Low Fat:** IBD can disrupt fat digestion and absorption, and poorly absorbed fats worsen diarrhea.

Additionally, soluble fiber aids in the treatment of IBD-related diarrhea. One kind of indigestible carbohydrate that slows down digestion is soluble fiber, which absorbs water and gels during digestion.

Good sources of soluble fiber include:

- Apples
- Barley
- Beans
- Beet pulp
- Carrots
- Chicory
- Green beans
- Jerusalem artichokes (sunchokes)
- Lentils
- Oats (choose gluten-free)
- Peas
- Psyllium
- Pumpkin
- Sweet potatoes

### *7.5.3.2. Prebiotics: Fiber that Feeds Beneficial Bacteria in the Intestines*

Furthermore, soluble fiber functions as a prebiotic. Prebiotics are indigestible carbohydrates (i. e. fiber) that feed probiotics and encourage the colon's beneficial bacteria to proliferate and/or activate. Prebiotics are not living things, in contrast to probiotics. Soluble fiber known as prebiotics makes its way from the small to the large intestine unaltered, where it feeds the beneficial bacteria in the colon. Fructooligosaccharides (FOS), the most prevalent prebiotic, are highly fermentable by a variety of bifidobacteria species, a class of bacteria linked to gut health, as well as by certain other beneficial microflora species (Swanson et al., 2002). Like FOS, inulin is a prebiotic. Whole grains, chicory root, asparagus, bananas, and Jerusalem artichokes are among the foods that contain prebiotics.

An additional class of indigestible carbohydrates that support GI health are mannan-oligosaccharides (MOS). The sugars mannose and glucose make up MOS, which are typically extracted from *Saccharomyces cerevisiae* yeast cell walls. Similar to FOS, MOS affect immune function by boosting the body's defenses locally, improving nutrient digestibility, and enhancing and safeguarding the health of the large intestine (Swanson et al., 2002; Yamka et al., 2006).

Some specific benefits of prebiotics include:

- Boost the immune system by assisting in the growth of good bacteria (remember that the gut contains the largest immune organ in the body).
- Help prevent diarrhea by keeping bad bacteria in the gut under control.
- Help prevent disease by aiding the immune system.
- Improve digestion.
- Improve nutrient absorption, especially of minerals.

Because they complement each other, prebiotics and probiotics are best given together. Synbiotics are foods and supplements that combine probiotics and prebiotics. Choose fermented dairy products made from sheep or goat milk rather than cow's milk as they are considered synbiotics. Examples of fermented dairy products are yogurt and kefir. Make sure the yogurt or kefir you buy says it contains live or active cultures; keep in mind, probiotics need to be living in order to function.

Prebiotics are also a common ingredient in commercial probiotic supplements. If you want to give your dog a fiber supplement to help with prebiotics, proceed with caution because too much soluble fiber can cause gas and loose stools. Gradually raise the dosage from a low starting point. If you experience negative symptoms like diarrhea or gas, cut back or stop (Straus, 2012).

Additionally, since they can lead to gas or bloating, we advise staying away from highly fermentable fiber sources like lactulose, guar gum, and citrus pectin. Keep in mind that insoluble fiber (e.g., cellulose), commonly referred to as roughage, can slow down the absorption of minerals and has a laxative effect by accelerating intestinal transit time. Since insoluble fiber cannot be fermented by the colon's friendly bacteria, it is not regarded as a prebiotic (Straus, 2012).

### *7.5.3.3. Probiotics: Essential for the Health of Your Dog's GI Tract*

The gut's trillions of beneficial bacteria perform a multitude of vital tasks, such as controlling digestion, generating and metabolizing vitamins and other trace minerals, preserving the integrity of the gut lining, and fending off harmful bacteria like *E. coli* and other infections while shielding the body from harm.

Abdominal cramps, surface erosions, ulcers, and bloating are just a few of the digestive disorders that can affect both humans and animals when the balance between good and bad bacteria is out of whack. You may be able to reverse your dog's gastrointestinal distress and restore optimal gut health by simply supplementing his diet with a high-quality probiotic.

### *7.5.3.4. GI Cleansing Diet*

The following GI cleansing diet is suitable for dogs with IBD. Note that this diet can also serve as a liver cleansing diet.

#### *7.5.3.4.1. Ingredients*

- White potato
- Sweet potato
- White fish (such as cod, halibut, sole, haddock, tilapia). You can substitute baked chicken or pork if your dog won't eat fish.

Once the potatoes are soft (e.g., bake in the skin (or take off the skin, chop, boil, or steam). Peel and mash very lightly. After cooked, mix half sweet potato and half white potato in a 50/50 ratio.

Fish should be baked, broiler-roasted, or steamed until cooked through. Note: Avoid using fish that are high in mercury, like swordfish and tuna. In a ratio of two thirds potato mixture to one third fish, combine the potato mixture with the cooked fish. Use parsley or a mixture of Italian herbs to season. When your dog starts to feel better later on, you can progressively add in finely chopped, gently cooked (e.g., steamed) veggies like green beans, carrots, zucchini, yellow squash, spinach, kale, and scrambled eggs, if your dog is able to eat these items.

#### *7.5.3.4.2. Supplements*

Include the supplements that are indicated in the sample diet, such as a premium probiotic. To help reduce inflammation, make sure to take an omega-3 fatty acid supplement. Adapt the dosage of supplements based on your dog's weight. Feed your child three to four small meals a day, or at least two if your schedule doesn't permit more. Your dog should receive enough food to meet their daily caloric needs. We also advise talking with your veterinarian about the possibility of supplementing these nutrients in amounts up to double the recommended daily allowance, as dogs with IBD also frequently experience excessive loss of potassium and both water- and fat-soluble vitamins (Guilford and Matz, 2003). Your dog can live a normal life and probably not require medication once he is on an anti-inflammatory diet that facilitates the healing of the colon.

### 7.5.4. Heart Disease

Dogs are fun-loving creatures with enormous hearts, as we all know, but canine heart disease is no laughing matter. Unfortunately, one in ten dogs have heart disease of some kind, according to the American Veterinary Medical Association. Thankfully, over 50 years of studying the management of cardiovascular disease in humans has led to the discovery of nutritional therapies that can help dogs as well. These therapies can reduce the need for prescription drugs, improve quality of life, and, in certain cases, even cure cardiovascular disease (Dove, 2001; Devi and Jani, 2009; Li et al., 2013).

Canine heart disease refers to many conditions involving dysfunction of the heart. Many of these conditions can benefit from nutritional therapies, including:

- Diseases of the heart valves (mitral regurgitation and valvular disease).
- Diseases of the heart ventricles (primary diastolic dysfunction).
- Diseases of the heart muscle tissue (cardiac hypertrophy, dilated and restrictive cardiomyopathy, congestive heart failure).
- Diseases resulting from abnormal electrical activity (cardiac arrhythmias, atrial fibrillation).
- Injury from ischemia (restricted blood supply to the tissues, causing a lack of oxygen and nutrients) and reperfusion (tissue damage

caused when blood supply returns to the tissue after ischemia, creating a condition in which the restoration of circulation results in inflammation and oxidative damage).

- Non-fatal heart attacks.

Here are some of the most well-researched nutrients to benefit your dog's heart health:

1. **Coenzyme Q10 (CoQ10)**

- Found in almost every cell in the body and is critical for a number of functions, including energy production.
- Antioxidant that can neutralize damaging free radicals implicated in a number of diseases, including degenerative heart disease.
- May provide enhanced antioxidant protection, help slow the progression of canine heart disease and improve overall heart health.

2. **L-Carnitine**

- Amino acid used by the body to turn fat into energy.
- In several breeds, a deficiency results in the development of cardiomyopathy (deterioration in the function of the heart muscle).
- Typically used in combination with taurine, has been shown to improve outcome of dogs with congestive heart disease.

3. **Magnesium**

- Mineral important to healthy functioning of the heart.
- Properties include antiarrhythmic (alleviates abnormal heart rhythms) and antithrombotic (reduces blood clots) activities.
- Deficiencies can cause conditions such as hypertension, coronary artery disease, congestive heart failure and cardiac arrhythmias.
- Deficiencies may correlate to the development of mitral valve prolapse in dogs.

4. **Omega-3 Fatty Acids**

- Benefits include antiarrhythmic, anti-inflammatory, hypotensive (promotes low blood pressure) and anti-vasopressor (lowers blood pressure) properties.
- Decrease the production of inflammatory cytokines interleukin-1 (IL-1) and tumor necrosis factor (TNF) in dogs with congestive heart failure (CHF). Inflammatory cytokines are a major cause of cardiac cachexia (loss of weight and lean body mass). Reduced IL-1 is correlated with increased survival in dogs with CHF.
- May prevent fatal arrhythmias in dogs with CHF.

- Low plasma concentrations of EPA and DHA are associated with heart failure in dogs.

5. **Taurine**

- An amino acid found in relatively high concentrations in heart tissue.
- Even though dogs' bodies are thought to make adequate amounts of taurine and it is not considered required in the diet, taurine deficiency is linked with dilated cardiomyopathy (DM) in several breeds, including Cocker Spaniels, Doberman Pinchers, English Setters, Golden Retrievers, Labrador Retriever, Portuguese Water Dogs and Saint Bernards.
- One study showed that taurine supplementation at 100 mg per kilogram of body weight daily over the course of a month improved the condition of dogs with congestive heart failure.

6. **Vitamin D**

- Plays an important role in cardiac function.
- In people, evidence indicates that vitamin D deficiency is associated with the development of congestive heart failure (CHF); a recent study conducted at Cornell University's College of Veterinary Medicine showed that vitamin D might play a similar role in dogs.
- The study of 82 companion dogs compared blood levels of vitamin D in dogs with CHF to healthy dogs.
- The researchers found that the dogs with CHF had lower blood levels of vitamin D than the healthy dogs. Low blood levels of vitamin D were also associated with poor survival rates in the dogs.
- The researchers concluded that improving vitamin D levels in some dogs with CHF might prove beneficial without causing toxicity.
- Consult with your veterinarian regarding appropriate levels of vitamin D supplementation to avoid toxicity.

7. **Vitamin E**

- Protects the myocardium (the muscular tissue of the heart) from damage brought about by free radicals.
- Lower plasma concentrations of vitamin E correlates with increased incidence and severity of heart disease in dogs.

To control cardiac cachexia, dogs with heart disease also need enough calories and protein. Dogs' bodies will use protein to make up for the calories they don't consume, which will cause them to lose lean body mass. Try adding a little raw honey or maple syrup to your dog's food to make it more palatable if he has anorexia (loss of appetite) as a result of medication side effects (Devi, 2009).

It is advisable to consult your veterinarian about nutritional therapies for canine heart disease, as your dog's age, breed, and specific medical condition will determine the recommended recommendations and dosages (Dove, 2001).

### 7.5.5. Liver Disease (Hepatopathy)

The liver is a complex organ involved in almost every biochemical process essential to life, such as:

- Filters toxins such as drugs, chemicals, bacteria and excess ammonia (a by-product of protein metabolism) from the blood.
- Metabolizes fats, carbohydrates and proteins.
- Processes raw materials.
- Removes waste.
- Rids the blood of old or damaged red blood cells.
- Stores, activates and degrades vitamins.
- Stores glycogen (the storage form of glucose) for energy.
- Synthesizes proteins essential to normal blood clotting.
- Produces bile to aid in digestion.

Hepatocytes, the cells that make up the liver, must efficiently perform thousands of chemical reactions per second. As one might expect, major issues can arise when the liver becomes sick and is unable to function correctly (Smith, 2009).

Typical signs of liver disease include:

- Anorexia (lack of appetite)
- Diarrhea
- Increased thirst
- Increased urination
- Lethargy
- Jaundice
- Orange urine
- Seizures (severe cases)
- Vomiting

Due to the liver's close ties to numerous bodily systems and the non-specific nature of liver disease symptoms, diagnosing liver disease can be challenging.

An additional issue is that the liver can function normally even when up to 80% of the organ is diseased; by the time a diagnosis is made, the damage may already be too great to reverse. The good news is that, more than any other organ, liver cells exhibit remarkable regenerative capabilities. According to Smith (2009), a full recovery is thus achievable if care is taken to identify damage as soon as possible and to start the right treatment.

Conventional therapy is often limited unless liver damage is caused by a known toxin that is treatable or a condition that calls for surgery. This means that your best bet for maximizing your dog's liver health may be complementary therapies like dietary changes and the addition of functional nutrients (Messonnier, 2007).

Here are recommendations to bolster the health of your dog's liver:

### *7.5.5.1. Eliminate Toxins*

Imagine the strain the liver experiences in filtering chemicals, pesticides, heavy metals, pharmaceuticals, anesthetic agents, and all the other impurities that we've previously talked about in detail and that are so prevalent in today's processed foods. The liver is the main organ that removes toxins from the blood.

The epidemic of liver disease in Labrador Retrievers since the mid-to-late 1990s serves as one example of how commercial pet foods and canine liver disease are related. Researchers came to the conclusion that an excess of copper in the dogs' diets was probably the cause of these illnesses after looking at the copper concentrations in their livers, a symptom that indicates liver disease. It's interesting to note that at the same time, AAFCO advised dog food producers to move away from cupric oxide and toward more bioavailable forms of copper, like cupric sulfate or other chelated forms (Johnston et al., 2013). The researchers came to the conclusion that, "We believe that the amount of copper in many commercial dog foods is excessive and that Labrador Retrievers, because of their popularity as family pets, may be functioning as sentinels for trends in hepatic [liver] copper concentrations in the general pet dog population," even though they were unable to conclusively prove that the higher incidences of liver disease were caused by the copper in the commercial pet foods (Johnston et al., 2013).

A wide array of prescription and non-prescription drugs used to treat dogs can also damage the liver, including:

- Anthelmintics (worming medication)
- Glucocorticoids (corticosteroids, such as cortisone)

- Ketaconazole (fungal treatment)
- Parasiticides (to rid the body of parasites)
- Phenobarbital (seizure/epilepsy medication)
- Rimadyl (arthritis treatment)
- Thiacetarsamide (heartworm treatment)
- Tylenol (acetaminophen)

Eliminate toxins from food and the environment, and make sure your veterinarian is informed about the use of any medications. This includes skipping vaccinations, which can strain and harm the immune system and liver (assuming your dog has already received all of his puppy shots).

### *7.5.5.2. Beware of this Liver Damaging Imitation "Vitamin"*

Never give Menadione Sodium Bisulfate (also known as Menadione Dimethyl-Pyrimidinol Bisulfate, Menadione Dimethyl-Pyrimidinol Bisulfite, Menadione Sodium Bisulfate Complex, Menadione Sodium Bisulfite, or Menadione Sodium Bisulfite Complex) to a dog with liver problems, even though natural vitamin K can help prevent bleeding disorders linked to some liver diseases (Smith, 2009). Synthetic vitamin K, known as menadione, is hazardous to liver cells and is responsible for several other health issues. Please be advised that Mephyton® and AquaMephyton®, brands of synthetic vitamin K1 (phytonadione) sold by Merck, are safe and effective for use in both humans and animals.

# REFERENCES


1. Adams, L. S., Kanaya, N., Phung, S., Liu, Z., & Chen, S. (2011). Whole blueberry powder modulates the growth and metastasis of MDA-MB-231 triple-negative breast tumors in nude mice. *Journal of Nutrition, 141*(10). https://doi.org/10.3945/jn.111.140178
2. American Diabetes Association. (2013). Gene therapy used in dogs to treat type 1 diabetes. Retrieved from http://www.diabetes.org/formedia/2013/gene-therapyused-in-dogs-to-treat-type1-diabetes.html (accessed on 7th October 2024).
3. Anand, P., Kunnumakara, A. B., Sundaram, C., Harikumar, K. B., Tharakan, S. T., Lai, O. S., & Aggarwal, B. B. (2008). Cancer is a preventable disease that requires major lifestyle changes. *Pharmaceutical Research, 25*(9), 2097–2116. https://doi.org/10.1007/s11095-008-9661-9
4. Armstrong, S. (2013). Foods as medicine. *Animal Wellness, 15*(1), 18–21.
5. Association for Pet Obesity Prevention. (2013). 2012 national pet obesity survey results. Retrieved from http://www.petobesityprevention.com/2012-national-petobesity-surveyresults/ (accessed on 7th October 2024).
6. Ayman, E. I., Abuzinadah, O. A., Baeshen, N. A., & Rahmy, T. R. (2012). Differential control of growth, apoptotic activity, and gene expression in human breast cancer cells by extracts derived from medicinal herbs *Zingiber officinale. Journal of Biomedicine and Biotechnology, 2012.* https://doi.org/10.1155/2012/614356
7. Banfield Pet Hospital. (2012). State of pet health 2012 report. Retrieved from www.stateofpethealth.com (accessed on 7th October 2024).
8. Banfield Pet Hospital. (2013). State of pet health 2013 report. Retrieved from http://www.stateofpethealth.com (accessed on 7th October 2024).
9. Bartges, J. (2013, May 29). Personal correspondence.
10. Basu, A., Du, M., Leyva, M. J., Sanchez, K., Betts, N. M., Wu, M., & Lyons, T. J. (2010). Blueberries decrease cardiovascular risk factors in obese men and women with metabolic syndrome. *Journal of Nutrition, 140*(9), 1582–1587.
11. Becker, K. (2013a). A single treatment cures diabetes—How to protect your pet now. Retrieved from http://healthypets.mercola.com/sites/healthypets/archive/2013/03/27/dogdiabetes-treatment.aspx (accessed on 7th October 2024).
12. Becker, K. (2013b). PBDE: Are you exposing your pet to this thyroid ravager? Retrieved from http://healthypets.mercola.com/sites/healthypets/archive/2013/10/25/pbdeexposure-aspx (accessed on 7th October 2024).

13. Becker, K. (2014). Vitamin D: Don't let your dog overdose on this popular human nutrient. Retrieved from http://healthypets.mercola.com/sites/healthypets/archive/2014/05/09/dogvitamin-d-levels.aspx?e_cid=20140509Z1_PetsNL_art_1&utm_source=petnl&utm_medium=email&utm_con (accessed on 7th October 2024).
14. Bensimoun, C. (2013). Brain food. *Animal Wellness, 15*(1), 32–36.
15. Beynen, A. C., & Legerstee, E. (2010). Influence of dietary beta-1,3/1,6-glucans on clinical signs of canine osteoarthritis in a double-blind, placebo-controlled trial. *American Journal of Animal and Veterinary Science, 5*(2), 97–101.
16. Boileau, C., Martel-Pelletier, J., Caron, J., Msika, P., Guillou, G. B., Baudouin, C., & Pelletier, J. P. (2009). Protective effects of total fraction of avocado/soybean unsaponifiables on the structural changes in experimental dog osteoarthritis: Inhibition of nitric oxide synthase and matrix metalloproteinase-13. *Arthritis Research & Therapy, 11*(2), R41.
17. Boler, B. V. (2011). Nutrition and canine behavior: How do proteins, lipids, carbohydrates and antioxidants in an animal's diet affect aggression, stress, and cognition? Retrieved from http://www.petfoodindustry.com/NutritionBehavior.html (accessed on 7th October 2024).
18. Bosch, G., Beerda, B., Hendriks, W. H., van der Poel, A. F. B., & Verstegen, M. W. A. (2007). Impact of nutrition on canine behavior: Current status and possible mechanisms. *Nutrition Research Reviews, 20*(1), 180–194.
19. Bottiglieri, T. (2002). S-adenosyl-L-methionine (SAMe): From the bench to the bedside—Molecular basis of a pleiotrophic molecule. *The American Journal of Clinical Nutrition, 76*(5), 1151S–1157S.
20. Brand-Miller, J. C., Holt, S. H., Pawlak, D. B., & McMillan, J. (2002). Glycemic index and obesity. *American Journal of Clinical Nutrition, 76*(1), 281S–285S.
21. Bren, L. (2006, September-October). Pain drugs for dogs: Be an informed pet owner. *FDA Consumer Magazine*, 7.
22. Broad Institute. (2013). Oncogene. Retrieved from http://www.broadinstitute.org/education/glossary/oncogene (accessed on 7th October 2024).
23. Brogan, K. (2013). Methylwho? Why you should know about methylation. Retrieved from http://www.greenmedinfo.com/blog/methylwho-why-you-should-knowaboutmethylation?utm_source=www.GreenMedInfo.com&utm_campaign=f3dd2bd89fGreenmedinfo&utmf3dd2bd89f-86768346 (accessed on 7th October 2024).

24. Brown, D. C., & Reetz, J. (2012). Single-agent polysaccharopeptide delays metastases and improves survival in naturally occurring hemangiosarcoma. *Evidence-Based Complementary and Alternative Medicine.* https://doi.org/10.1155/2012/384301
25. Brown, S. A., Brown, C. A., Crowell, W. A., Barsanti, J. A., Kang, C. W., Allen, T., Cowell, C., & Finco, D. R. (2000). Effects of dietary polyunsaturated fatty acid supplementation in early renal insufficiency in dogs. *Journal of Laboratory and Clinical Medicine, 135*(3), 275–286. https://doi.org/10.1067/mlc.2000.105178
26. Buchanan, B. B., & Frick, O. L. (2002). The dog as a model for food allergy. *Annals of the New York Academy of Sciences, 964*, 173–183.
27. Bugianesi, E., McCullough, A. J., & Marchesini, G. (2005). Insulin resistance: A metabolic pathway to chronic liver disease. *Hepatology, 42*(5), 987–1000. https://doi.org/10.1002/hep.20920
28. Byers, C. G., Wilson, C. C., Stephens, M. B., Goodie, J., Netting, F. E., & Olsen, C. (2011, April). Obesity in dogs, Part 1: Exploring the causes and consequences of canine obesity. *DVM 360 Magazine*. Retrieved from http://veterinarymedicine.dvm360.com/vetmed/Nutrition/Exploring-thecauses-and-consequences-ofcanineob/ArticleStandard/Article/detail/715423 (accessed on 7th October 2024).
29. Byers, C. G., Wilson, C. C., Stephens, M. B., Goodie, J., Netting, F. E., & Olsen, C. (2011, April). Obesity in dogs, Part 2: Treating excess weight with a multiple modality approach. *DVM 360 Magazine*. Retrieved from http://veterinarymedicine.dvm360.com/vetmed/Nutrition/Treating-excessweight-with-a-multimodalityapproach/ArticleStandard/Article/detail/715431 (accessed on 7th October 2024).
30. Canine Epilepsy Guardian Angels. (2011). Canine epilepsy. Retrieved from http://www.canine-epilepsy-guardian-angels.com/default.htm (accessed on 7th October 2024).
31. Carr, A. (2013). Ketogenic diet, zonisamide ease seizures. Retrieved from http://www.veterinarypracticenews.com/vet-practice-newscolumns/abstract/ketogenicdiet-zonisamide-ease-seizures.aspx (accessed on 7th October 2024).
32. Case, L. P., Daristotle, L., Hayek, M., & Rausch, M. F. (2011). *Canine and feline nutrition: A resource for companion animal professionals* (3rd ed.). Maryland Heights, MO: Mosby Elsevier.
33. Cave, N. J., & Marks, S. L. (2004). Evaluation of the immunogenicity of dietary proteins in cats and the influence of the canning process. *American Journal of Veterinary Research, 65*(10), 1427–1433.

34. Celleno, L., Tolaini, M. V., D'Amore, A., Perricone, N. V., & Preuss, H. G. (2007). A dietary supplement containing standardized *Phaseolus vulgaris* extract influences body composition of overweight men and women. *International Journal of Medical Sciences, 4*(1), 45–52.

35. Cerundolo, R., Michel, K. E., Court, M. H., Shrestha, B., Refsal, K. R., Oliver, J. W., & Shofer, F. S. (2009). Effects of dietary soy isoflavones on health, steroidogenesis, and thyroid gland function in dogs. *American Journal of Veterinary Research, 70*(3), 353–360.

36. Challacombe, S. J. (1987). The induction of secretory IgA responses. In *Food allergy and intolerance* (pp. 214–220). Eastbourne, England: W. B. Saunders.

37. Coates, J. (2013). Dietary therapy for canine epilepsy? Retrieved from http://www.petmd.com/blogs/nutritionnuggets/jcoates/2013/june/dietarytherapy-for-canineepilepsy-30491#.UsBsfRaAGFJ (accessed on 7th October 2024).

38. Cotman, C. W., Head, E., Muggenburg, B. A., Zicker, S. S., & Milgram, N. W. (2002). Brain aging in the canine: A diet enriched in antioxidants reduces cognitive dysfunction. *Neurobiology of Aging, 23*(5), 809–818.

39. Davidson, S. J., & Joshi, D. K. (2006). In vivo measurement of flatulence and nutrient digestibility in dogs fed poultry by-product meal, conventional soybean meal, and low-oligosaccharide low-phytate soybean meal. *American Journal of Veterinary Research, 67*(1), 88–94.

40. Davis, C. D., & Uthus, E. O. (2004). DNA methylation, cancer susceptibility, and nutrient interactions. *Experimental Biology and Medicine (Maywood), 229*(10), 988–995.

41. Day, M. J. (2005). The canine model of dietary hypersensitivity. *Proceedings of the Nutrition Society, 64*(4), 458–464.

42. Deckelbaum, R. J., Worgall, T. S., & Teo, S. (2006). N-3 fatty acids and gene expression. *American Journal of Clinical Nutrition, 83*(6), S1520–S1525.

43. Demerath, E. W., Guan, W., Pankow, J. S., Grove, M. L., North, K. E., Fornage, M., & Boerwinkle, E. A. (2013). Genome-wide methylation study of body mass index (BMI) in African American adults: Preliminary data from the ARIC study. *Circulation, 127*, A052.

44. DeNapoli, J. S., Dodman, N. H., Shuster, L., Rand, W. M., & Gross, K. L. (2000). Effect of dietary protein content and tryptophan supplementation on dominance aggression, territorial aggression, and hyperactivity in dogs. *Journal of the American Veterinary Medical Association, 217*(4), 504–508.

45. Devi, S., & Jani, R. G. (2009). Review on nutritional management of cardiac disorders in canines. *Veterinary World, 2*(12), 482–485.

46. Divisi, D., Di Tommaso, S., Salvemini, S., Garramone, M., & Crisci, R. (2006). Diet and cancer. *Acta Bio Medica, 77*(2), 118–123.

47. Dodds, W. J. (2001). Vaccination protocols for dogs predisposed to vaccine reactions. *Journal of the American Animal Hospital Association, 38*(1), 1–4.

48. Dodds, W. J. (2014). Food intolerance: Diagnostic testing & dietary management. *Journal of the American Holistic Veterinary Medical Association, 36*, 36–42 (Summer issue).

49. Dodds, W. J., & Lassin, A. J. (2013). Reduce inflammation, improve detoxification, and promote repair. Retrieved from www.hemopet.org (accessed on 7th October 2024).

50. Dodman, N. H., Reisner, I., Shuster, L., Rand, W., Luescher, U. A., Robinson, I., & Houpt, K. A. (1996). Effect of dietary protein on behavior in dogs. *Journal of the American Veterinary Medical Association, 208*(3), 376–379.

51. Donaldson, M. (2004). Nutrition and cancer: A review of the evidence for an anticancer diet. *Nutrition Journal, 3*(19). https://doi.org/10.1186/1475-2891-3-19

52. Dotinga, R. (2013). Gene variants may play role in obesity. Retrieved from http://www.14news.com/story/22612476/gene-variants-may-playrole-in-obesity (accessed on 7th October 2024).

53. Dove, S. (2001). Nutritional therapy in the treatment of heart disease in dogs. *Alternative Medicine Review, 6*, S-38–S45.

54. Drake, V. J. (2009). Carotenoids. Retrieved from http://lpi.oregonstate.edu/infocenter/phytochemicals/carotenoids/ (accessed on 7th October 2024).

55. Durham University. (2012). Modern dog breeds genetically disconnected from ancient ancestors. *ScienceDaily*. Retrieved from http://www.sciencedaily.com/releases/2012/05/120521163845.htm (accessed on 7th October 2024).

56. Ehrlich, S. D. (2011). Milk thistle. Retrieved from http://umm.edu/health/medical/altmed/herb/milk-thistle (accessed on 7th October 2024).

57. Eisele, I. S., Wood, I. J., German, A. J., Hunter, L. J., & Trayhurn, P. (2005). Adipokine gene expression in dog adipose tissues and dog white adipocytes differentiated in primary culture. *Hormone and Metabolic Research, 37*(8), 474–481.

58. Endocrine Today. (2011). Fat linked to DNA methylation may help explain obesity-related diseases. *Endocrine Today*. Retrieved from http://www.healio.com/endocrinology/obesity/news/print/endocrinetoday/%7Bf0484521-8015-4f71-a5dd-383d0236a57d%7D/fat-linked-todna-methylation-may-help-explain-obesityrelated-diseases (accessed on 7th October 2024).

59. Environmental Working Group. (2007). 7 ways to reduce your exposure to PBDE flame retardants. Retrieved from http://www.ewg.org/enviroblog/2007/09/7-waysreduce-your-exposurepbde-flame-retardants (accessed on 7th October 2024).

60. Environmental Working Group. (2009). Dog food comparison shows high fluoride levels. Retrieved from http://www.ewg.org/research/dogfood-comparison-showshigh-fluoride-levels (accessed on 7th October 2024).

61. Environmental Working Group. (2009)(a). Dog food comparison shows high fluoride levels: Fluoride and osteosarcoma. Retrieved from http://www.ewg.org/research/dog-food-comparison-shows-high-fluoridelevels/fluoride-and-osteosarcoma (accessed on 7th October 2024).

62. Environmental Working Group. (2009)(b). Dog food contaminated with levels of fluoride above EPA's legal limit for humans. Retrieved from http://www.ewg.org/ (accessed on 7th October 2024).

63. Esteller, M. (2008). Epigenetics in cancer. *New England Journal of Medicine, 358*(11), 1148–1159. https://doi.org/10.1056/NEJMra072067

64. Evert, A. (2013). Vitamin A. Retrieved from http://www.nlm.nih.gov/medlineplus/ency/article/002400.htm (accessed on 7th October 2024).

65. Fekete, S. G., & Brown, D. L. (2007). Veterinary aspects and perspectives of nutrigenomics: A critical review. *Acta Veterinaria Hungarica, 55*(2), 229–239.

66. Fiore, K. (2013). Obesity tied to DNA regulation. Retrieved from http://www.medpagetoday.com/Genetics/GeneralGenetics/38082 (accessed on 7th October 2024).

67. Food and Agriculture Organization of the United Nations & World Health Organization (FAO/WHO). (2001, October 1-4). Health and nutritional properties of probiotics in food including powder milk with live lactic acid bacteria [Report of a joint FAO/WHO expert consultation]. Retrieved from http://www.who.int/foodsafety/publications/fs_management/en/probiotics.pdf (accessed on 7th October 2024).

68. Foster, A. P., Knowles, T. G., Hotson-Moore, A., Cousins, P. D., Day, M. J., & Hall, E. J. (2003). Serum IgE and IgG responses to food antigens in normal and atopic dogs, and dogs with gastrointestinal disease. *Veterinary Immunology and Immunopathology, 92*(1-2), 113–124.

69. Foti, S. B., & Roskams, A. J. (2011). A tale of two epiphenomena: The complex interplay of epigenetics and epilepsy. In *Underlying mechanisms of epilepsy*. https://doi.org/10.5772/24529

70. Fougère, B. (2012). Herbs for cancer [Newsletter]. *Holistic Veterinary Society of the New Zealand Veterinary Association.*

71. Franks, P. W., & Ling, C. (2010). Epigenetics and obesity: The devil is in the details. *BMC Medicine, 8*, 88. https://doi.org/10.1186/1741-7015-8-88

72. Fritsch, D., Allen, T. A., Dodd, C. E., Jewell, D. E., Sixby, K. A., Leventhal, P. S., & Hahn, K. A. (2010). A multi-center study of the effect of a therapeutic food supplemented with fish oil omega-3 fatty acids on the carprofen dosage in dogs with osteoarthritis. *Journal of the American Veterinary Medical Association, 236*(5), 535–539.

73. Gantar, M., Dhandayuthapani, S., & Rathinavelu, A. (2012). Phycocyanin induces apoptosis and enhances the effect of topotecan on prostate cell line LNCaP. *Journal of Medicinal Food, 15*(12), 1091–1095. https://doi.org/10.1089/jmf.2012.0123

74. German, A. J. (2006). The growing problem of obesity in dogs and cats. *Journal of Nutrition, 136*(7), 1940S–1946S.

75. Gilmore, T. (n.d.). NF-kB transcription factors. Retrieved from http://www.bu.edu/nf-kb/ (accessed on 7th October 2024).

76. Glickman, L. T., Raghavan, M., Knapp, D. W., Bonney, P. L., & Dawson, M. H. (2004). Herbicide exposure and the risk of transitional cell carcinoma of the urinary bladder in Scottish Terriers. *Journal of the American Veterinary Medical Association, 224*(8), 1290–1297.

77. Glickman, L. T., Schofer, F. S., McKee, L. J., Reif, J. S., & Goldschmidt, M. H. (1989). Epidemiologic study of insecticide exposures, obesity, and risk of bladder cancer in household dogs. *Journal of Toxicology and Environmental Health, 28*(4), 407–414.

78. Gosslau, A., Jao, D. L., Huang, M. T., Ho, C. T., Evans, D., Rawson, N. E., & Chen, K. Y. (2011). Effects of black tea polyphenol theaflavin -2 on apoptotic and inflammatory pathways in vitro and in vivo. *Molecular Nutrition & Food Research, 55*(5), 198–208.

79. Guilford, W. G., & Mataz, M. E. (2003). The nutritional management of gastrointestinal disorders in companion animals. *New Zealand Veterinary Journal, 51*(6), 284–291.

80. Hardy, T. M., & Tollefsbol, T. O. (2011). Epigenetic diet: Impact on the epigenome and cancer. *Epigenomics, 3*(4), 503–518. https://doi.org/10.2217/epi.11.71

81. Harvard School of Public Health (HSPH). (2013). Genes are not destiny. Retrieved from http://www.hsph.harvard.edu/obesity-prevention-source/obesity-causes/genesand-obesity/ (accessed on 7th October 2024).

82. Head, E., Rofina, J., & Zicker, S. (2008). Oxidative stress, aging and central nervous system disease in the canine model of human brain aging. *Veterinary Clinics of North America: Small Animal Practice, 38*(1), 167–178.

83. Heinze, C. R., Gomez, F. C., & Freeman, L. M. (2012). Timely topics in nutrition: Assessment of commercial diets and recipes for home-prepared diets recommended for dogs with cancer. *Journal of the American Veterinary Medical Association, 241*(11), 1453–1460.

84. Howard, B. (2012). Banfield sees bump in fat pets. *DVM360, 43*(6), 1–43. Retrieved from http://veterinarynews.dvm360.com/dvm/Veterinary+news/Banfieldseesbump-in-obese-and-overweightpets/ArticleStandard/Article/detail/775267 (accessed on 7th October 2024).

85. Howell, A. B. (2012). Health benefits of cranberries—it's not just about UTIs anymore! *Dietitians in Integrative & Functional Medicine, 14*(3), 49–51.

86. Hu, W. T., Murray, J. A., Greenaway, M. C., Parisi, J. E., & Josephs, K. A. (2006). Cognitive impairment and celiac disease. *Archives of Neurology*, 63(10), 1440–1446. https://doi.org/10.1001/archneur.63.10.1440

87. Huskisson, E. C. (2008). Glucosamine and chondroitin for osteoarthritis. *Journal of International Medical Research, 36*(6), 1161–1179.

88. Hyman, M. (2010). Glutathione: The mother of all anti-oxidants. Retrieved from http://www.huffingtonpost.com/dr-markhyman/glutathione-the-motherof_b_530494.html (accessed on 7th October 2024).

89. Hyman, M. (2011, February 8). Maximizing methylation: The key to healthy aging [Web log post]. Retrieved from http://drhyman.com/blog/2011/02/08/maximizingmethylation-the-key-tohealthy-aging-2/ (accessed on 7th October 2024).

90. Hyman, M. (2012). How hidden food sensitivities make you fat. Retrieved from http://drhyman.com/blog/2012/02/22/how-hidden-foodsensitivities-make-youfat/ (accessed on 7th October 2024).

91. Hyman, M. (2013). Three hidden ways wheat makes you fat. Retrieved from http://drhyman.com/blog/2012/02/13/three-hidden-ways-wheatmakes-you-fat/ (accessed on 7th October 2024).

92. Iwasa, M., Kobayashi, Y., Mifuji-Moroka, R., Hara, N., Miyachi, H., Sugimoto, R., … Takei, Y. (2013). Branched-chain amino acid

supplementation reduces oxidative stress and prolongs survival in rats with advanced liver cirrhosis. *PLoS ONE, 8*(7), e70309. https://doi.org/10.1371/journal.pone.0070309

93. Jeffers, J. G., Shanley, K. J., & Meyer, E. K. (1991). Diagnostic testing of dogs for food hypersensitivity. *Journal of the American Veterinary Medical Association, 198*(2), 245–250.
94. Jensen, P. (2009). *The ethology of domestic animals* (2nd ed.). Cambridge, MA: CAB International.
95. Johnston, A. N., Center, S. A., McDonough, S. P., Wakshlag, J. J., & Warner, K. L. (2013). Hepatic copper concentrations in Labrador Retrievers with and without chronic hepatitis: 72 cases (1980–2010). *Journal of the American Veterinary Medical Association, 242*(3), 372–380.
96. Joseph, J. A., Shukitt-Hale, B., Denisova, N. A., Bielinski, D., Martin, A., McEwen, J. J., & Bickford, P. C. (1999). Reversals of age-related declines in neuronal signal transduction, cognitive, and motor behavioral deficits with blueberry, spinach, or strawberry dietary supplementation. *Journal of Neuroscience, 19*(18), 8114–8121.
97. Kealy, R. D., Lawler, D. F., Ballam, J. M., Mantz, S. L., Biery, D. N., Greeley, E. H., … Stowe, H. D. (2002). Effects of diet restriction on life span and age-related changes in dogs. *Journal of the American Veterinary Medical Association, 220*(9), 1315–1320.
98. Kilpeläinen, T. O., Qi, L., Brage, S., Sharp, S. J., Sonestedt, E., Demerath, E., … Loos, R. J. F. (2011). Physical activity attenuates the influence of FTO variants on obesity risk: A meta-analysis of 218,166 adults and 19,268 children. *PLoS Medicine, 8*(11), e1001116. https://doi.org/10.1371/journal.pmed.1001116
99. Kiyono, H., Kweon, M. N., Hiroi, T., & Takahashi, I. (2001). The mucosal immune system: From specialized immune defense to inflammation and allergy. *Acta Odontologica Scandinavica, 59*(3), 145–153.
100. Kobow, K., & Blümcke, I. (2011). The methylation hypothesis: Do epigenetic chromatin modifications play a role in epileptogenesis? *Epilepsia, 52*(1), 15–19. https://doi.org/10.1111/j.1528-1167.2011.03145.x
101. Koonce, R. C., & Braverman, J. T. (2013). Obesity and osteoarthritis: More than just wear and tear. *Journal of the American Academy of Orthopaedic Surgeons, 21*(3), 161–169. https://doi.org/10.5435/JAAOS-21-03-161
102. Kraft, S. C., Rothberg, R. M., & Kramer, C. M. (1967). Gastric output and circulating anti-BSA in adults. *Clinical and Experimental Immunology, 2*(4), 321–326.

103. Kraus, M. S., Rassnick, K. M., Wakshlag, J. J., Gelzer, A. R., Waxman, A. S., Struble, A. M., & Refsal, K. R. (2014). Relation of vitamin D status to congestive heart failure and cardiovascular events in dogs. *Journal of Veterinary Internal Medicine, 28*(1), 109–115. https://doi.org/10.1111/jvim.12239

104. Kressler, C. (2012). RHR: Pioneering researcher Alessio Fasano M.D. on gluten, autoimmunity & leaky gut. Retrieved from http://chriskresser.com/pioneering-researcher-alessio-fasano-m-d-on-gluten-autoimmunity-leaky-gut (accessed on 7th October 2024).

105. Krikorian, R., Shidler, M. D., Nash, T. A., Kalt, W., Vinqvist-Tymchuk, M. R., Shukitt-Hale, B., & Joseph, J. A. (2010). Blueberry supplementation improves memory in older adults. *Journal of Agricultural and Food Chemistry, 58*(7), 3996–4000.

106. Laflamme, D. P. (2004). Osteoarthritis and diet: Joined at the hip. *Nestle Purina Clinical Edge*, 10–12.

107. Laflamme, D. P. (2012). Nutritional care for aging cats and dogs. *Veterinary Clinics of North America: Small Animal Practice, 42*(4), 769–791.

108. Landsberg, G. M., DePorter, T., & Araujo, J. A. (2011). Clinical signs and management of anxiety, sleeplessness, and cognitive dysfunction in the senior pet. *Veterinary Clinics of North America: Small Animal Practice, 41*(3), 565–590.

109. Lavebratt, C., Almgren, M., & Ekström, T. J. (2012). Review: Epigenetic regulation in obesity. *International Journal of Obesity, 36*(6), 757–765. https://doi.org/10.1038/ijo.2011.178

110. Lawenda, B. D. (2013). Is there any other anti-cancer botanical compound as exciting as curcumin? Retrieved from http://www.integrativeoncologyessentials.com/2013/03/is-there-any-other-anti-cancer-botanicalcompound-as-exciting-ascurcumin/ (accessed on 7th October 2024).

111. Lawler, D. F., Larson, B. T., Ballam, J. M., Smith, G. K., Biery, D. N., Evans, R. H., … Kealy, R. D. (2007). Diet restriction and aging in the dog: Major observations over two decades. *British Journal of Nutrition, 98*(3), 497–503.

112. Lee, Y. H., & Wong, D. T. (2009). Saliva: An emerging biofluid for early detection of disease. *American Journal of Dentistry, 22*(4), 421–428.

113. Lefebvre, S. L., Yang, M., Wang, M., Elliott, D. A., Buff, P. R., & Lund, E. M. (2013). Effect of age at gonadectomy on the probability of dogs becoming overweight. *Journal of the American Veterinary Medical Association, 243*(2), 236–243.

114. Li, G.-R., Wang, H.-B., Qin, G.-W., Jin, M.-W., Tang, Q., & Sun, H.-Y. (2013). Acaetin, a natural flavone, selectively inhibits human atrial repolarization potassium currents and prevents atrial fibrillation in dogs. *Circulation, 117*(21), 2449–2457.
115. Li, W. (2010). Can we eat to starve cancer? [Video]. Retrieved from http://www.youtube.com/watch?v=B9bDZ5-zPtY (accessed on 7th October 2024).
116. Lieber, C. S. (2002). S-adenosyl-L-methionine: Its role in the treatment of liver disorders. *American Journal of Clinical Nutrition, 76*(5), 1183S–1187S.
117. Lila, M. A. (2004). Anthocyanins and human health: An in vitro investigative approach. *Journal of Biomedical Biotechnology, 2004*(5), 306–313.
118. Lim, E. L., Hollingsworth, K. G., Aribisala, B. S., Chen, M. J., Mathers, J. C., & Taylor, R. (2011). Reversal of type 2 diabetes: Normalisation of beta cell function in association with decreased pancreas and liver triacylglycerol. *Diabetologia, 54*(10), 2506–2514. https://doi.org/10.1007/s00125-011-2204-7
119. Link, A., Balaguer, F., Shen, Y., Lozano, J. J., Leung, H. C., Boland, C. R., & Goel, A. (2013). Curcumin modulates DNA methylation in colorectal cancer cells. *PLoS One, 8*(2), e57709. https://doi.org/10.1371/journal.pone.0057709
120. Linus Pauling Institute. (2013). The bioavailability of different forms of vitamin C (ascorbic acid). Retrieved from http://lpi.oregonstate.edu/infocenter/vitamins/vitaminC/vitCform.html (accessed on 7th October 2024).
121. Lippiello, L., Nardo, J. V., Harlan, R., & Chiou, T. (2008). Metabolic effects of avocado/soy unsaponifiables on articular chondrocytes. *Evidence-Based Complementary and Alternative Medicine, 5*(2), 191–197. https://doi.org/10.1093/ecam/nem132
122. Logar, D. B., Komadina, R., Prezelj, J., Ostanek, B., Trošt, Z., & Marc, J. (2007). Expression of bone resorption genes in osteoarthritis and in osteoporosis. *Journal of Bone and Mineral Metabolism, 25*(4), 219–225.
123. Lu, P., Mamiya, T., Lu, L. L., Mouri, A., Zou, L. B., Nagai, T., ... Nabeshima, T. (2009). Silibinin prevents amyloid β peptide-induced memory impairment and oxidative stress in mice. *British Journal of Pharmacology, 157*(7), 1270–1277.
124. Magee, E. (2007). The super-veggies: Cruciferous vegetables. Retrieved from http://www.webmd.com/food-recipes/features/super-veggies-cruciferous-vegetables (accessed on 7th October 2024).

125. Manteca, X. (2011). Nutrition and behavior in senior dogs. *Topics in Companion Animal Medicine, 26*(1), 33–36.

126. Marchesini, G., Moscatiello, S., Di Domizio, S., & Forlani, G. (2008). Obesity-associated liver disease. *The Journal of Clinical Endocrinology & Metabolism, 93*(11), s74–s80. https://doi.org/10.1210/jc.2008-1399

127. Mason, S. (2013). UCLA researchers find link between intestinal bacteria and white blood cell cancer. Retrieved from http://newsroom.ucla.edu/portal/ucla/ucla-researchers-find-link-between-245945.aspx (accessed on 7th October 2024).

128. Medical College of Georgia at Georgia Regents University. (2013, March 21). Scientists identify gene that is consistently altered in obese individuals. *ScienceDaily*. Retrieved from http://www.sciencedaily.com/releases/2013/03/130321133114.htm (accessed on 7th October 2024).

129. MedlinePlus, U.S. National Library of Medicine. (2012). Fish oil. Retrieved from http://www.nlm.nih.gov/medlineplus/druginfo/natural/993.html (accessed on 7th October 2024).

130. MedlinePlus. (2013). Vitamin D. Retrieved from http://www.nlm.nih.gov/medlineplus/druginfo/natural/929.html (accessed on 7th October 2024).

131. Mercola, J. (2010). Is this the new silver bullet for cancer? Retrieved from http://articles.mercola.com/sites/articles/archive/2010/03/11/is-vitamin-d-the-silver-bullet-for-cancer.aspx (accessed on 7th October 2024).

132. Mercola, J. (2012). A daily 900 mg dose of this fat helped reverse memory loss. Retrieved from http://articles.mercola.com/sites/articles/archive/2012/02/06/without-krill-oil-your-brain-could-degenerate.aspx (accessed on 7th October 2024).

133. Mercola, J., & Hofmekler, L. (2010). This one antioxidant keeps all other antioxidants performing at peak levels. Retrieved from http://articles.mercola.com/sites/articles/archive/2010/04/10/can-you-use-food-to-increase-glutathione-instead-of-supplements.aspx (accessed on 7th October 2024).

134. Mercola.com. (2012). This new drug appears to cause cancer cells to self-destruct. Retrieved from http://articles.mercola.com/sites/articles/archive/2012/08/04/dca-and-turmeric-on-cancer.aspx (accessed on 7th October 2024).

135. Messonnier, S. (2006). *The natural vet's guide to preventing and treating cancer in dogs*. Novato, CA: New World Library.

136. Messonnier, S. (2007). Liver disease. Retrieved from http://www.petcarenaturally.com/articles/liver-disease.php (accessed on 7th October 2024).

137. Middleton, R. P., & Hannah, S. S. (2010). Osteoarthritis and its origins: Disease development at the cellular and molecular level. *Nestle Purina Clinical Edge*, 6–9.
138. Milgram, N. W., Zicker, S. C., Head, E., Muggenburg, B. A., Murphey, H., Ikeda-Douglas, C. J., & Cotman, C. W. (2002). Dietary enrichment counteracts age-associated cognitive dysfunction in canines. *Neurobiology of Aging, 23*(5), 737–745.
139. Miller, C. S., Foley, J. D., Bailey, A. L., Campbell, C. L., Humphries, R. L., Christodoulides, N., & McDevitt, J. T. (2010). Current developments in salivary diagnostics. *Biomarker Medicine, 4*(2), 171–189.
140. Moreau, M., Dupuis, J., Bonneau, N. H., & Lécuyer, M. (2004). Clinical evaluation of a powder of quality elk velvet antler for the treatment of osteoarthrosis in dogs. *The Canadian Veterinary Journal, 45*(2), 133–139.
141. Moreau, M., Troncy, E., Del Castillo, J. R., Bédard, C., Gauvin, D., & Lussier, B. (2012). Effects of feeding a high omega-3 fatty acids diet in dogs with naturally occurring osteoarthritis. *Journal of Animal Physiology and Animal Nutrition, 97*(5), 830–837. https://doi.org/10.1111/j.1439-0396.2012.01325.x
142. Murata, N., Murakami, K., Ozawa, Y., Kinoshita, N., Irie, K., Shirasawa, T., & Shimizu, T. (2010). Silymarin attenuated the amyloid β plaque burden and improved behavioral abnormalities in an Alzheimer's disease mouse model. *Bioscience, Biotechnology, and Biochemistry, 74*(11), 2299–2306.
143. Murray, L. (2012, August 28). [Personal correspondence].
144. Nahleh, Z., Bhattie, N. S., & Mal, M. (2011). How to reduce your cancer risk: Mechanisms and myths. *International Journal of General Medicine, 4*, 277–287. https://doi.org/10.2147/IJGM.S18657.
145. Narayanapillai, S., Agarwal, C., Deep, G., & Agarwal, R. (2013). Silibinin inhibits ultraviolet B radiation-induced DNA damage and apoptosis by enhancing interleukin-12 expression in JB6 cells and SKH-1 hairless mouse skin. *Molecular Carcinogenesis*. https://doi.org/10.1002/mc.22000.
146. National Kidney and Urologic Diseases Information Clearinghouse (NKUDIC), National Institute of Diabetes and Digestive and Kidney Diseases (NIDDK), U.S. Department of Health and Human Services. (2012). The kidneys and how they work. Retrieved from http://kidney.niddk.nih.gov/kudiseases/pubs/yourkidneys/ (accessed on 7th October 2024).
147. National Library of Medicine. (2013). Pyridoxine-dependent epilepsy. Retrieved from http://ghr.nlm.nih.gov/condition/pyridoxine-dependent-epilepsy (accessed on 7th October 2024).

148. National Research Council of the National Academy of Sciences. (2006). *Nutrient requirements of dogs and cats*. Washington, DC: NRC Press.

149. Neto, C. C. (2011). Cranberries: Ripe for more cancer research. *Journal of the Science of Food and Agriculture, 91*(13), 2303–2307.

150. Norris, S. (2013). You are what you eat—Nutrigenomics and obesity. *Veterinary Nursing Journal, 28*(6), 190–191. https://doi.org/10.1111/vnj.2013.28.issue-6

151. O'Grady, M. (2002). DCM in Doberman Pinschers: Lessons learned in the first decade of study. *ACVIM Abstr. 2002*. Retrieved from www.DobermanData.com (accessed on 7th October 2024).

152. Oates, L. (2009). Medicinal mushrooms. *CM: The Journal of Complementary Medicine, 8*(3), 44–48.

153. Paige, C. (2007). *The good behavior book for dogs: The most annoying dog behaviors... solved!*. Beverly, MA: Quarry Books.

154. Pan, Y., Larson, B., Araujo, J. A., Lau, W., de Rivera, C., Santana, R., ... Milgram, N. W. (2010). Dietary supplementation with medium-chain TAG has long-lasting cognition-enhancing effects in aged dogs. *British Journal of Nutrition, 103*(12), 1746–1754. https://doi.org/10.1017/S0007114509993866

155. Perlmutter, D. (n.d.). New study links gluten sensitivity to brain failure [Web log post]. Retrieved from http://www.vanguardneurologist.com/new-study-links-gluten-sensitivity-to-brain-failure/ (accessed on 7th October 2024).

156. Perricone, N. (2010). *Forever young: The science of nutrigenomics for glowing, wrinkle-free skin and radiant health at every age*. New York: Atria Books.

157. Pollack, P. (2013). Is there a systemic link between obesity and OA? Retrieved from http://www.aaos.org/news/aaosnow/mar13/clinical10.asp (accessed on 7th October 2024).

158. Pollard, B., Guilford, W. G., Ankenbauer-Perkins, K. L., & Hedderley, D. (2006). Clinical efficacy and tolerance of an extract of green-lipped mussel (*Perna canaliculus*) in dogs presumptively diagnosed with degenerative joint disease. *New Zealand Veterinary Journal, 54*(3), 114–118.

159. Pons, L. (2006). Pterostilbene's healthy potential: Berry compound may inhibit breast cancer and heart disease. Retrieved from http://www.ars.usda.gov/is/AR/archive/nov06/health1106.htm (accessed on 7th October 2024).

160. Puotinen, C. J. (2010). Alternatives to canine surgery. *Whole Dog Journal, 13*(2), 13–23.

161. Raghavan, M., Knapp, D. W., Bonney, P. L., Dawson, M. H., & Glickman, L. T. (2005). Evaluation of the effect of dietary vegetable consumption on reducing the risk of transitional cell carcinoma of the urinary bladder in Scottish Terriers. *Journal of the American Veterinary Medical Association, 227*(1), 94–100.

162. Raghavan, M., Knapp, D. W., Dawson, M. H., Bonney, P. L., & Glickman, L. T. (2004). Topical flea and tick pesticides and the risk of transitional cell carcinoma of the urinary bladder in Scottish Terriers. *Journal of the American Veterinary Medical Association, 225*(3), 389–394.

163. Rakyan, V. K., Beyan, H., Down, T. A., Hawa, M. I., Maslau, S., Aden, D., ... Leslie, R. D. (2011). Identification of type 1 diabetes–associated DNA methylation variable positions that precede disease diagnosis. *PLoS Genetics, 7*(9), e1002300. https://doi.org/10.1371/journal.pgen.1002300

164. Reddy, C. M., Bhat, V. B., Kiranmai, G., Reddy, M. N., Reddanna, P., & Madyastha, K. M. (2000). Selective inhibition of cyclooxygenase-2 by C-phycocyanin, a biliprotein from *Spirulina platensis*. *Biochemical and Biophysical Research Communications, 277*(3), 599–603.

165. Rialland, P., Bichot, S., Lussier, B., Moreau, M., Beaudry, F., Del Castillo, J. R., & Troncy, E. (2013). Effect of a diet enriched with green-lipped mussel on pain behavior and functioning in dogs with clinical osteoarthritis. *Canadian Journal of Veterinary Research, 77*(1), 66–74.

166. Ricci, R., Granato, A., Vascellari, M., Boscarato, M., Palagiano, C., Andrighetto, I., ... Mutinelli, F. (2013). Identification of undeclared sources of animal origin in canine dry foods used in dietary elimination trials. *Journal of Animal Physiology and Animal Nutrition, 97*(1), 32–38.

167. Rinkinen, M., Teppo, A. M., Harmoinen, J., & Westermark, E. (2003). Relationship between canine mucosal and serum immunoglobulin A (IgA) concentrations: Serum IgA does not assess duodenal secretory IgA. *Microbiology and Immunology, 47*(2), 155 159.

168. Robinson, L. E., & Reeves, S. (2013). Review of sIgA's major role as a first line of immune defense and new indications regarding inflammation and gut health. Retrieved from http://www.embriahealth.com (accessed on 7th October 2024).

169. Robinson, N. G. (2008). Why the buzz about turmeric? Retrieved from http://csuvets.colostate.edu/pain/Articlespdf/TheBuzzAboutTurmeric.pdf (accessed on 7th October 2024).

170. Romay, C. H., González, R., Ledón, N., Remirez, D., & Rimbau, V. (2003). C-phycocyanin: A biliprotein with antioxidant, anti-inflammatory and neuroprotective effects. *Current Protein and Peptide Science, 4*(3), 207–216.

171. Roudebush, P., Schoenherr, W. D., & Delaney, S. J. (2008). Timely topics in nutrition: An evidence-based review of the use of nutraceuticals and dietary supplementation for the management of obese and overweight pets. *Journal of the American Veterinary Medical Association, 232*(11), 1646–1655.

172. Roush, J. K., Cross, A. R., Renberg, W. C., Dodd, C. E., Sixby, K. A., Fritsch, D. A., ... Hahn, K. A. (2010). Effects of dietary supplementation with fish oil omega-3 fatty acids on weight-bearing in dogs with osteoarthritis. *Journal of the American Veterinary Medical Association, 236*(1), 67–73.

173. Roush, J. K., Dodd, C. E., Fritsch, D. A., Allen, T. A., Jewell, D. E., Schoenherr, W. D., ... Hahn, K. A. (2010). A multi-center veterinary practice assessment of the effects of omega-3 fatty acids on osteoarthritis in dogs. *Journal of the American Veterinary Medical Association, 236*(1), 59–66.

174. Sanchez, C., Deberg, M. A., Bellahcène, A., Castronovo, V., Msika, P., Delcour, J. P., ... Henrotin, Y. E. (2008). Phenotypic characterization of osteoblasts from the sclerotic zones of osteoarthritic subchondral bone. *Arthritis and Rheumatism, 58*(2), 442–455. https://doi.org/10.1002/art.23159

175. Sanghi, D., Avasthi, S., Srivastava, R. N., & Singh, A. (2009). Nutritional factors and osteoarthritis: A review article. *Internet Journal of Medical Update, 4*(1), 42–53.

176. Schachter, S. C. (2006). Nutritional deficiencies. Retrieved from http://www.epilepsy.com/epilepsy/provoke_nutrition (accessed on 7th October 2024).

177. Schaffer, J. E. (2003). Lipotoxicity: When tissues overeat. *Current Opinion in Lipidology, 14*(3), 281–287.

178. Serini, S., Fasano, E., Piccioni, E., Cittadini, A. R. M., & Calviello, G. (2012). Apoptosis as a mechanism involved in the anticancer effect of dietary n-3 polyunsaturated fatty acids. In *Novel Apoptotic Regulators in Carcinogenesis*. Dordrecht, Netherlands: Springer Science+Business Media.

179. Shanmuganayagam, D., Beahm, M. R., Osman, H. E., Krueger, C. G., Reed, J. D., & Folts, J. D. (2002). Grape seed and grape skin extracts elicit a greater antiplatelet effect when used in combination than when used individually in dogs and humans. *Journal of Nutrition, 132*(12), 3592–3598.

180. Sjøgren, K. (2013). Seven new genetic causes of obesity identified. Retrieved from http://sciencenordic.com/seven-new-genetic-causes-obesity-identified (accessed on 7th October 2024).

181. Smith, J. E., Rowan, N. J., & Sullivan, R. (2002). Medicinal mushrooms: A rapidly developing area of biotechnology for cancer therapy and other bioactivities. *Biotechnology Letters, 24*(22), 1839–1845.

182. Smith, L. (2009). Canine liver disease foundation. Retrieved from http://canineliverdiseasefoundation.org/?cat=5 (accessed on 7th October 2024).

183. Stafstrom, C. E. (2004). Dietary approaches to epilepsy treatment: Old and new options on the menu. *Epilepsy Currents, 4*(6), 215–222. https://doi.org/10.1111/j.1535-7597.2004.46001.x

184. Straus, M. (2007). Canine arthritis treatment: Many effective tools can help in the fight against canine arthritis. *Whole Dog Journal, 10*(3).

185. Straus, M. (2012). How prebiotics improve your dog's digestion and nutrient absorption: Special ingredients support the friendly bacteria that aid digestive health. *Whole Dog Journal*. Retrieved from http://www.whole-dog-journal.com/issues/15_4/features/Prebiotics-and-Canine-Digestive-Health_20496-1.html (accessed on 7th October 2024).

186. Straus, M. (2012). Senior supplements: These nutraceuticals may offer hope for treating canine cognitive dysfunction. *Whole Dog Journal, 15*(11).

187. Straus, M. (2013). Dishing on diets: Veterinary nutritionists conclude that only they can properly formulate a homemade diet. *Whole Dog Journal, 16*(9), 3–5.

188. Swanson, K. S. (2007). Using nutritional genomics to study canine obesity and diabetes. Retrieved from http://en.engormix.com (accessed on 7th October 2024).

189. Swanson, K. S., Grieshop, C. M., Flickinger, E. A., Bauer, L. L., Healy, H. P., Dawson, K. A., ... Fahey Jr., G. C. (2002). Supplemental fructooligosaccharides and mannanoligosaccharides influence immune function, ileal and total tract nutrient digestibilities, microbial populations and concentrations of protein catabolites in the large bowel of dogs. *Journal of Nutrition, 132*(5), 980–989.

190. Swanson, K. S., Schook, L. B., & Fahey, G. C. (2003). Nutritional genomics: Implications for companion animals. *Journal of Nutrition, 133*(10), 3033–3040.

191. Therkleson, T. (2013). Topical ginger treatment with a compress or patch for osteoarthritis symptoms. *Journal of Holistic Nursing*. [Epub ahead of print]. https://doi.org/10.1177/0898010113514063

192. Thomas, W. B. (2011). Does the ketogenic diet work for dogs? Retrieved from http://www.canine-epilepsy.com/KetogenicDiet.html (accessed on 7th October 2024).

193. Tizard, I., & Ni, Y. (1998). Use of serologic testing to assess immune status of companion animals. *Journal of the American Veterinary Medical Association, 213*, 54–60.

194. Tremayne, J. (2010). Using supplements to fight cancer. *Veterinary Practice News, 22*(7).

195. Twark, L., & Dodds, W. J. (2000). Clinical application of serum parvovirus and distemper virus antibody titers for determining revaccination strategies in healthy dogs. *Journal of the American Veterinary Medical Association, 217*(10), 1021–1024.

196. Tyagi, A., Agarwal, C., Dwyer-Nield, L. D., Singh, R. P., Malkinson, A. M., & Agarwal, R. (2011). Silibinin modulates TNF-α and IFN-γ mediated signaling to regulate COX2 and iNOS expression in tumorigenic mouse lung epithelial LM2 cells. *Molecular Carcinogenesis*. https://doi.org/10.1002/mc.20851

197. Tyagi, A., Singh, A., Bhardwaj, P., Sahu, S., Yadav, A. P., & Kori, M. L. (2012). Punicalagins – A large polyphenol compound found in pomegranates: A therapeutic review. *Academic Journal of Plant Sciences, 5*(2), 45–49.

198. U.S. Food and Drug Administration (USFDA). (2004). Letter updating the phosphatidylserine and cognitive function and dementia qualified health claim [Public letter]. Retrieved from http://www.fda.gov/food/ingredientspackaginglabeling/labelingnutrition/ucm072993.htm (accessed on 7th October 2024).

199. University of Colorado Denver. (2011). Milk thistle extract stops lung cancer in mice, study shows. Retrieved from http://www.sciencedaily.com/releases/2011/11/111115145236.htm (accessed on 7th October 2024).

200. VCA Animal Hospitals. (n.d.). Glycosaminoglycans. Retrieved from http://www.vcahospitals.com/main/pet-health-information/article/animal-health/glycosaminoglycans-gags/468 (accessed on 7th October 2024).

201. Veerman, J. L. (2011). On the futility of screening for genes that make you fat. *PloS Medicine, 8*(11), e1001114. https://doi.org/10.1371/journal.pmed.1001114

202. Venier, M., & Hites, R. A. (2011). Flame retardants in the serum of pet dogs and in their food. *Environmental Science & Technology, 45*(10), 4602–4608. https://doi.org/10.1021/es1043529

203. VetInfo. (2012). Fatty liver disease in dogs. Retrieved from http://www.vetinfo.com/fatty-liver-disease-in-dogs.html#b (accessed on 7th October 2024).

204. Vojdani, A. (2009). Detection of IgE, IgG, IgA and IgM antibodies against raw and processed food antigens. *Nutrition and Metabolism, 6*(22), 1–15.

205. Waldron, M. (2004). The role of fatty acids in the management of osteoarthritis. *Nestle Purina Clinical Edge*, 14–16.

206. Walker, A. K. (2013, July 24). Obesity can lead to liver disease. *The Baltimore Sun*. Retrieved from http://www.baltimoresun.com/ (accessed on 7th October 2024).

207. Ward, E. (2007). Weight reduction in dogs – General information. Retrieved from http://www.petobesityprevention.com/wp-content/uploads/2010/05/Weight_Reduction_in_Dogs_General_Information (accessed on 7th October 2024).

208. Wasser, S. P. (2011). Current findings, future trends, and unsolved problems in studies of medicinal mushrooms. *Applied Microbiology and Biotechnology, 89*(5), 1323–1332. https://doi.org/10.1007/s00253-010-3067-4

209. WebMD. (2012). Vitamins and supplements lifestyle guide: Vitamin B-12 (Cobalamin). Retrieved from http://www.webmd.com/vitamins-and-supplements/lifestyleguide-11/supplement-guide-vitamin-b12 (accessed on 7th October 2024).

210. Weese, J. S., & Arroyo, L. (2003). Bacteriological evaluation of dog and cat diets that claim to contain probiotics. *Canadian Veterinary Journal, 44*(3), 212–215.

211. Weil, A. (2013). Condition care guide: Diarrhea. Retrieved from http://www.drweil.com/drw/u/ART00344/diarrhea.html (accessed on 7th October 2024).

212. Wein, H. (2012). Understanding insulin sensitivity and diabetes. Retrieved from http://www.nih.gov/researchmatters/april2012/04092012insulin.htm (accessed on 7th October 2024).

213. Wilken, R., Veena, M. S., Marilene, B., Wang, M. B., & Srivatsan, E. S. (2011). Curcumin: A review of anti-cancer properties and therapeutic activity in head and neck squamous cell carcinoma. *Molecular Cancer, 10*(12). https://doi.org/10.1186/1476-4598-10-12

214. Wilson-Robles, H. (2013). Second-hand smoke and your pet [Web log post]. Retrieved from http://vetmed.tamu.edu/news/pet-talk/second-hand-smoke-and-your-pet#.UgqtgRaAEII (accessed on 7th October 2024).

215. Wilson, L. (2013). Epilepsy and seizures. Retrieved from http://drlwilson.com/Articles/epilepsy.htm (accessed on 7th October 2024).

216. Wolf, D. (2009). *Superfoods: The food and medicine of the future.* Berkeley, CA: North Atlantic Books.

217. Wong, M. (2013, December 4). [Personal correspondence].

218. Woodruff Health Sciences. (2012). Immune system compensates for 'leaky gut' in IBD susceptibility. Retrieved from http://news.emory.edu/stories/2012/09/immune_system_compensates_for_leaky_gut/ (accessed on 7th October 2024).

219. Yamka, R. M., Frantz, N. Z., & Friesen, K. G. (2007). Effects of 3 canine weight loss foods on body composition and obesity markers. *International Journal of Applied Research in Veterinary Medicine, 5*(3), 125–132.

220. Yamka, R. M., Friesen, K. G., Gao, X., & Al-Murrani, S. (2007). Identification of genes related to obesity in dogs. *The FASEB Journal, 21*(28), 4.

221. Yamka, R. M., Harmon, D. L., Schoenherr, W. D., Khoo, C., & Gross, K. L. (2002). Supplemental fructooligosaccharides and mannanoligosaccharides influence immune function, ileal and total tract nutrient digestibilities, microbial populations and concentrations of protein catabolites in the large bowel of dogs. *Journal of Nutrition, 132*(5), 980–989.

222. Yuen, A. W., Sander, J. W., Fluegel, D., Patsalos, P. N., Bell, G. S., Johnson, T., & Koepp, M. J. (2005). Omega-3 fatty acid supplementation in patients with chronic epilepsy: A randomized trial. *Epilepsy Behavior, 7*(2), 253–258.

223. Yurko-Mauro, K., McCarthy, D., Rom, D., Nelson, E. B., Ryan, A. S., Blackwell, A., & Stedman, M. (2010). Beneficial effects of docosahexaenoic acid on cognition in age-related cognitive decline. *Alzheimer's & Dementia: The Journal of the Alzheimer's Association, 6*(6), 456–464. https://doi.org/10.1016/j.jalz.2010.01.013

224. Zhang, R., Fang, H., Chen, Y., Shen, J., Lu, H., Zeng, C., ... Zhao, Q. (2012). Gene expression analyses of subchondral bone in early experimental osteoarthritis by microarray. *PLoS ONE, 7*(2), e32356. https://doi.org/10.1371/journal.pone.0032356

225. Zicker, S. C., Jewell, D. E., Yamka, R. M., & Milgram, N. W. (2012). Evaluation of cognitive learning, memory, psychomotor, immunologic, and retinal functions in healthy puppies fed foods fortified with docosahexaenoic acid-rich fish oil from 8 to 52 weeks of age. *Journal of the American Veterinary Medical Association, 241*(5), 583–594. https://doi.org/10.2460/javma.241.5.583

*Chapter* 8

# NUTRITIONAL MANAGEMENT OF EQUINE DISEASES

## Contents

# INTRODUCTION

Nutrigenomics, the study of how nutrition influences gene expression, offers a promising approach to managing equine diseases. By understanding the interaction between an individual horse's genetic makeup and its diet, nutrigenomics allows for the development of targeted nutritional strategies aimed at preventing and managing a range of diseases common in horses. This personalized approach to nutrition is especially valuable in equine medicine, where traditional management practices often rely on generalized dietary recommendations that may not address specific genetic predispositions.

Horses are prone to various diseases, including metabolic disorders like equine metabolic syndrome (EMS), laminitis, and gastric ulcers. Nutrigenomics provides insights into how specific nutrients can influence the expression of genes related to these conditions, offering new avenues for prevention and treatment. For example, horses with a genetic predisposition to EMS may benefit from a diet low in non-structural carbohydrates, which can help regulate insulin sensitivity and reduce the risk of laminitis. Similarly, nutrigenomic approaches can be used to formulate diets that support gut health, potentially reducing the incidence of ulcers by modulating the expression of genes involved in the digestive process.

The integration of nutrigenomics into equine disease management also has implications for improving overall horse health and performance. By tailoring diets to an individual horse's genetic profile, it is possible to enhance nutrient absorption, optimize energy metabolism, and support immune function, all of which contribute to better health outcomes. As research in equine nutrigenomics continues to evolve, it holds the potential to revolutionize the way we approach the prevention and management of diseases in horses, leading to more effective and personalized care.

## 8.1. MINIATURE HORSES AND PONIES

The feeding of miniature horses and ponies—two of the tiniest members of the genus, species, and subspecies *Equus Ferus* caballus—will be covered in this chapter. With the obvious exception that because they are smaller, they should be fed similarly to light breed horses and ponies in terms of total nutrients on a body weight basis. Both pony and tiny breeds may benefit from many of the standard feeding and husbandry techniques used with other breeds.

The American Horse Association recognized the American Miniature horse as a single breed in the late 1970s, but many miniature horse breeders believe there are multiple distinct breeds of the breed (e.g., Dartmoor Pony, Micro Mini, Australian Miniature Pony, Falabella, Miniature Toy Horse, and so forth).

These breeds can be linked to seventeenth-century European royal families. These horses are currently kept as service, show, and pet animals. Miniature horses are defined as being under 97 cm (38 in) in height at the withers and typically live 25–35 years; it is likely that all of them carry some dwarfism-related genes. Numerous miniature horses that are not used as guides can weigh up to 90 kg, but those used as service animals should weigh between 24 and 45 kg (55 and 100 lbs) and be no taller than 66 cm (26 in).

### 8.1.1. General Feeding of Miniature Horses

In contrast to ponies, the dietary needs of miniature horses have not been the subject of many rigorous scientific investigations. Because they are an "easy keeper" breed, minis shouldn't be overfed in order to avoid obesity. Similar to other horse breeds, maintenance energy requirements are typically satisfied by feeding 1–2 kg of high-quality dry forage or 1–8% of the body weight in dry matter derived from good-quality forages each day. Miniature horses can be used to graze or "mow" lawns, or they can be fed small amounts of concentrates and grass or hay. Body condition score (BCS) should be used to determine when to add supplemental grain to the diet. The body condition scoring guidelines for miniature horses are the same as those for other horse breeds. Maintaining a BCS of 5–6/9, where ribs are easily palpable but not visible, and where there are no noticeable fat deposits on the neck, shoulders, withers, or base of the tail, is ideal for miniature horses. The body condition score should be used to assess whether or not energy intake is adequate. Caretakers should think about improving the quantity and/or quality of forages or gradually introducing a small amount (0.25–0.5 lb/day) of concentrate, when BCS is less than 5/9. Overestimating body weight and underestimating concentrate offered are common errors made when concentrate feeding. Both errors carry the risk of obesity and perhaps one of several metabolic disorders (e.g., hyperlipemia, laminitis, equine metabolic syndrome, etc. observed in obese small horses. It is advisable to advise owners of miniature horses to acquire scales so they can accurately weigh their feed and prevent overfeeding. Fishermen use inexpensive, easily obtainable scales to weigh their catch. For other equines as well, having access to clean, fresh water is essential to ensuring proper feed intake, reducing the risk of colic, and maintaining general health. The general recommendations for the percentage of diet that should be consumed in water, energy, protein, minerals, and vitamins depend on factors like age, growth, and production status (e.g., They are comparable to other breeds and use miniature horses (early, mid, late gestation, or lactation). Although they may be more vulnerable to enteroliths and hyperlipemia, miniature horses are prone to many of the nutritionally related conditions observed in other breeds.

## 8.1.2. Pony Feeding

Horses under 147 cm (14.2 hands, or 58 in) at the withers are considered ponies. In comparison to other horse breeds, pony breeds usually have shorter heads with wider foreheads, thicker necks, wider barrels, and shorter legs. Pony breeds are utilized as working, show, and pet animals. Numerous unique ponies with diverse applications exist, which aid in identifying appropriate feeding regimens (e.g., carriage horses as opposed to hunter/jumper horses).

Ponies' availability and size have made them useful for numerous equine nutrition research studies. As a result, there is a wealth of information available that addresses pony nutrition specifically. For ponies, feeding guidelines designed for other light breeds typically apply. By the time they are 12 months old, many pony breeds will have reached 75% of their mature weight, whereas thoroughbred horses will only have reached roughly 69% of their mature weight. Because of their quicker growth rate than other breeds, ponies require different feeding procedures. Most pony breeds are easier to care for than other horse breeds because they were chosen and developed in environments with rough terrain, sparse or poor-quality pasture.

Most ponies, with the possible exception of those that are growing, working, and nursing, will seldom need concentrates and will quickly gain weight. Numerous metabolic disorders, including hyperlipemia and equine metabolic syndrome, are more common in ponies. In Shetland ponies, increased fat supplementation with soybean oil at 10% of dry matter intake was linked to glucose intolerance. In comparison to other horse breeds, ponies seem to have a higher voluntary intake.

In a particular study, ponies were fed 30.9 kg of alfalfa hay for every 100 kg of body weight, or 30.9% of their body weight in dry matter. Caretakers ought to exercise caution and reserve the use of high-energy foods for specific situations. To reduce obesity, the BCS should be continuously monitored when a concentrate or concentrates are provided.

Regarding small equines, ponies ought to be fed premium fodder, accounting for 1.0–1.8 percent of their body weight in dry matter each day. As with other breeds, body condition scores for ponies should be estimated, and dietary adjustments should be made to keep ideal body condition levels around 5–6/9. Horses kept at maintenance (not gaining weight, not losing it) can typically get by on grass pastures or hay alone, but horses used for light work may need a concentrate that makes up 20% of their diet. Ponies should only be denied food under strict supervision and when medically necessary. Hyperlipemia is the result of prolonged low energy intake, which is made worse by underlying medical conditions like obesity, pregnancy, and/or illness.

## 8.2. THE ASSOCIATION BETWEEN NUTRITION AND COLIC

Horses are meant to be grazing animals that are continuously fed high-fiber, low-carbohydrate diets, with the fermentation of the hindgut facilitating this process. The digestive system of horses was evolved to be refined for the digestion of fiber; modern horse husbandry either ignores this fact or is unable to provide a lifestyle for the needs of the digestive system. For the majority of modern horses, stall confinement, intermittent feedings, and large boluses of cereal-based feeds are the norm. Although horses can adapt to some degree, it is not surprising that these adaptations occasionally fail to allow horses to coexist with humans. Although gastrointestinal pain is frequently linked to nutrition, the multifactorial nature of the issue—which includes feed types, feed quality, and differences in feeding techniques—makes it challenging to determine the precise epidemiological contribution of diet to colic.

In veterinary medicine, low record keeping, small group sizes, and budgetary constraints frequently make it difficult to determine the cause and effect of a condition. But since the information we learn will directly affect the treatments offered, it is crucial that we approach all of it objectively and critically. Integrating objective research and clinical knowledge with a grading system for published literature is known as evidence-based medicine. It recognizes that not all evidence is created equal and that applying research to clinical patients requires careful thought. To provide statistical evidence linking nutrition and colic, only studies with a level of evidence of grade 2 or higher were included in this chapter. The veterinary professional is advised to consider the advantages and disadvantages of the published literature when evaluating the evidence so that they can base their decisions only on the most important data. An analysis of the data pertaining to the most popular feeds given to horses, such as concentrates, dried forages, and grasses, is the first step in connecting feeds to colic.

### 8.2.1. Feeds and Colic: Pastures

The horse's evolutionary purpose was to feed on grasses. In particular, the large colon evolved to ferment these grasses into energy-producing short chain fatty acids (acetate, propionate, and butyrate). Even though grasses can store a lot of carbohydrates in the form of fructans and starches, which could disturb the delicate microbial populations, constant grazing should enable the bacteria to adapt to changes in carbohydrate content gradually. On the other hand, colic and laminitis have been linked to pastures with high levels of fermentable carbohydrates, which may account for the apparent seasonality of colic episodes. Although grazing on pasture has historically been thought to protect against colic, other factors such as weather, water availability, feed intake rate, stocking density, pasture quality, and supplementation may have an impact.

In a case control observational study involving 364 horses, pasture access was found to reduce the likelihood of colic. The study also found that horses with no pasture turnout or that had recently reduced paddock size or time at pasture had a three-fold increased risk of colic (95% CI 1.4–6.6, P=0.007). Stall confinement was found to be associated with a higher risk of colic (OR=9.30, 95% CI 1.68–51.40, P=0.011) in a different case control study conducted in the UK. The highest chance of developing colic was linked to 24-hour stabling (OR=35.2). Enterolithiasis, which was found to occur 2.8 (95% CI 1.06–7.59, P=0.04) to 4.0 (95% CI 1.3–12.2, P=0.02) times more frequently in horses that spent less than 50% of their time outdoors, may also be predisposed to certain types of colic if there is a lack of grazing activities.

**Table 8.1.** Summary of Findings of Studies Investigating Factors Affecting the Microbiota of Horses

| **Factor Influencing the Equine Microbiota** | **References** | **Key Findings** |
|---|---|---|
| Diet | Julliand et al. (2017); Garber et al. (2020); Hesta et al. (2020) | Diet alters the composition of the horse's gut microbiome. |
| | Muhonen et al. (2008); Muhonen et al. (2009); Fernandes et al. (2014) | The gut microbiome of horses is altered by abrupt changes in diet. |
| | Daly et al. (2012); Zhu et al. (2021); Warzecha et al. (2017) | Non-structural carbohydrate (concentrate) based diets promote the growth of lactic acid-producing bacteria (e.g., *Bacillus-Lactobacillus-Streptococcus* group). Fiber-based diets are considered optimal for horses because they promote the growth of bacteria associated with good intestinal health (e.g., Ruminococcaceae, Lachnospiraceae, *Fibrobacter*). |
| | Destrez et al. (2015); Bulmer et al. (2019) | Non-structural carbohydrate-based diets have been suggested to alter equine behavior (e.g., "pace-change" behavior) and reactivity compared to high-fiber diets. |
| Supplements | Proudman et al. (2014) | Six weeks of amylase supplementation did not change microbiota richness. |

| | | |
|---|---|---|
| | Proudman et al. (2015) | Amylase-rich malt extract increased the relative abundance of lactate-utilizing bacteria. |
| Age | Husso et al. (2020) | The colonization of the newborn foals originated from the maternal vaginal and fecal microbiota, as well as environmental bacteria. |
| | Quercia et al. (2019); Garber et al. (2020) | Until weaning, diet (milk) the dam's udder, skin, hair, and feces influence the foal's microbiota. |
| | Lidenberg et al. (2019) | At approximately 60-days-of-age, the dominant bacteria found in the foal feces are similar to those found in adult horses, but the overall composition remains significantly different until yearling. |
| | Dougal et al. (2014) | Similar bacterial communities were found between adult and aged horses. However, a reduction in bacterial diversity was observed in older animals. |
| Climate | Salem et al. (2018) | Seasonal differences in the microbiota composition were observed throughout the year. |
| Social interactions | Antwis et al. (2018) | The composition of the gut microbiota may be influenced by group interactions and social behavior. |
| Geographic location | Ang et al. (2022); Arnold et al. (2021); Ayoub et al. (2022); Kaiser-Thom et al. (2021) | The fecal and skin microbiota of horses tended to differ based on geographic location and habitat, suggesting that environment and management might influence the microbiota composition. |
| Exercise | Pagan et al. (1998) | Exercise improved feed digestibility in horses. |
| | Almeida et al. (2016) | Exercise significantly changed the microbiota of unfit horses. A training program was also associated with specific changes in the gut bacteria. |
| | Janabi et al. (2017) | No changes were found in the microbiota of standardbred racehorses after 12 weeks of intensive training. |
| | Plancade et al. (2019) | No associations were found between gut microbiota and blood biochemical and metabolic markers in horses competing in endurance races. |
| | Górniak et al. (2021) | There was a significant increase in Firmicutes and Bacteroidetes after exercise. |
| Transportation | Schoster et al. (2016) Perry et al. (2018); Faubladier et al. (2013) Mach et al. (2020) | There is evidence that transportation can decrease diversity as well as the abundance of certain bacterial taxa (e.g., Clostridia, Rickettsiales, Streptococci, Bacteroidetes). Those changes are likely associated with the stress caused by transportation. |

| | | |
|---|---|---|
| Antibiotics and other medications | Barr et al. (2013); Harlow et al. (2013); Di Pietro et al. (2021) | The administration of antibiotics can lead to antibiotic-induced colitis, which allows for the overgrowth of pathogens and leads to a decrease in the richness and diversity of species. |
| | Costa et al. (2015); Liepman et al. (2022); Gomez et al. (2023) | Each drug principle has a distinct effect on the composition of the intestinal microbiota. |
| | Costa et al. (2015); Liepman et al. (2015); Di Pietro et al. (2023) | Most of the microbial changes caused by antibiotic administration were considered minimal 30 days after treatment, but some individuals might bear a less resilient microbiota. |
| | Whitfield-Cargile et al. (2018) | After administration of nonsteroidal anti-inflammatory drugs, transient dysbiosis was observed in the feces of healthy horses. |
| | Schoster et al. (2016) | Pre-surgery medication and general anesthesia affected the bacterial community. However, most of the changes in the microbiota could be attributed to transport and fasting. |
| | Kunz et al. (2019); Sirois (2013) | Studies on the effects of helminths and anthelmintics on gut microbiota are conflicting. |
| Diseases | Arnold et al. 2021); Arnold et al. (2021); Zakia et al. (2023); Costa et al. (2012); Ayoub et al. (2022); McKinney et al. (2020); McKinney et al. (2021); Arroyo et al. (2020) | Many diseases have been associated with changes in the gut microbiota, including colitis, equine metabolic syndrome, and colic. In general, decreased diversity and altered bacterial composition have been reported. Many studies show a decrease in the relative abundance of the Firmicutes phyla in diarrheic horses and an increase in other taxa, such as *Lactobacillus*, *Pseudomonas*, *Streptococcus*, *Enterococcus*, and Enterobacteriaceae. |
| | Uzal et al. (2022); Zakia et al. (2022); Zakia et al. (2023) | Unconventional Clostridia species have been suggested to cause diarrhea in horses, as *Clostridium sensu stricto* was increased three-fold in the cecal content of horses with diarrhea. Another study found no significant difference in the prevalence of *Clostridium innocuum*. |

*Source: Boucher L, Leduc L, Leclère M, Costa MC. (2024). Current understanding of equine gut dysbiosis and microbiota manipulation techniques: comparison with current knowledge in other species. Animals, 14(5):758. https://doi.org/10.3390/ani14050758.*

Reeves et al. (1996) found that having access to pasture did not lower the risk of colic, however the risk of colic more than doubled if there was no water in the paddock (OR=2.2, 95% CI 1.2–4.3). It is challenging to distinguish any positive effects of grazing activities from the timing and patterns of ingestion, the horse's activity level, and the effects of exercise on intestinal motility, even with this published evidence (Table 8.1).

## 8.2.2. Feeds and Colic: Dried Forages

Because of the limitations of space and resources, hay frequently makes up a significant portion of a modern horse's diet. Even though the diet of a horse fed dried forages is somewhat consistent, variations in the source, the kind of hay fed, and even the preservation method can cause variability from batch to batch. High concentrations of hemicellulose, cellulose, and lignin in low-quality forages raise the possibility of colic linked to impaction. Feeding hay from round bales may also raise the risk by 2.5 (95% CI 1.1–5.6, P=0.028). This increase is probably caused by the preservation techniques used, the horse's exposure to the weather before and during feeding, and the unrestricted amounts of hay that are available.

Colic has been observed to be frequently caused by sudden changes in the type or batch of hay. While a feed change in the previous two weeks was significantly associated with colic (OR=5.0, 95% CI 2.6–9.7, P< 0.001), a change in hay increased the risk even further (OR=9.8, 95% 1.2–81.5, P=0.035), according to a case control study of horses suffering from colic in Texas. An additional study conducted in 2001 confirmed this (OR=4.9, 95% 2.1–11.4, P< 0.001). Changes in diet were found to be associated with an increase in colic incidence of 2.1 times (95% confidence interval [CI] 1.2–3.8, P = 0.01) on the eastern seaboard of the United States, specifically with changes in hay consumption.

The particular variety of hay may also be a significant factor in the likelihood of colic. Coastal Bermudagrass hay has been linked to ileal impactions in multiple retrospective case series and has been implicated as a cause of colic in one cohort study (OR=1.65, 95% CI 1.01–2.7, P=0.045). 18–20 A retrospective case control study by Little and Blikslager confirmed that feeding horses coastal Bermudagrass hay increased the risk of ileal impaction by 4.4% CI 2.1–9.1) compared to non-colic controls; the risk of medical colic increased by 5.7% CI 2.4–13.6, and the risk of surgical colic increased by 2.7% CI 1.2–6.5). The risk of colic in general was not increased by feeding Coastal Bermudagrass alone, and the risk of ileal impactions was not decreased by diluting it with other hays, according to this study. Alfalfa hay has been linked to an alkalinizing effect on the ingesta of the colon, which leads to changes in the microflora and a decrease in the production of volatile fatty acids, creating an environment that is favorable for the development of enteroliths. According to one study, eating more than 50% alfalfa in the diet raised the risk of developing enterolithiasis (OR=4.2, 95% CI 1.3–12.9, P=0.01). This result was validated by two more studies, supporting the recommendation that horses at risk for enterolithiasis not eat alfalfa hay. Alfalfa's alkalinizing effect on colon contents may be attributed, in part, to its high magnesium content. Furthermore, a high protein content may increase ammonia and reduce the absorption of magnesium, causing minerals to precipitate and form enteroliths. On the other hand, if fed as more than 50% of the diet, grass hays have been shown to have a protective effect and may be helpful in preventing enteroliths.

## 8.2.3. Feeds and Colic: Concentrates

Since carbohydrates have been shown to have a significant impact on the flora of the equine gastrointestinal tract, they are frequently identified as the dietary cause of colic in horses. This includes grains. Grain content differs significantly from forage in terms of the type of carbohydrates present, even though grains contain a far higher concentration of carbohydrates overall. Carbohydrates are generally classified into two groups: fermentable carbohydrates, which are broken down by bacterial populations into short chain fatty acids (acetate, propionate, and butyrate) in the large intestine and cecum, and rapidly hydrolyzable carbohydrates, which are mainly broken down in the small intestine (starches, hexoses, disaccharides, and some oligosaccharides). Greater amounts of quickly hydrolyzable carbohydrates are found in grains.

In the small intestine, hydrolyzable carbohydrates are typically broken down into oligosaccharides by pancreatic α-amylase, which are then further broken down into glucose for absorption by brush-border enzymes. Although enterocytes can improve their transport of glucose by adapting to high starch diets, this adaptation is dependent on the presence of amylase-produced monosaccharides, which stimulate this adaptive process. Nonetheless, the horse's total amylase activity is quite variable and low when compared to other species. It has been suggested that the rate-limiting step in the digestion of carbohydrates in horses is an insufficient amount of amylase.

This conclusion is supported by research, which shows that starch intakes greater than 0.4 percent of body weight in a single meal can permit hydrolyzable carbohydrates to enter the horse's hindgut. Saccharolytic bacteria in the cecum and colon ferment carbohydrates to lactic acid, which lowers colonic pH and, consequently, the survival of bacteria needed for fiber fermentation (*Clostridiaceae*, *Spirochaetaceae*, and *Fibrobacter* spp.). A horse's susceptibility to colic can be increased by changes in the microbial environment, which can also affect the digestion of fermentable carbohydrates, decrease the synthesis of volatile fatty acids, dehydrate the ingesta, lengthen transit times, and cause the large intestine to distention gas.

Concentrates and colic in clinical patients showed varying associations; in one study, there was a significant effect (OR=2.6, 95% CI 0.9–7.2, P=0.064), while in another, there was no effect. In the study by Hudson et al., feeding more than 2.7 kg of oats daily raised the risk of colic. (OR=5.9, 95% CI 1.6–22, P=0.009), earlier research revealed that any whole grain (oats, barley, etc. was linked to a higher risk of colic if fed in quantities more than 2.5 kg each meal (OR=4.8, 95% CI 1–46.6, P = 0.01). Only whole corn was found to be a risk factor in another study. Grain pelleting may raise the chance of colic, although the results of the studies were unclear. Nonetheless, it was demonstrated statistically that variations in the concentrate given increased the likelihood of colic (OR=3.6, 95% CI 1.6–5.4, P< 0.001). To avoid negative effects on digestion, current concentrate recommendations suggest feeding less than 0.2 percent of body weight per meal.

### 8.2.4. General Practices to Prevent Colic

As feeding changes are the most common factor linked to equine colic, maintaining a consistent diet should be the main focus of colic prevention. The pH of ingesta and the bacterial populations necessary for efficient digestion can be affected by changes in the type of feed (forages, concentrates, and pasture turnout), the quantity supplied, and the frequency of feeding. As a result, the horse needs to receive at least 1.5 percent of its body weight, or at least 60% of its diet, from forage sources. To minimize the amount of acid produced in the stomach and colon, concentrates should be limited and given in three or more feedings per day. Vegetable oil, strained sugar beet pulp, or soy hulls are suitable substitutes for starch-based feeds If extra energy sources are needed. In order to allow for the adjustment of the microbial population, small intestinal enzymes, and glucose transporters necessary for the digestion of carbohydrates, any change in the horse's diet should finally take place over a period of 7 to 10 days.

## 8.3. NUTRITIONAL PLANS FOR HORSES WITH COLIC

### 8.3.1. Identifying Nutritional Status

Assessing the horse's current nutritional status and body condition score (BCS) is the first step in choosing a nutrition plan for a colic-stricken horse. It is thought that horses with good BCS can fast for two to three days without experiencing any negative consequences. On the other hand, horses with comorbidities that interfere with their ability to gain and maintain weight or with emaciated BCS should receive nutritional support right away. Growth, sepsis, endotoxemia, lactation, gestation in the third trimester, abdominal or thoracic abscesses, diarrhea, bowel resection, intestinal ischemia, or malabsorptive disorders are examples of concurrent conditions that may suggest the need for supplements. Although underweight horses are easily recognized as needing assistance, over conditioned horses may also require early nutritional intervention because of their susceptibility to hyperlipemia and intolerance to feed deprivation.

The initial examination can pose challenges in determining the overall nutritional status, frequently because of incomplete or unreliable dietary intake history, recent weight loss, length of illness, and recent dietary modifications. For humans, a number of scales have been proposed to assess baseline nutritional deficiencies. These include measures of body weight, height, protein markers, dietary history, nutrient intake, gastrointestinal function, and co-occurring conditions. But since these instruments were developed with accuracy in mind, based on particular patient populations, their findings cannot be generalized to other groups, let alone different species.

Furthermore, these scales require the clinician to select an outcome, which can be either identifying the risk of malnutrition or recognizing established malnutrition. Regretfully, veterinarians typically prefer to recognize both risks in equines. Although there isn't a validated tool for use with equine patients just yet, information about the patient's nutritional requirements can be found through a thorough history, physical examination, lab results, and feed analysis. The veterinarian should be able to identify horses that are already malnourished and those that might need extra nutritional support to avoid deficiencies with sufficient investigation.

First and foremost, the vet should get a thorough history that includes signaling, training schedule, and present housing. Medical records can be used to determine changes in physical condition, but in order to minimize bias, it is usually necessary to conduct a thorough interview with the owner. It is important to record any relevant medical history, such as prior colic surgery and current medication regimen. Not only should prescribed drugs be kept on file, but also complementary and alternative treatments such as probiotics, enzyme supplements, vitamins, minerals, and herbal remedies. It is important to talk about any changes to the diet, housing, appetite, attitude, or fecal production.

The next step is for the patient to undergo a comprehensive examination, which will help the clinician better understand the patient's current nutritional status. When assessing changes during a hospital stay, body weight should, if at all feasible, be objectively measured using a scale to establish an admission baseline. Body weight is also helpful for calculating nutritional needs, medication dosages, and feeding management adjustments. To avoid sudden changes brought on by water intake and gastrointestinal fill, body weight should be measured repeatedly at a regular time, ideally before feeding in the morning. When evaluating changes in body weight, it's crucial to take feed intake, body cavity effusions, and hydration state into account at the time of weighing. For instance, depending on one's level of hydration, body weight can vary by 5–12%, and certain procedures, like large colon lavages, can remove 50–70 liters of food at once. In situations where a scale is unavailable, the following formula can be used to estimate body weight based on the horse's length and heart girth measurements.

$$\text{Weight in pounds} = (\text{heart girth}[\text{inches}])^2 \times \text{body length}[\text{inches}]/330$$

or

$$\text{Weight in kilograms} = (\text{heart girth}[\text{cm}])^2 \times \text{body length}[\text{cm}]/11{,}877$$

The circumference of the horse's thorax at the highest point of the withers, directly behind the triceps, is known as the heart girth. It is best to measure the circumference of the heart while exhaling. Measuring from the point of the shoulder to the tuber ischii yields the horse's length. Using an inelastic tape is crucial for optimal accuracy. The horse should be standing level and square, and the tape shouldn't be pulled too tightly (Table 8.2).

**Table 8.2.** Diet Composition of Buryatian Horse Fecal Samples from Microhistological Analyses

| **Summer** | | **Winter** | |
|---|---|---|---|
| **Species** | **Mean%** | **Species** | **Mean%** |
| Total graminoids: | 41.9 ± 11.7 | Total graminoids: | 7 ± 5.3 |
| *Poa botryoides* | 14.3 | *Poa botryoides* | 2.9 |
| *Agropyron cristatum* | 11.1 | *Carex pediformis* | 2.4 |
| *Carex pediformis* | 8.9 | *Agropyron cristatum* | 1.2 |
| *Stipa capillata* L. | 3.0 | *Koeleria cristata* | 0.3 |
| *Achnatherum splendens* | 1.7 | *Achnatherum splendens* | 0.2 |
| *Koeleria cristata* | 1.4 | | |
| *Carex* sp. | 1.1 | | |
| *Carex duriuscula* C. A. Meyer 1831 | 0.8 | | |

| | | | |
|---|---|---|---|
| Total forbs: | 57.7 ± 11.8 | Total forbs: | 92.6 ± 5.9 |
| *Lappula myosotis* | 8.9 | *Schizonepeta multifida* | 51.43 |
| *Silene repens* | 8.5 | *Aster alpines* | 9.4 |
| *Aster alpines* | 7.8 | *Artemisia sieversiana* | 8.7 |
| Unknown grass 6 | 4.3 | *Heteropappus altaicus* | 6.8 |
| *Potentilla acaulis* | 3.9 | *Potentilla acaulis* | 6.1 |
| *Allium altaicum* | 2.6 | *Heteropappus biennis* | 2.5 |
| *Thalictrum etaloideum* L. | 2.5 | *Lappula myosotis* | 2.4 |
| Unknown grass 2 | 2.3 | *Cymbaria dahurica* | 1.4 |
| *Heteropappus biennis* | 2.3 | | |
| *Serratula centauroides L. s. str.* | 1.7 | | |
| *Achnatherum splendens* | 1.7 | | |
| *Pulsatilla turczaninovii* | 1.5 | | |
| *Artemisia dracunculus* L. | 1.3 | | |
| *Saussurea salicifolia* L. | 1.1 | | |
| *Phlomis tuberosa* L. | 1.0 | | |
| *Nonea rossica* | 1.0 | | |
| *Potentilla longifolia* | 0.8 | | |
| *Schizonepeta multifida* | 0.8 | | |
| *Artemisia frigida* Wil | 0.7 | | |
| *Axyris amaranthoides* L. | 0.6 | | |
| *Leontopodium leontopodoides* | 0.4 | | |
| Unknown grass 9 | 0.4 | | |
| *Hypecoum (Chiazospermum) erectum* | 0.3 | | |
| Unknown grass 4 | 0.3 | | |
| Unknown grass 5 | 0.3 | | |
| *Astragalus adsurgens* Pallas | 0.3 | | |
| *Artemisia gmelinii* Webb ex Stechm. | 0.3 | | |
| *Descurainia sophia* (L.) Webb. et Berth. | 0.2 | | |
| *Potentilla bifurca* | 0.2 | | |
| *Allium anisopodium* Ledeb | 0.2 | | |

| | | | |
|---|---|---|---|
| *Pulsatilla multifida* | 0.2 | | |
| *Cymbaria dahurica* | 0.2 | | |
| *Heteropappus altaicus* | 0.1 | | |
| *Artemisia sieversiana* | 0.1 | | |
| *Saposhnikovia divaricata* (Turcz.) Hiroe | 0.1 | | |
| *Thalictrum squarrosum* Stephan ex Willd. | 0.1 | | |
| *Galium verum* L. | 0.1 | | |
| *Leonurus* sp. | 0.1 | | |
| Total shrubs: | 0.4 ± 1.3 | Total shrubs: | 0.43 ± 1.3 |
| *Kochia prostrata* Sschra-ber | 0.4 | *Kochia prostrata Sschra-ber* | 0.43 |
| *Spiraea* sp. | 0.1 | | |

*Source: Zaitseva S, Dagurova O, Radnagurueva A, Kozlova A, Izotova A, Krylova A, Noskov S, Begmatov S, Patutina E, Barkhutova DD. (2023). Fecal microbiota and diet composition of buryatian horses grazing warm- and cold-season grass pastures. Microorganisms.; 11(8):1947. https://doi.org/10.3390/microorganisms11081947.*

This formula determines the mature body weight and is independent of feed intake or state of hydration. Although the calculation has proven to be accurate for smaller horses, horses taller than 15 hands should use it with some caution. The ideal way to estimate weight is to take three measurements of the body length and heart circumference, then average the results to lower error.

Although body weight is an objective indicator of nutritional status, it cannot assess over-conditioning or provide information about the horse's nutritional history in terms of chronic catabolic conditions. It is advised to use a BCS system to provide an individualized evaluation of the horse's lean muscle mass and adipose reserves. A nine-point system that works well for light breeds has been developed and modified for use with warm-blood horses. This scale is useful for determining whether a horse's weight is appropriate for its frame and provides a BCS for the amount of fat noted at six discrete areas (neck, withers, behind the shoulders, over the ribs, along the spine, and around the tail head). Ultrasonography has confirmed the positive correlation between body condition score (BCS) and body fat percentage by observing a strong association between BCS and the amount of fat deposited at the tailhead, thirteenth rib, and withers.

Human nutritional status has been assessed using biochemical markers of plasma protein, such as albumin, transferrin, and retinol binding protein, presuming that decreased hepatic synthesis is the cause of decreased plasma protein levels. But since they are acute phase proteins, the majority of these conventional protein markers have not been proven to work for critically ill patients. Particularly albumin is impacted by a variety of disease-related changes. Reduced production by the liver as a result of illness or a negative acute phase response, dilution, increased renal, vascular, or gastrointestinal permeability, and overhydration all lower blood albumin concentration. Since albumin has a half-life of 19–21 days in horses, changes in serum concentration won't be noticed right away. Consistent with findings in critically ill humans, albumin has not been demonstrated to be sensitive to nutritional status or supplemental feeding in horses. Following small intestinal resections, albumin levels considerably rose in horses that were not fed as opposed to those that received parenteral nutrition; this increase was probably caused by a dilutional effect ($P=0.044$). Thus, the most accurate indicators of nutritional status continue to be medical examination and past medical records.

## 8.3.2. Nutritional Requirements of Horses with Colic

Through feeding trials, the National Research Council (NRC) and others have published the dietary requirements for healthy horses based on mature body weight. Although daily requirements have been established for various life stages and activities, it is crucial to understand that due to individual differences in nutrient absorption, losses, and metabolism, each patient with colic will differ from these published requirements. Furthermore, these guidelines only include the minimal amounts of each nutrient; as a result, if fed according to standard, horses with poor digestion, absorption, or metabolism may not receive enough of these nutrients. Although useful, it is challenging to apply these recommendations to horses that have colic, and additional considerations such as medical history, physical examination, and BCS assessment are crucial in identifying dietary requirements.

It is believed that horses in good condition can endure a brief fast. The main organs of donkeys and mules that were denied food for ten days did not suffer any harm. The actual amount of time a horse can survive without food may be considerably greater because past research on humans has shown that starvation does not negatively impact organ systems for up to 30 days. Three things can impact a patient with colic's metabolism and energy needs: (1) stall rest, which lowers needs; (2) illness processes like surgery or sepsis, which may raise needs; and (3) starvation, which can lower needs by reducing energy lost through fermentation and digestive processes. Disease and starvation have hard to measure effects. The aim of the equine clinician caring for horses with colic is to calculate the resting energy requirements (RER) or resting energy expenditure by utilizing the stalled patient's requirements in a thermoneutral environment. Since indirect calorimetry does not rely on patient population-specific metabolic formulas, it is the best method for estimating RER in critically ill human patients. Using closed circuit spirometry to measure oxygen uptake and carbon dioxide excretion, this method determines each patient's specific caloric needs. The patient's inspired oxygen concentration, variations in ventilation, sudden pH shifts, sympathomemetic drugs, leaks or resistance in the spirometry apparatus, and pain or anxiety can all cause measurements to be inaccurate. Accurate measurements are challenging due to these factors and the dearth of spirometers in veterinary practices.

The RER of normal, stall rested horses using indirect calorimetry can be estimated by the following calculation:

$$\text{RER} = 21.28\,\text{kcal} \times \text{body weight}(\text{kg}) + 975$$

Using this equation, the calculated RER for a 500 kg horse is 11,615 kcal. This figure would be a great place to start when figuring out feeding schedules for horses that are admitted to the hospital with severe colic. Nonetheless, the majority of horses that experience colic recover fast and don't significantly alter their daily routine or activities. As a result, a different formula known as the daily digestible energy requirement (DER) might be more suitable. It takes into account the extra energy needed to maintain normothermia in a variety of settings in addition to low activity levels (i.e., strolling, pasture).

$$\text{DER horses} < 600\,\text{kg}(\text{Mcal}) = 1.4 + (0.03 \times \text{BW}[\text{kg}])$$

$$\text{DER horses} > 600\,\text{kg}(\text{Mcal}) = 1.82 + (\text{BW}[\text{kg}] \times 0.0383) - (0.000015 \times \text{BW}[\text{kg}])$$

DER is equivalent to 16,400 kcal per day for a 500 kg horse. When calculating the diet for refeeding, the clinician should take the severity of the illness into account to determine whether RER or DDE is more appropriate.

Stress factors or multipliers have been developed to quantify the effect of disease and illness on the energy expenditure of the critically ill. These factors estimate the increase or decrease in basal metabolic rate. 89 It's crucial to keep in mind that these factors were created using arbitrary study groups, so their applicability has lately been called into question. When extrapolating to an equine patient, it is important to take into account the multitude of factors that may impact a critical patient's metabolic rate, in addition to the fact that these multipliers were developed for human use.

There are no energy expenditure multipliers available for equine patients, particularly after gastrointestinal surgery; however, energy expenditure during surgical procedures in humans is estimated to be 23–31 kcal/kg/day, or about 1–24% greater than resting energy expenditure. Based on human studies, the

RER of post-operative colic patients in horses has been estimated to be one and a third times the RER. Closed circuit spirometry has been utilized more recently to offer a more precise estimate. Under general anesthesia, a standardized abdominal exploratory procedure using a ventral midline incision led to a daily increase in Mcal/kg of 1.0, which is equivalent to a 1.1-fold increase in RER. This study suggests that energy expenditure multipliers need not be used for adjustments.

### 8.3.3. Routes for Feeding Horses Recovering from Colic

In equines, gas colic or mild impaction are common short-term disease processes that cause colic. These ailments allow for a speedy return to oral feedings and swift resolution with little medical intervention. However, a period of feed withholding during the acute recovery period may be necessary for severe impactions and certain surgical conditions, such as small intestinal resections. Veterinarians should ideally give enteral or parenteral nutritional support to any horse that has been denied feed for more than four to five days. In actuality, financial constraints or the unanticipated course of a patient's illness frequently put clinicians in situations where they must forgo supplements for longer than intended.

The best way to nourish a horse that has had colic will differ based on several factors, such as the reason behind the colic, the horse's appetite, the surgical procedures that were done, the length of the illness, and the owner's financial situation. When it comes to gastrointestinal diseases, it is crucial to take care of the main issue before allowing enteral feeding. Though every program has its advantages and disadvantages, the ultimate objective is to progressively restore voluntary intake. For horses recuperating from colic, there exist three fundamental feeding techniques: (1) voluntary intake; (2) supportive enteral nutrition; and (3) supportive parenteral nutrition.

#### *8.3.3.1. Voluntary Intake*

The hypothesis that horses may tolerate short fasts without negative consequences has been validated by clinical cases. If feed is given and after a gradual reintroduction >85% of the calculated RER can be achieved, then no additional supportive nutrition is needed. Feed should be reintroduced at 25% of the RER and increased by 25% each day, if tolerated, in horses that have experienced prolonged anorexia until the amount supplied exceeds 85% of the estimated energy requirements. Reintroduction can happen much faster in horses whose disease progresses more quickly. Once the colic symptoms have subsided, the transit time has returned to normal with no indication of positive nasogastric reflux, and the gastrointestinal borborygmi are satisfactory, the feeds should be resumed. Horses are typically first tested with small amounts of water to ensure proper progressive motility. Feed is introduced if oral fluids are accepted without causing colic. To ensure compliance and avoid feed changes during early convalescence, the veterinarian should provide a feed that the owners can easily obtain and will provide the energy needed for work. Refeeding should not involve these feeds if dietary factors—like the kind of hay fed—were a contributing factor to colic. In order to prevent additional disruptions in the gastrointestinal microflora, it is also advised to refrain from feeding grains or other concentrates for the first two weeks following the colic episode. If the horse's level of athletic performance depends on concentrates, grains should be gradually added back in at a rate of 0.1 percent of body weight (or 0.5 kg per 500 kg of body weight) per feeding.

Rarely do horses recuperating from colic refuse voluntarily to eat, and they frequently make an effort to eat even when they are experiencing excruciating stomach pain. Feeds can occasionally be rejected for a variety of reasons, including novelty, consistency, palatability, or boredom. To find out which food the horse prefers, a buffet-style arrangement of various feeds should be offered to these horses. Although some feedstuffs might not be the best (i.e., sweet feed), the items the horse chooses may enable the clinician to increase intake by adding tiny amounts of less appetizing but healthier feeds. Adding molasses or pancake syrup to the feed to sweeten it is another way to increase the palatability of the feed. While the use of medications to increase horse appetite is debatable, diazepam (0.02–0.03 mg/kg, IV) has been shown to improve things. Acepromazine (0.05 mg/kg, IV) also had an impact in this investigation, albeit a less noticeable one. Anabolic steroids and cyproheptadine (0–1–2 mg/kg, PO, q24 h) are also effective appetite stimulants.

Because grass is the ideal feed for reintroduction, it makes sense for the horse's natural physiology. The horse can be gradually returned to eating by grazing for 20 to 30 minutes at a time. This also enables the veterinarian to keep an eye out for any untreated gastrointestinal issues. Furthermore, the exercise it offers will increase motility in the gastrointestinal tract. According to studies on the digestible energy gained from grazing over time, this level of grazing will yield between 1.2 and 1.8 kcal/kg per feeding. Thus, to provide about 25% of the RER, a 500 kg horse can be grazed three to five times a day at first. After two to three days, full feeding should be achieved, and grazing should be increased as tolerated.

Since pastures are frequently scarce, hay can be offered as a substitute for pasture. Initially, small amounts of hay ought to be provided. The majority of sources advise beginning with 0.5 kg (1 lb) of high-quality hay given 4–6 times daily. First cutting hays are preferred in situations where feed analysis is not available because hay quality can vary greatly. When fed six times a day, this conservative feeding rate will provide roughly half of the RER for a 500 kg horse. To guarantee proper ingesta passage, these meals may first be smaller due to worries about strictures or edema decreasing lumen size. For a 500 kg horse, feeding 0.25 kg of hay will cover about 25% of its daily requirements. Any horse that has been anorexic for longer than four days should introduce feed gradually. Since different grasses, growing climates, and baling techniques can significantly affect the weight of the hay per volume, bale weight, and amount of hay per flake, weighing the hay will aid in monitoring intake.

Forages are similar in general in that they are high in fiber and will encourage the hindgut to produce volatile fatty acids. The amount of protein in different kinds of hay can differ dramatically. For this reason, horses fed low-protein hay (i.e., Coastal Bermudagrass), particularly if the horse is undernourished. Additionally, when refeeding colicky horses, serum concentrations should be closely monitored because horses that have been denied feed frequently lack calcium and potassium.

Horses with dental disease, metabolic or respiratory problems, limited gastrointestinal luminal size, or those living in places with limited access to appropriate feedstuffs may benefit from forages that are cubed or pelleted. These feeds can take the place of pasture or hay in the diet, and soaking can help to minimize dust and lower the risk of esophageal blockage. Because of the benefits of long stem fiber for the gut flora, alfalfa cubes or chopped hay are better than pelleted forages. A minimum of 1–2.5% of the body weight in hay is needed for upkeep (Table 8.3).

**Table 8.3.** Essential Oils Used in Horses' Diets and Dosages Customarily Used

| Type of Essential Oil | Dose | Effect |
|---|---|---|
| Roman chamomile Marjoram | Apply 20–30 drops of pure oil (dilute with V-6 vehicular oil if phenol-rich oils are used). | Can be placed in a capsule and mixed with food. It helps to combat nerve problems and anxiety in horses. |
| Mint Cypress | Essential oil drops on the coat approximately 10 cm in size. | Therapeutic application for horses. |
| Valeriana | 20–30 pure drops per application. | Relieves anxiety in animals; the calming effect lasts from days to weeks. |

| | | |
|---|---|---|
| Marjoram | Application of essential oil drops on the coat, approximately 10 cm in size. | Therapeutic application for horses. |
| Gauteria | Apply diluted in a vegetable oil such as almond oil or even arnica, which is also an anti-inflammatory. | It is a powerful anti-inflammatory and analgesic, but contraindicated during pregnancy and lactation. |
| Rosemary chemotype camphor | Apply diluted in a vegetable oil such as almond oil or even arnica, which is also anti-inflammatory. | Muscle relaxant and analgesic, ideal for muscle contractures, torticollis, and other types of pain. |
| BIOBRON is a food supplement based on a synergistic blend of essential oils. | 20 mL/day per 100 kg of live weight for 5 days (there are also more concentrated versions). | It helps animals recover due to essential oils' expectorant and fluidizing properties. Stimulates food consumption and general health thanks to plant extracts and vitamin C. |
| Soybean or sunflower oil | 20% or 450 mL daily of the total diet. | Contributes to the proper growth of the foal. Enhances absorption of vitamins A, D, E, K, and linoleic acid. Prevents colic. |
| Corn oil | 90 mL in a single meal; after 3 days, 90 mL in each meal and continue to increase gradually until reaching the desired level. | Maintains reserves of muscle glycogen and glycemia levels for a longer time. |

Source: Elghandour MMMY, Maggiolino A, García EIC, Sánchez-Aparicio P, De Palo P, Ponce-Covarrubias JL, Pliego AB, Salem AZM. (2023). Effects of microencapsulated essential oils on equine health: nutrition, metabolism and methane emission. Life, 13(2):455. https://doi.org/10.3390/life13020455.

Complete pelleted rations are another option for recovering colic-stricken horses instead of forages. In cases where large colon fiber digestibility has been compromised by resection or ischemia, or when recovering from more complex causes of colic, such as an anastomosis with questionable lumen size, complete feeds are frequently preferred. Although there are a lot of complete feeds available, any one of them can be fed as the only ration if necessary. Complete feeds offer enough fiber to promote mucosal healing and are nutritionally balanced for maintenance. To maintain healthy hindgut microbial populations, hay, chopped hay, or hay cubes are suggested as a source of long stem fiber once the horse has fully recovered from the colic episode.

Separated into meals fed every 3–6 hours, pelleted feeds can be introduced at a rate of 25–50% of the total daily ration digestible energy (DE), increasing by 25% each day until a full ration is reached. For instance, 6.1 kg (13.4 lb) of a complete senior feed would be needed daily for maintenance on a 500 kg horse. Starting at 25% of the daily energy requirements, 1.5 kg (3.3 lb) of feed would be needed, divided into feedings of 0.2 kg (0.5 lb) every three hours. To increase water intake and improve motility through the intestinal lumen, pelleted rations can be soaked in water. However, because of its texture or temperature, some horses might not accept wet feed. It is essential to customize the diet to the unique requirements and inclinations of every horse. Once the horse is back on its full ration, further dietary modifications for performance, maintenance, and malnourishment can be addressed.

Given their nutrient profiles, complete manufactured feeds and forages are ideal, but some horses may not want to eat them. Further feed options could be chosen in accordance with clinician preference or availability. For horses recovering from colic, common feedstuffs include soy hulls, rice bran, vegetable oils, and beet pulp.

Beet pulp is frequently used as a substitute source of fiber; however, because it is deficient in certain essential vitamins and minerals, it should not be used as the only source of forage. Although the exact maximum that can be fed is unknown, it is thought to be very nearly 1% of body weight each day. Even though none of these feeds are nutritionally complete on their own, they can give the horse the variety it needs to pique its interest and appetite. Additionally, they can be added to pelleted and fully foraged diets to boost protein, energy, and palatability.

When reintroducing feed to a horse, it is crucial to take its digestive system into account. Specifically, in the small intestine, lipolysis occurs before fats are absorbed. Therefore, the capacity to convert dietary fats to energy may be diminished by small intestinal disease. Giving horses with metabolic disorders feeds that might raise their glycemic index is another issue that needs to be taken seriously. Beet pulp, alfalfa, rice bran, and soy hulls are generally safe, but sweet feeds, corn, oats, and other grains should be avoided. However, because beet pulp with molasses has a glycemic index that is similar to oats, care must be taken when preparing some feeds. Concerns regarding wheat bran's energy content have been raised by some medical professionals because it is believed to have a laxative effect by drawing water into the intestinal lumen. The glycemic index of wheat bran is intermediate, so feeding large amounts of it to horses with metabolic diseases or those experiencing colonic flora disruptions may be problematic as it can worsen diarrhea. Furthermore, when fed in excess and without additional calcium, the inverted calcium-phosphorus rations of wheat and rice bran can result in nutritional secondary hyperparathyroidism. For a 500 kg horse, it is currently advised to restrict these feeds to less than one pound per day.

### *8.3.3.2. Supportive Enteral Nutrition*

The best way to give nutrition to horses recovering from colic is through voluntary intake. As long as it is fed nice meals, horses almost never refuse to eat. Oral intake should not be permitted in certain situations, though. This most frequently happens as a result of illnesses or injuries to the esophagus or mouth, such as esophageal strictures and pharyngeal trauma, which cause dysphagia, or the inability to swallow. In these cases, nutrients can be given enterally through a nasogastric tube or esophagostomy.

Because enteral nutrition promotes mucosal healing and maintains the gastrointestinal barrier against endotoxins and enteric pathogens, it is favored over parenteral methods of nutrition. A complete and balanced oral diet supplies nutrients that are chosen by various enterocytes along the gastrointestinal tract to preserve this barrier, according to research done on other species. A source of fiber should be included in enteral diets for horses, as this is particularly crucial for gastrointestinal activity, mucosal blood flow, cell growth, and nutrient absorption. Given its beneficial effects on mortality, length of hospital stay, infection, and metabolic response to illness in critically ill human patients, enteral nutrition should be given as soon as possible.

A nasogastric tube inserted into the stomach (roughly at the 12th rib) can be used to give horses extra enteral nutrition. For diets with blenderized components, a 14 French nasogastric tube is recommended; however, a smaller 18F feeding tube can be used for liquid diets. Feeding tubes can be inserted at the site of injury or obstruction via an esophagostomy aboral or nasogastric intubation. For larger tubes, accurate placement can be achieved by aspirating or palpating the gastrointestinal contents; for smaller tubes, proper positioning needs to be ensured by endoscopy or radiography. Once inserted, the tube is kept patent by being flushed with warm water before and after each feeding, capped when not in use, and taped to the halter.

The horse has received three different types of diets through feeding tubes. First comes a commercially available complete pellet diet that has been blended and can be given via a 14F nasogastric tube. To prevent clogging the tube, it is advised to stay away from any feed that contains whole grains, such as oats. Investing in a blender can help reduce particle size even more. To make a slurry meal, combine about 0.5 kg (1 lb) of pellets with 2–4 l of water. Vegetable oil can be added to provide extra calories if needed; one cup (240 ml) will provide one point nine Mcal. These blenderized diets are more affordable than other enteral diets, despite taking more time.

Feeding a commercial diet in powdered form (Welsolve Wel - Gel®, Purina Mills, St. Louis, MO) in an effort to lessen particle size issues brought on by pellet grinding. This is a comprehensive formula that meets all of the NRC's requirements for protein, vitamins, minerals, and fiber to support fermentation in the hindgut. For a maximum of five days, this diet can be given via nasogastric tube at a rate of 0.3 percent of bodyweight each day. To meet energy requirements, horses requiring prolonged enteral feeding may require more calories, and the formula may need to include molasses, dextrose, or vegetable oil.

A secondary option for enteral feeding in horses is component diets. Here, choosing whole feeds that will supply the necessary nutrients satisfies the dietary requirements. Since these diets are frequently alfalfa pellet-based, a larger nasogastric tube will still be needed for administration. The most often used formula, which consists of alfalfa, casein, dextrose, and electrolytes, was released by Naylor. Although a clinician can customize a diet more precisely to meet the needs of a given patient, component diets are more difficult to combine and frequently lack essential vitamins and minerals, making them unsuitable for long-term use.

Additionally offered is a more comprehensive commercial enteral diet for horses (Critical Care MealsTM, MD's Choice, Louisville, TN). The known amounts of vital nutrients in this diet and the flexibility to customize its composition to each patient's needs are its advantages. Lactase, glutamine, L-arginine, vitamins, trace minerals, and antioxidants are additional ingredients.

Adult horses have been fed a low residue elemental liquid feed (Vital® HN and Osmolite®; Abbott Nutrition, Columbus, OH) as the third enteral diet type. These low-fat diets consist of individual peptides, carbohydrates from sucrose and/or maltodextrin, medium-chain triglycerides, lecithin, and oils from safflower, canola, and medium-chain triglycerides. These meals were created as a full enteral feed to be given via an 8F feeding tube to patients with gastrointestinal disorders in humans. The nutrient profiles of herbivores are most likely incomplete, and neither diet includes a fiber source. Low volume diarrhea is therefore frequently associated with these diets. An extra drawback is the increased expense. When starting an enteral diet, the usual protocol is to start at 25% of the target on the first day and increase by 25% every day until the dietary goal is reached. The number of feedings per day can be decreased to make it easier for caregivers to handle if the horse is responding well to these little meals. Nonetheless, considering the stomach capacity of a 500 kg horse, a single feeding should never go over 8 liters. Every feeding, horses receiving enteral nutrition should have their stomach reflux checked and their diet modified as necessary.

Enteral feeding complications are frequently the consequence of diet plans that are high in carbohydrates and low in fiber. This is frequently observed with low residue, elemental diets that may give horses diarrhea. Colic, ileus, gastrointestinal stasis, and laminitis are further issues. Nasogastric intubation can result in aspiration, pharyngitis, rhinitis, and esophageal or pharyngeal perforation as direct complications. Diverticula, laryngeal hemiplegia, mediastinitis, and local infection are possible side effects of esophagostomy. When gastrointestinal tract function is compromised in horses with cardiovascular shock, hypovolemia, or gastrointestinal obstruction, enteral nutrition is not advised. Parenteral nutrition is advised for these patients.

Enteral nutrition, particularly elemental diets, requires close observation and routine physical examinations of the horse. To check if the horse is receiving an adequate amount of feed, the feeding tube should be flushed on a regular basis and the horse weighed every day. The horse should be gradually weaned off of enteral feedings after it has recovered. The horse can begin voluntary feedings if fewer feedings are given and the tube is removed in between. It is crucial to keep in mind that, in order to avoid mediastinitis, an esophagostomy tube must be left in place for 10–14 days after healing has taken place.

### 8.3.3.3. Parenteral Nutrition

For horses that cannot receive enteral feedings, parenteral nutrition is advised. Infectious causes, ileus, persistent anorexia, and gastrointestinal blockages are frequently the cause of colic in horses. When a horse is expected to go without food for longer than five days, parenteral feeding should be taken into consideration. Parenteral nutrition shouldn't be put off in horses with low BCS (<3/9) or in situations where there are higher metabolic demands, such as during pregnancy or lactation, as this could lessen the body's capacity to make up for even brief feed shortages. Because their reduced gastrointestinal perfusion hinders normal digestion and absorption, horses in circulatory shock are also good candidates for parenteral nutrition.

Parenteral nutrition may even raise infection and mortality rates, according to some data, and there is currently no evidence from human retrospective studies supporting the early institution of IV nutritional support. Parenteral nutrition, on the other hand, increases survival in malnourished patients (less than 90% of ideal body weight); for this reason, supplementation is recommended without delay in these cases. In fact, parenteral nutrition did not significantly alter the clinical outcome when compared directly to enteral nutrition. Parenteral nutrition should only be given to patients in adequate body condition if enteral nutrition is not available or if the issue lasts longer than seven days, according to current guidelines from the American Society for Parenteral and Enteral Nutrition and the Society of Critical Care Medicine. However, based on strict guidelines to prevent malnutrition, the European Society for Parenteral and Enteral Nutrition recommends immediate supplementation if the patient is not expected to take food enterally within three days.

Although there is less research in veterinary medicine, it seems to support the more traditional human practice guidelines. There is no proof that parenteral nutrition given to horses right after surgery reduces hospital stays, increases survival rates, or lowers treatment costs. One study found that horses receiving parenteral nutrition had higher death rates; however, the study may have been biased by choosing to administer parenteral nutrition only to the most severe cases. Notwithstanding this evidence, certain positive outcomes have been documented in equines, such as noteworthy enhancements in triglyceride concentration and decreased hyperbilirubinemia in post-operative intestinal resections receiving parenteral nutrition, potentially indicating a better nutritional state in these equines. If enteral intake falls short of 75% of the total caloric requirement, intravenous nutrition can also be utilized as a supplement.

Parenteral nutrition components include reduced catabolic muscle scavenging, amino acid supplementation for protein synthesis, and supplements of carbohydrates and/or fats. Depending on each patient's needs, additional vitamins, minerals, and electrolytes may be added. It is unknown what the ideal nutrient composition is for parenteral nutrition in horses. To avoid complications like hepatic lipidosis, hyperglycemia, and sepsis, it is crucial to provide these ingredients in the right amounts and ratios.

An energy source, often 50% dextrose, is the first part of parenteral nutrition. Although 3.4 kcal is provided by one gram of dextrose, dextrose solutions alone may not be able to provide the patient enough energy due to their high osmolality and low caloric density. Dextrose typically provides 40–60% of the non-protein calories in parenteral nutrition, and equine patients seem to tolerate this rate well.

It's critical to keep in mind that carbs are not a necessary food. Numerous substances, such as lactate, glycerol, and amino acids, can be converted into glucose by the liver, kidneys, muscles, and gastrointestinal tract. Human basal glucose requirements are set at 2 g/kg/day 125 to provide easily accessible energy to tissues that are highly dependent on glucose, like the brain, and tissues lacking mitochondrial metabolism, like red blood cells, leukocytes, and the renal medulla. Exogenous glucose supplementation can lower muscle protein catabolism, but high doses in parenteral diets can cause oxidative damage and inflammation. Tight glucose control has been recommended for treating over supplementation or glucose intolerance in humans, but it frequently results in hypoglycemia, which raises the risk of death. The current advice is to simply use prudent parenteral supplementation to prevent hyperglycemia.

One substitute for carbs that can help meet some of the energy requirements for metabolism is lipids. They might also help to lessen variations in serum glucose levels. A 20% lipid solution yields 2 kcal/ml, while one gram of lipid provides 9 kcal/g. Glycerol is added to lipid solutions containing long-chain fatty acids, which are typically made of soy or safflower oils and are iso-osmolar. Up to 30–40% of the non-protein calories can be initially obtained from lipids, with the potential to supply up to 60% of the total calories. Adding lipids has the benefit of lowering the solution's dextrose content, which lowers the tonicity and lowers the chance of thrombophlebitis. Reduced dextrose concentrations also lessen the chance of hypokalemia, osmotic diuresis, and hepatic lipidosis.

Lipids, on the other hand, are very expensive, and the long-chain fatty acids they contain may inhibit T-cell function, neutrophils, and the reticuloendothelial system, thus raising the risk of infection. Because endotoxins inhibit lipoprotein lipase production and activity, which can lower lipid clearance, it is crucial to closely monitor patients for lipemia and hypertriglyceridemia. Any patient who has elevated serum triglycerides or is at risk for hyperlipemia should avoid lipids.

The last component frequently added to parenteral nutrition for horses is precursors for protein synthesis. These are supplied by solutions of amino acids with different concentrations, ranging from 8.5 to 10.0%. The purpose of amino acid supplementation is to prevent the breakdown of lean muscle, which is needed for energy during times of scarcity of fats and carbohydrates, and to promote the synthesis of new proteins. It is challenging to determine the ideal protein needs for horses because, in the face of severe illness, whole-body nitrogen balance is not a valid indicator of sufficient protein synthesis.

Enteral protein requirements in a healthy horse have been found to be about 40 times the daily DER in Mcal. Around 0.9 g of protein per kilogram of body weight per day is provided by this level of protein supplementation, assuming that 70% of the total intake is absorbed and digested. Parenteral nutrition requires 0–0.8 g/kg/day of protein because parenteral proteins are closer to 100% available due to their bypass of the digestive process. 68,125 To improve nitrogen balance in severely stressed patients, it is currently recommended to increase protein supplementation to 0–10.5 g/kg/day without increasing metabolic demands in critically ill humans. A clinician must still make an educated guess because the protein requirements of a horse with colic are currently unknown. Parenterally supplied protein supplementation in horses has ranged from 0.68–2.0 g/kg/day.

Many factors that limit the ability to increase or replace lean muscle further impede the anabolic effects of protein supplementation in critically ill patients. One factor that has a positive correlation with muscle protein synthesis is physical activity. Convalescence inevitably results in inactivity, which lessens the anabolic effect that amino acid supplements typically provide. Even with sufficient nutritional support, catecholamines and inflammatory mediators also reduce insulin activity and speed up catabolism. The computation of protein requirements for parenteral nutrition is made more difficult by these metabolic factors.

To avoid using protein for energy or storing it as fat, the amount of protein supplementation must also be balanced against the energy provided by carbs and fats. The ideal ratio between non-protein energy (kcal) and nitrogen (g) is 1:100–1:200 kcal/g. Since 16% of nitrogen is found in protein, 0.16 times the total amount of protein equals grams of nitrogen. Hepatic insufficiency generally results in lower protein requirements; however, nitrogen requirements rise by 25–30% in permissive hypoalimentation and may even be higher in patients who are malnourished.

It is crucial to keep in mind that micronutrients like vitamins, minerals, and trace elements are also included in a comprehensive parenteral nutrition formula. The NRC has established the nutritional requirements for healthy horses, but it is challenging to apply these recommendations to critically ill patients. Nutrient deficiencies that preexist could make things more difficult. Micronutrient requirements cannot be measured, but energy and protein requirements can be tracked using indirect calorimetry. Any horse receiving parenteral nutrition for longer than three days should receive supplements.

### *8.3.3.3.1. Calculation of Parenteral Nutrition*

The parenteral nutrition required for a horse with colic can be calculated using the previously discussed resting energy requirement, multiplied by a factor for surgery (if desired), and extra calories added to account for malnutrition in the past. Preventing muscle wasting and maintaining body weight until enteral feeding is resumed are the objectives of the nutrition formulation. Energy expenditure is frequently overestimated in calculations, which can result in hyperlipemia, hypercapnea, and hyperglycemia. Hypoalimentation, however, can also have detrimental effects, such as an elevated risk of infection and an elevated mortality rate. In human medicine, the current recommendation is permissive hypoalimentation for critically ill patients, which provides 33–65% of targeted calories (9–18 kcal/kg/day) to prevent complications from overfeeding or underfeeding. This tactic has been demonstrated to increase survival and lower sepsis while avoiding excess calories and providing essential energy and protein. Equine parenteral nutrition should be calculated by multiplying the total resting energy requirement by 33–65%, based on these permissive hypoalimentation principles. This will result in an initial rate of 7.7–15.0 kcal/kg/day. The horses may be weaned off parenteral nutrition and put on oral feeds based on their recuperation, or the rate may be increased to RER, DER, or even higher depending on the requirements of each patient. To assess whether the parenteral nutrition given to horses is sufficient, it's critical to keep an eye on changes in body weight and disease state.

Physicians who have access to compounding pharmacies can precisely calculate the number of calories, protein, and energy-supplying components that their patients need in their prescriptions. A step-by-step algorithm is offered to enable complete parenteral nutrition (TPN) prescriptions. In actual practice, the majority of clinicians mix their own using easily accessible products. These consist of solutions containing 50% dextrose, 8.5– 10.0%% amino acid solutions, and 10–20% lipid solutions. Currently, parenteral nutrition is administered to horses in three different ways.

- **Dextrose Solutions:** The first is a parenteral solution that contains only dextrose. Although dextrose can give critically ill patients short-term energy, it is not a complete food and won't stop their muscles from continuing to deteriorate. However, due to its lower cost, dextrose solutions are frequently used in horses. Given the higher risk of complications associated with higher (hypertonic: 600 mOsm/L) concentrations infused into a peripheral vein, dextrose infusions, if chosen, should be diluted to a concentration of 5% or less. To achieve the required concentration, one method is to mix a balanced electrolyte solution with a hypertonic dextrose solution. Although rates ranging from 0.5–2.0 mg/kg/min have been reported, horses generally tolerate 1.7 mg/kg/min. However, this rate may not be sufficient for certain horses as it only provides 35% of the RER, which is the minimum rate for permissive hypoalimentation. Administering dextrose at excessive rates can cause hyperglycemia, hypovolemia, and hyperosmolality. Additionally, there may be a noticeable dilutional effect from large volumes of isotonic fluids. Parenteral nutrition should only be administered for brief periods of time—less than three days—using dextrose solutions.
- **Amino Acid and Dextrose Solutions:** For parenteral nutrition in

horses, a more physiological solution is a mixture of protein and carbohydrate solutions. This approach combines the delivery of the daily RER with a solution of amino acids that lowers the catabolism of lean muscle and 50% dextrose. The concentration of amino acid solutions varies greatly, but 8.5–10.0% is the most commonly used range for horses. The target range for protein supplementation is 0.6–2.0 g/kg/day. Lower levels of protein supplementation should be used if liver disease or renal failure are diagnosed; on the other hand, higher levels may be needed in cases of severe sepsis or burns. To avoid using the amino acids for energy, it's critical to make sure there are enough calories after determining the total amount of protein needed. To produce a balanced formula, the non-protein energy to nitrogen ratio calculation should guarantee a ratio between 100 and 150:1.

These solutions are frequently supplied in a dual chamber bag, which breaks an internal membrane to allow a combination of the dextrose and amino acids, simplifying parenteral nutrition. In equine medicine, the most widely used solution is made up of 8.5% amino acids and 50% dextrose in a 1:1 ratio. Following mixing, 1 liter of a 25% dextrose solution in a 4.25% amino acid mixture is produced, yielding a total of 1.002 kcal/ml. The ratio of non-protein energy to nitrogen is 200:1, and the solution's osmolality is 780 mOsm/L. Initial target infusion rates for a 500 kg horse are about 343 ml/h, based on 60% of the horse's resting energy needs. It is advised to begin at about 25% of this rate, or 85 milliliters per hour, and increase for the first 24 hours as tolerated.

Lipid-free formulations have the advantage of being less expensive and reducing infection-related complications due to the removal of immunosuppressive long-chain fatty acids. On the other hand, hyperglycemia, which has been shown to raise mortality in both human and equine patients, may become more likely due to the high concentration of dextrose. Furthermore, because of the solution's osmolarity, the formula needs to be diluted before being administered or combined with the infusion of a crystalloid solution at the same time.

- **Amino Acid, Dextrose, and Lipid Solutions:** A combination of lipids, carbohydrates, and amino acids is the typical nutrient combination used for equine TPN. Lipids must supply a minimum of 8% of total energy needs to guarantee sufficient fatty acid supplementation. It is more costly and requires more work to calculate, but it has several benefits. The final solution's increased osmolarity comes first. Lipid solutions that are 20% pure and have an osmolarity of 260 mOsm/L are iso-osmotic. This lowers the final solution's osmolarity, which may lower the chance of thrombophlebitis. Additionally, lipid solutions include linoleic acid, an essential fatty acid that can gradually be lost in horses that aren't fed.

Any horse who is at risk for hyperlipemia or has evidence of it, which would worsen with the administration of additional lipids, is contraindicated. Lipid dosages should not exceed 2.5 grams per kilogram per day. It's critical to keep in mind that certain electrolytes (calcium and magnesium), sunlight, pH, and storage conditions can all destabilize lipids. Amino acid solutions, however, provide some lipid protection. The lipid solution must be mixed with the amino acid and dextrose solutions before combining TPN. Yellow droplets or streaks that suggest stearation indicate that the mixture has become unstable and should be thrown out.

### *8.3.3.3.2. Administration of Parenteral Nutrition*

Parenteral nutritionation in horses can be administered via peripheral veins, such as the lateral thoracic and jugular. It is best to use a specialized multilumen catheter made of silicone or polypropylene with separate injection ports for fluids and medications. Even though it's not ideal, delivery via a single-lumen catheter has worked as long as a Y port is utilized to supply fluids and medications independently. It is crucial to keep the lines sterile at all times. To lower the risk of infection, triple antibiotic ointment can be applied to all connections, gloves should be worn when handling them, and all ports should be cleaned before use. To lower the chance of bacterial colonization, all lines should be changed every 24 to 48 hours, particularly when lipids are being used.

Though newer products provide both glucose and amino acids in separate chambers inside one bag, which allows for mixing without contamination after breaking the internal seal, solutions should always be mixed under sterile conditions in a laminar flow hood. If lipids are required, they must still be added in a sterile environment. Since lipids are destabilized by pH and several electrolytes, additional additives, such as vitamins and minerals, must be compatible with the solution, as guaranteed by the manufacturer. Because certain vitamins, amino acids, and lipids are light-sensitive, it's critical to shield the solution from the sun's rays both before and after administration. Lipid embolization may occur from the oxidation of fatty acids, which can form big droplets that are impossible for the endothelium to remove. After mixing, the solution can be kept in the fridge between 2 and 8 degrees Celsius for up to 24 hours before using; however, it must be used right away after being taken out of the fridge.

When administering parenteral nutrition, an infusion pump works best to guarantee precise and continuous infusions. To minimize costs and prevent complications from overeating, parenteral nutrition is usually given as 60–85% of the horse's total daily resting energy requirements. Parenteral nutrition should generally be started at 25% of this decreased rate and increased by 25% every 3–6 hours if the patient is responding well to the solution. Serum chemistries, body weight, and changes in the patient's disease status will indicate whether additional increases are required once the patient is receiving the full infusion rate. After enteral feeding has begun, parenteral nutrition should be stopped in a similar phased manner.

Extensive observation is necessary to guarantee appropriate levels of serum glucose, triglycerides, electrolytes, and organ function, both during the introduction phase and afterward. Initial blood glucose readings should be taken every 4 hours, and after that, every 6 hours at the full rate of parenteral nutrition. Even though stringent glycemic control has been recommended, while receiving parenteral infusions, it is reasonable to anticipate maintaining blood glucose levels between 110 and 150 g/dl. The horse's glucose tolerance can also be ascertained by looking at its urine. Urine may show glucose if the renal threshold (serum glucose of 200–220 mg/dl) is exceeded. The infusion rate should be lowered and rate increases should be made more gradually if hyperglycemia develops. If the hyperglycemia doesn't go away, the amount of dextrose in the solution can be lowered or an insulin infusion can be started.

Since hyperlipemia is known to occur without noticeably changing the clarity of the serum, serum triglycerides should also be checked at least once daily. In spite of protein supplementation, blood urea nitrogen (BUN) and serum creatinine levels should be monitored every day to guarantee appropriate hydration and renal function. Determining whether amino acid supplementation is adequate may also be aided by serum creatinine and BUN concentrations. BUN may fall if protein intake is insufficient; on the other hand, it may rise if supplementation is excessive. A normal creatinine concentration combined with an elevated BUN can suggest muscle catabolism and protein deamination for energy. Lastly, a daily evaluation of liver function is necessary.

#### *8.3.3.3.3. Complications*

Hyperglycemia is the most frequent side effect of parenteral nutrition in horses, and it can be made worse by endotoxin, cortisol, catecholamines, and the concentration of the glucose solution. Because of glucose intolerance and rebound hypoglycemia, high glucose administration rates (1.91–7.41 mg/kg/min) must be introduced and discontinued gradually; in contrast, slower rates (1.39 mg/kg/min) can be given and stopped more quickly. To preserve normoglycemia, a continuous rate infusion of normal insulin at 0.04–0.1 IU/kg/h can be added if necessary. Serum glucose levels must be checked frequently because severe illness and sepsis can worsen glucose intolerance. The patient's inability to keep their serum glucose levels within normal ranges could be the first indication that something is wrong with them.

Hyperinsulinemia can also result from hyperglycemia. Although glucose given by parents is known to raise insulin levels, enteral glucose has a greater effect. Hyperinsulinemia may help the horse by improving glucose and triglyceride absorption and having anabolic effects on skeletal muscle, even though glucose intolerance may occur.

While complications from intravenous catheters are not unusual in humans, it appears that this is not as common in veterinary patients. It is still important to keep a close eye out for any symptoms of thrombosis or infection in catheters, such as heat, swelling, and loss of patency. Although adult horses typically leave their catheters exposed, a bandage should be taken into consideration because parenteral nutrition increases the risk of catheter-related sepsis.

TPN side effects include hyperlipidemia and hypertriglyceridemia. The most vulnerable animals are ponies, donkeys, and miniature horses, whose triglyceride and serum biochemistries should be regularly checked. Furthermore, due to increased adipose tissue catabolism by catecholamines, inhibition of hepatic metabolism of chylomicrons, and suppression of lipoprotein lipase activity by inflammatory mediators, horses with sepsis or endotoxemia are more likely to develop hyperlipidemia.

One horse receiving parenteral nutrition has been reported to have hypermagnesemia leading to seizures. Other research, meanwhile, has not detected any appreciable variations in electrolytes. Human cases of hypermagnesemia have also been documented, presenting with bradycardia, apnea, hypotension, and loss of muscle tone. Horses receiving parenteral nutrition as well as additional electrolyte solutions need to have their electrolytes regularly monitored.

Amino acid solutions can result in uremia and hyperammonemia, whereas dextrose and amino acid solutions can cause phlebitis, acidosis, and osmotic diuresis. Furthermore, duodenal motility, migrating motor complexes, and small intestinal transit time may all be shortened by elevated amino acid concentrations. In horses receiving parenteral nutrition, it has been observed that 5% dextrose solutions increase the frequency and volume of gastric reflux.

# 8.4. DIETS FOR SPECIFIC DISEASES

## 8.4.1. Uncomplicated Colic

Most cases of colic are not diagnosed with a specific cause. The main suspects are mild impactions or vagotonia, or gas colic. In these cases, a physical examination is usually unremarkable, and advanced diagnostics are not always necessary. Treatment, which consists of mineral oil, oral fluids, and non-steroidal anti-inflammatory drugs, is generally effective, and most patients recover completely in 12 to 24 hours.

Feeding should be resumed in cases of simple colic as soon as normal fecal production and borborygmi are observed. For the first 12 to 24 hours of refeeding, high-quality forages should be given in tiny amounts (0.5 kg or 1.1 lb) every 4 to 6 hours. Feedings can be gradually increased to normal rations over the course of 24 to 48 hours if there are no new symptoms of discomfort. While forages are the ideal diet for horses, there are other options as well, such as hay cubes, alfalfa pellets, complete pellets, or hand grazing for 15–20 minutes at a time. Mashing food is a common and acceptable way to improve feed's palatability and water intake. For a further 10–14 days, grains should be avoided to prevent further microbial disruption. The kind of feed the horse will be given again after it heals should be attempted to be reintroduced by the clinician. If the owner wants to go back to the original diet, adjustments should be made gradually over 7–10 days if the feed that the horse is reintroduced to differs from their usual diet.

The majority of simple colic cases don't come back. Nonetheless, a thorough examination of the diet, surroundings, and feeding schedule is necessary to determine any risk factors that might have contributed to the colic. Preventive actions can be suggested if an issue is found, such as a recent trip or a change in the feed supplier (Zimmel, 2002; Shirazi - Beechey, 2008).

## 8.4.2. Equine Gastric Ulcer Syndrome

The term "equine gastric ulcer syndrome" (EGUS) refers to a group of medical conditions that cause ulceration of the glandular and squamous portions of the stomach, as well as colic symptoms. The cause of gastric ulceration is an imbalance between the protective (mucus, bicarbonate, mucosal blood flow, prostaglandin E2, epidermal growth factor, gastroduodenal motility) and ulcer-promoting (bile acids, pepsin, and hydrogen chloride) factors. High concentrate diets, low roughage or low-quality fiber diets, meal feeding, and fasting are dietary factors that may aggravate EGUS. The main dietary recommendations for EGUS prevention are to feed animals more frequently, feed fewer starches, and feed forages that are of higher quality. The main defenses against the acid the horse's stomach produces are saliva and food that has been swallowed. For this reason, pasture turnout is frequently cited as the most effective way to manage and avoid stomach ulcers. Though horses naturally tend to eat less at certain times of the day, this may not completely eradicate ulcers. Regardless of their housing, it has been proposed that these foraging patterns are what cause a circadian pattern to the pH of the proximal stomach in horses. This could be the cause of gastric ulcer diagnoses in horses, even in those kept only on pasture. Since intermittent feeding or feed withholding is an experimental model used to consistently produce gastric ulcers, free choice forages should still be provided to horses even though we have no control over their voluntary intake.

Different factors have been linked to altered feed intake and gastric ulceration: stall confinement and meal feeding. According to a study, feeding meals more than six hours apart raised the risk of non-glandular ulceration, indicating that having constant access to forages could be essential to lowering this risk. The findings of a second study, which discovered ulcers in 75% of horses fed twice daily and 579.9% of horses fed three times daily, corroborated this. These results support the recommendation that horses not on pasture be continuously fed alfalfa or high-quality grass hay to preserve the non-glandular mucosa. Compared to grass hays alone, alfalfa may offer superior protection due to its high protein and calcium content. Alfalfa may have a direct buffering effect due to its high protein content or by lowering the secretion of gastrin. Even in cases where concentrates like barley, oats, or pelleted rations were fed to the alfalfa, the pH effect remained constant. These results suggest that alfalfa should be the preferred feed for horses who have been diagnosed with or are at risk of developing gastric ulcers.

Because grains have been linked to the development of gastric ulcers, it is advisable to minimize their intake in horses suffering from EGUS. The majority of horse cereals contain soluble carbohydrates, which the stomach ferments into volatile fatty acids. Because these fatty acids can enter mucosal cells at low pH levels, they cause cell disruption and swelling.

It is also known that concentrate diets stimulate the release of gastrin, the hormone that secretes acid. A pH of less than 4 exacerbates the effects of volatile fatty acids, although gastric acid can be disruptive on its own. Furthermore, it has been demonstrated that grains postpone gastrin secretion, raising the possibility of unintended acid secretion following ingesta. Since the non- glandular mucosa is more susceptible to acid exposure when the stomach is empty, feeding grains may increase the risk of ulceration.

To lower volatile fatty acids below the threshold for mucosal damage, grains should only be fed in amounts less than 0.5 kg/100 kg body weight and no less frequently than every 6 hours. Because a starch intake between 1–2 g/kg/meal was linked to a 2 × 6 increase in the likelihood of ulcers (95% CI 1.3–5.2; P=0.006) and a starch intake greater than 2 g/kg bodyweight per meal increased the likelihood by a 3 × 2 multiplier (95% CI 1.3–7.7; P=0.009), the grains chosen should be low in starch. There should always be hay available, preferably alfalfa. Vegetable oils and other fat supplements can be used in place of concentrates to give you energy. Lipids may be helpful in the treatment and prevention of glandular ulceration by lowering gastric acid and raising prostaglandin synthesis, even though they were not able to prevent non-glandular ulceration.

Although there is no greater risk of gastric ulcers in horses with colic than in the general population, the possibility of ulcer formation may increase if the horses are likely to be denied feed because of their illness. 188 Equine illnesses that could necessitate extended periods without oral nutrition (i.e., proximal jejunitis and duodenitis) are especially vulnerable. This supports the claim that, in order to lower the risk of iatrogenic complications from feed withholding, horses with colic should be fed as soon as their medical conditions permit.

### 8.4.3. Duodenitis/Proximal Jejunitis

Proximal jejunitis (DPJ), also called anterior enteritis, is a type of inflammation-related disease that causes copious nasogastric reflux, gastrointestinal ileus, and hypersecretion in the proximal small intestine. You can find a complete explanation of the course of this disease, its diagnosis, and its treatment elsewhere. Although a number of infectious organisms have been linked to the problem, including Salmonella species, Clostridium difficile, and *Clostridium perfringens*, the precise cause is still unknown. Necropsy findings of lesions consistent with DPJ have also led to a possible connection with aflatoxicosis or fusariotoxicosis. For this reason, it's critical to keep a close eye out for mold in concentrates and other feeds that contain corn.

Because high concentrate diets contain large amounts of hydrolysable carbohydrates that ferment in the stomach and possibly the small intestine, it was previously believed that these diets increased the risk of DPJ. 197 A case control study conducted in Texas found a modest odds ratio of 1.3–1.62 (P< 0.001), supporting the statistical relationship between grains and DPJ. The study fed horses with DPJ a median of 4.1 kg concentrate per day, while the control group received 2.7 kg.

In this same study, there was a significant increase in the likelihood that horses with DPJ had access to pasture grazing, with an odds ratio of 3.5–4.0 (P<0.0005). Due to this association with pasture grazing, DPJ and equine dysautonomia, also known as equine grass sickness, which is another ileus cause in horses connected to grazing pastures, have been compared. Similar to DPJ, mycotoxins and Clostridium spp. are suggested causes of dysautomonia., indicating that both syndromes have an infectious cause.

For DPJ, medical therapy entails prokinetics, anti-inflammatory drugs, plasma or colloids to support oncotic pressure, and IV fluids to maintain hydration. If no other cause for the obstruction of the small intestine can be identified or if medical treatment proves ineffective, surgery ought to be taken into consideration. If surgery is chosen, an incomplete bypass or small intestine decompression into the cecum may be carried out. If aggressive treatment is pursued and the disease is allowed to run its course, there is a good chance that both medical and surgical treatments will result in a full recovery.

The severity of ileus in horses with DPJ typically prevents them from receiving enteral feed, resulting in large amounts of gastric reflux. Furthermore, prolonged cachexia can lead to severe protein loss because of the breakdown of the intestinal mucosal barrier and protein catabolism. Parenteral nutrition should be considered for any horse diagnosed as malnourished (low BCS), for horses with increased demands (pregnancy, lactation), and for any horse with DPJ that refluxes for longer than 48 hours, depending on the BCS at the time of diagnosis. Parenteral nutrition should be administered until the horse is willingly consuming at least 60% of its energy needs while at rest.

Once progressive motility is verified by an absence of net gastric reflux and ultrasonographic confirmation of small intestinal contractions, enteral feeding can start. To confirm progressive motility, water is initially given in small amounts (1–2 liters every 2–4 hours). Hay, softened complete feeds, or alfalfa pellets can be gradually introduced if water is accepted without any problems. To begin, 25% of the daily ration should be divided into meals every 2–4 hours. In two to three days, the horse should be returned to a full diet. Vital signs, hunger, disposition, and recurrent ultrasonography exams will evaluate how the feedings progress during reintroduction. Avoid using concentrates or fresh grasses that have a high level of fermentable carbohydrates. Grain consumption should be restricted to meals weighing less than 0.5 kg/100 kg of body weight and more than four feedings per day during the first two weeks following recovery. Gaining back the weight lost during the illness might necessitate eating forages high in protein, such as alfalfa or vegetable oils (Zimmel, 2002; Shirazi - Beechey, 2008).

### 8.4.4. Small Intestinal Strangulation

Both the intestinal lumen and the blood supply are blocked by small intestinal strangulations. Numerous conditions, such as mesenteric lipomas, mesenteric rents, intussusceptions, and epiploic foramen entrapment, can result in strangulating lesions; however, none of these conditions are directly linked to dietary factors. Clinical indicators of strangulation include peritoneal fluid changes, small intestinal distention on rectal and ultrasonography examinations, intense pain that is not relieved by medication, and varying degrees of nasogastric reflux based on the length and location of the obstruction. An exploratory laparotomy confirms the definitive diagnosis of the strangulating lesion, and if the bowel is considered non-viable, resection and anastomosis may be carried out.

The small intestine is highly susceptible to distention brought on by obstruction and the manipulation needed during surgery, both of which can lead to post-operative ileus. After surgery, enteral feeding is not feasible if ileus develops. Parenteral feeding is therefore frequently necessary to preserve bodily condition and offer nourishing nourishment to aid in healing. The owner's financial concerns, coexisting medical conditions, BCS at presentation, and other factors all influence when early parenteral nutrition should be started. Although there is evidence of better nutritional status in post-operatively fed horses receiving parenteral nutrition, including decreased serum triglycerides and total bilirubin levels, the advantages of parenteral nutrition for surgical colic cases remain unclear, with the potential drawback being higher client costs. Since there isn't enough proof, the majority of adult horses are just kept off feed until their ileus goes away.

After surgery, horses undergoing enterotomies or small intestinal resections should be given oral feed as soon as possible to avoid post-operative ileus. Traditionally, oral supplementation was started only after defecation and the presence of borborygmi, but modern medicine suggests enteral feeding even sooner. In that sense, veterinary medicine is evolving as well.

Freeman et al., suggest, rather than depending on defecation or intestinal sounds, reintroducing feed on a predetermined schedule following surgery to any horse that does not exhibit symptoms of colic or nasogastric reflux. This plan calls for giving horses small amounts of water 12–18 hours after surgery, then alfalfa hay in handfuls, which is gradually increased every 2–4 hours for the next 24 hours. Over the course of one or two days, the horse is to be reintroduced to at least 75% of the full feed. When severe mucosal edema is suspected or anastomosis function is doubtful, horses may benefit from alternatives to hay, such as softened complete feed, alfalfa pellets, or grass provided as small meals. To assess whether feeding is proceeding satisfactorily, factors such as attitude, packed cell volume, serum total solid concentration, progressive gastrointestinal motility without nasogastric reflux, and vital signs are taken into consideration. Controlled small feedings improve gastrointestinal motility and lower expenses and complications, such as gastric ulcers.

A malabsorptive disorder characterized by strangulations with resections that leave more than 60% of the small intestine intact is known as short bowel syndrome. However, depending on individual cases where horses have survived with less (12.3–15.0 feet), the length of the small intestine required for adequate digestion and absorption may vary. The ileum is the special location for the absorption of fat-soluble vitamins, while the small intestine serves as the main site for the absorption of minerals, proteins, and carbohydrates. But the large intestine seems to be able to make up for the lack of absorption of B vitamins, proteins, and carbohydrates after surgery. Furthermore, the small intestine can undergo compensatory adaptive hyperplasia, which enlarges the remaining intestine's length, diameter, and surface area in order to improve digestion efficiency.

Short bowel syndrome is characterized by diarrhea, hepatic damage, loss of appetite, and weight loss. In order to support the production of volatile fatty acids, the objective, if identified following surgery, is to optimize large bowel fermentation using a highly digestible fiber source, such as alfalfa, beet pulp, or soybean hulls. Supplementing the diet with fat-soluble vitamins may be necessary if the ileum or more than 75% of the small intestine has been removed. In cases where ileal resection has occurred, supplementation with rice bran or corn oil may be necessary to provide extra energy to maintain body weight. Horses with distal small intestine resections up to 70% of the time have also been successfully supported by complete pelleted feeds given in small, frequent meals. However, because grains cannot properly process starches, they should be avoided if a large resection has been done.

### 8.4.5. Ileal Impaction (Non-Strangulating Small Intestinal Obstruction)

In the Southeast of the US, ileal impactions are most frequently seen in mature horses. The excessive non-fermentable fiber and fine stem character of low-quality coastal Bermudagrass hay are associated with this obstruction. These factors may be related to harvesting maturity, soil composition, curing, and storage. *Anoplocephala perfoliata* tapeworm infection is another risk factor frequently mentioned for ileal impactions. These tapeworms localize at the ileocecal junction and can cause inflammation-induced blockage of the ileocecal valve.

Depending on the length and degree of the blockage, moderate to severe colic symptoms can be observed as clinical indicators of ileal impaction. When the disease is still in its early stages, transrectal palpation may reveal the impaction as a firm tubular structure entering the right side of the abdomen's cecum. The small intestine oral to the ileum fills with ingesta as the obstruction deepens and becomes complete, and multiple loops may be palpable per rectum and observed on ultrasound examination. It is possible to get gastric reflux and ileus results. Currently, medical therapy that includes sedation, analgesics, gastric decompression, and intravenous fluids is advised. Surgery is advised if protein concentration and white blood cell count rise and medical therapy is not working. Serial abdominocenteses can be used to track the effectiveness of medical therapy. During surgery, direct intraluminal infusion of carboxymethylcellulose or balanced electrolyte solutions is frequently used in conjunction with external massage to reduce the impaction into the cecum. An enterotomy might be helpful in some circumstances to lessen the trauma brought on by outside manipulation. Risks associated with any surgical procedure used to relieve an ileal impaction include adhesions and post-operative ileus.

Parenteral nutrition is rarely necessary for supportive care because the majority of horses will recover from ileal impactions with medical therapy in less than 12 hours. Whether they have undergone surgical or medical management, horses should be fed according to the small intestinal strangulation schedule during the acute phase of recovery. Gradually resume feeding as soon as the impaction seems to have resolved. Clinical resolution would be defined as the absence of pain, good intestinal motility on ultrasonography examination, no palpable impaction, and no small intestinal distention in addition to the absence of significant nasogastric reflux. However, if feeding is done too aggressively, the ileum may become edematous or spasm due to inflammation from the impaction or surgical manipulation, increasing the likelihood of recurrence. Maintaining the use of non-steroidal anti-inflammatory drugs during refeeding has been suggested by some clinicians to lower this risk, and hay should not be given until 48 hours after the clinical signs have resolved. Based on firsthand observations, it is recommended to treat large intestinal impactions before refeeding because they frequently indicate a coexisting condition. After healing, it is advised not to feed coastal Bermudagrass hay because ileal impactions have been conclusively linked to it. Complete pelleted diets or pasture are suitable substitutes for forage If alternate hay sources are unavailable. If feeding coastal Bermudagrass hay is necessary, it ought to be done so gradually rather than in the form of free-choice round bales. It is advised to use first cutting, high-quality hays and to make sure there is water intake. Furthermore, feeding concentrates or a pelleted diet can lengthen the gastrointestinal transit period and potentially lower the chance of reimpaction (Zimmel, 2002; Shirazi Beechey, 2008).

## 8.4.6. Ascending (Large) Colon Impactions

The most common cause of colic in horses is ascending colon impactions, which can be brought on by a variety of external factors such as stress, poor feed quality, reduced water intake, or dietary modifications that cause the colon ingesta to become dehydrated. Usually, obstruction happens where the large colon narrows, which includes the diaphragmatic and sternal flexures, the pelvic flexure, and the left ventral colon. Mild to moderate colic accompanied by a decrease in fecal output are common clinical signs. Fecal balls that are firm and covered in mucus may be felt during transrectal palpation. If an impaction in the large colon is felt, the diagnosis is verified. Fluids, laxatives, and the prudent use of analgesics and anti-spasmotics are the mainstays of therapy. Feeding should be stopped while the horse receives treatment. If the impaction does not go away quickly, the abdomen becomes more swollen, or the pain becomes intolerable, surgery is recommended. During surgery, lavage via a pelvic flexure enterotomy is typically used to relieve the impaction.

Feeding shouldn't be postponed for horses receiving medical therapy alone after the impaction has cleared. The absence of pain, normal fecal production, no abnormalities on transrectal palpation, and normal gastrointestinal transit time as indicated by the passage of mineral oil are all indicators of resolution. Good quality forages, preferably alfalfa hay, should be given in small amounts 4–6 times a day in the beginning (0.5 kg or 1.1 lb), and then gradually increased over 24–48 hours to normal rations. In order to lower the risk of gastric ulcers, free choice hay might be preferred following the successful reintroduction of feed. Depending on the preferences of the patient and the clinician, complete pellets, alfalfa pellets, or grazing, beginning at 15-minute intervals, can be used as alternatives to hay during the acute phase. After feed is reintroduced, the horse should be back on a full ration in 1–3 days (1.5–2.5% of body weight in forages). After the colic has cleared up, grains should be avoided for ten to fourteen days in order to prevent undigested starches from causing additional microbial population disruptions. Reintroducing concentrates should be done gradually, beginning with less than 1 kg (2.2 lbs) per 500 kg horse twice a day and working your way up to 0.5–0.75 kg increments per day if necessary.

The feeding schedule for a resolved medically treated impaction should be the same if the horse has had a celiotomy to relieve the impaction. Once fully recovered from anesthesia, most horses can resume feeding, typically 3–6 hours after going back to their stall. A 12-to 24-hour feeding delay is preferred by certain medical professionals following an enterotomy. The colon lumen is sufficiently large at the pelvic flexure to prevent any distention by feed from stressing the suture line, but a secure closure and appropriate surgical technique won't prevent the reintroduction of feed. Furthermore, the large colon flora is disrupted by enterotomy and lavage; this should be restored as soon as possible with a nutritious substrate for fermentation. Grazing or grass hay are both advised in the early post-operative phase following an enterotomy because grass hay has been linked to a lower risk of post-operative diarrhea in horses. Because of the potential for salmonellosis and the inevitable changes in gastrointestinal flora, it is crucial to keep hospital biosecurity in mind when grazing horses that have experienced colic. Enterotomy was also found to increase the risk an additional 1.5 times ($P = 0.042$), making it especially important post-operatively for large intestinal problems, which double the risk of diarrhea over other types of intestinal lesions ($P < 0.006$).

The key to preventing impactions in horses that are prone to them is to provide high-quality fodder and promote sufficient water intake. Avoid late-season, high-fiber hay cuttings as well as forages known to be of lower quality, such as straw and coastal Bermudagrass. Diets based on alfalfa or green grass may have a laxative effect. Horses with dental disease will benefit from a diet that is more easily digested overall when fed complete pelleted feeds. Beet pulp, alfalfa pellets, or hay or alfalfa cubes can all offer a highly digestible source of fiber with shorter fiber lengths to enhance digestibility. Although the exact benefit of routine dental prophylaxis is still unknown, good occlusion will help to shorten fibers and may lower the chance of impactions. Up to eight tablespoons of salt (NaCl or KCl) can be added to the feed to promote water intake. Along with a bucket of plain water, interesting substitutes like electrolyte water, molasses-flavored water, or other flavorings can be added.

### 8.4.7. Sand Impactions

A particular kind of intestinal blockage brought on by consuming sand—typically from the environment—is known as a sand impaction. In particular, feeding in sandy arenas or on the ground has been linked to risk. Nonetheless, some horses—particularly foals—may purposefully consume sand if they have pica. Additionally, some areas' native sandy soils (i.e., Florida, Georgia, and Arizona) offer plenty of chances for inadvertent consumption during grazing, particularly on pastures with short cropping periods.

The irritated nature of sand can cause debridement of the mucosal lining, diarrhea, and signs of endotoxemia due to bacterial translocation. The clinical signs are often similar to feed impactions of the ascending colon. Sand impactions can happen in the small intestine, ascending and small colon, stomach, cecum, and other parts of the digestive tract. On the other hand, during surgery, horses most frequently have sand impactions of the dorsal ascending colon discovered. Large amounts of sand or persistent impacts that weaken the intestinal wall's integrity can cause peritonitis, or rupture of the stomach or large intestine.

A history of exposure and the identification of sand in the ingesta are frequently used to make the diagnosis of sand. Sand can be felt in the manure by feeling it "gritty" or as a palpable impaction. Determining exposure to sand may be made easier by fecal sedimentation. It is carried out by identifying sand sediment in the glove's fingers after dissolving three to four fecal balls in water inside a palpation sleeve. Fecal floatation, however, does not allow for the measurement of the amount of sand present in the colon and has a poor correlation with the presence of sand in the intestines because of a high percentage of false negatives. It is typical to observe a small amount of sand during a fecal sedimentation test in regions with sandy soil. Abdominal radiographs provide a better representation of the amount of sand present as a mineralized opacity; however, in larger horses, its sensitivity is low, particularly for sand impactions outside of the ascending colon. Sand accumulations range in grade from 0 to 4, and a sizable portion of them are longer than 5 cm and press the large colon up against the ventral body wall. While radiography can be used to determine the volume, ultrasound is not a suitable substitute for it when determining the presence of sand. While some horses may exhibit characteristic "waves on a beach" sounds caudal to the xyphoid process, auscultation is not sensitive for sand obstructions, particularly if the colic is reducing large intestinal motility. Treatment for colic resulting from sand impaction has focused on the use of psyllium fiber (0.5–1.0 g/kg), osmotic laxatives, and fluid therapy. Mucilage, which is found in psyllium husks, is thought to have laxative properties because it absorbs water. Psyllium needs to be given via a nasogastric tube as soon as possible because it gels when combined with water. An alternative is to combine 2 liters of mineral oil with psyllium to suspend it and make nasogastric tube administration simpler. The psyllium-oil mixture can be dosed and immediately followed by the administration of 2–4 liters of water.

The literature reports varying responses to psyllium treatment, even in the presence of additional laxatives or mineral oil. When sand was given via nasogastric tube at a dose of 0.5 g/horse, one study found no difference between treatment with psyllium (0.5 g/kg), wheat bran (1 g/kg), or a 1:1 mineral oil/water mixture (8 g/kg) compared with no treatment at all. Pseudocellium dosages up to 1.0 mg/kg did not work on horses that had surgery to insert 10 mg/kg of sand into their cecum. 240 A different study found that combining 0.5 kg of psyllium with 2 liters of mineral oil resulted in twice as much sand being cleared as using mineral oil alone. 241 However, only 51% of the sand was cleared following 5 days of treatment. When psyllium failed to resolve radiographic signs of impaction in horses, further treatment with magnesium sulfate (1 g/kg for 3 days) was successful.

Surgical candidates are horses that do not respond to medication or analgesics; during surgery, large colon displacements or volvulus are frequently discovered. The colon is meticulously exteriorized, a pelvic flexure enterotomy is carried out, and the sand is siphoned from the site of impaction—typically the right dorsal colon—through a ventral midline celiotomy. A thorough examination of the gastrointestinal tract is necessary to rule out transmural damage, vascular compromise, and pressure necrosis. For both medical and surgical treatment of sand impactions, short-term survival is good, and long-term survival is outstanding.

The feeding schedule for a horse receiving surgery and an enterotomy to treat sand impaction should resemble the one prescribed for surgically resolving feed impactions. It is crucial to understand, though, that horses treated with an enterotomy and lavage for ascending colon conditions are more susceptible to diarrhea. Because sand colic irritates the mucosa, horses are particularly susceptible to developing the condition. Administration of di-tri-octahedral smectite (0.5 kg/500 kg in 4 l water, q24h) for 3 days post-operatively may reduce diarrhea, based on positive results in clinical cases. Because di-tri-octahedral smectite neutralizes clostridial and Bacteroides spp., it may reduce damage to the colonic mucosa. poisons. Through the control of intestinal secretions and gastrointestinal flora, it may also have a long-lasting protective effect. This is a significant advantage for recuperating equines after surgery, since an enterotomy disturbs the mucous layer and the intestinal flora. The amount of sand that is important for causing clinical disease is unknown, and the amount may vary depending on the horse. Given that sand is abrasive, it makes sense to believe that any amount of it could affect the gastrointestinal tract's integrity and function. Prolonged exposure could be even more harmful.

Although sand impactions can be treated, it is preferable to avoid ingesting more sand in the first place. It is frequently impossible to take a horse out of its natural habitat, but it is possible to try limiting their exposure to sand. Feed should not be given to horses on the ground; instead, feed should be kept from spilling onto sandy soils by using big tubs or rubber matting.

To promote better growth, pastures shouldn't be overgrazed, rested, or reseeded. Forage types with a deeper root system and a lower propensity to be pulled from the ground while the horse grazes should be reseeded in pastures. Psyllium can be taken as a preventive (0.25 kg once daily for one week every month), either as a commercial pellet or as a generic powder. Psyllium could be administered daily to horses that have been exposed continuously, but there is some worry that the psyllium's effectiveness may be diminished if the gastrointestinal flora gets used to the new fiber source and ferments it. The best way to remove sand from the digestive system is to feed hay at a rate of up to 2.5 percent of body weight each day. Nearly 95% of swallowed sand can be eliminated by the bulk alone; in a controlled experiment, it outperformed psyllium, mineral oil, and wheat bran (Zimmel, 2002; Shirazi - Beechey, 2008).

### 8.4.8. Enteroliths and Fecaliths

Enteroliths are intestinal concretions that develop when minerals, mainly ammonium phosphate salts, magnesium, and nitrogen, are deposited around a foreign material nidus. Additionally, around fibrous, undigested feed materials known as fecaliths, concretions may form. These mineralized masses are thought to originate in the right dorsal colon ampulla. As they move aboral into the constricted transverse or descending colon, they may obstruct the passage completely or partially. If the enterolith is able to move orally back into the right dorsal colon, symptoms of discomfort in the abdomen may subside. These horses frequently have a history of intermittent and recurrent colic.

Sand, pebbles, rope, fabric, hair, or any other foreign material can act as the center for the formation of calculus. The cause of enterolithiasis appears to be multifactorial, according to recent research. All things considered, however, ultimately lead to an intestinal environment that encourages mineral deposition because of its basic pH and the minerals—magnesium, nitrogen, phosphorus, calcium, sulfur, sodium, and potassium—that are present.

Enteroliths have been linked to genetics; specific breeds, such as Arabians, Morgans, American Saddlebreds, American Miniature Horses, and donkeys, are predisposed to the condition. Siblings exhibit an increased predilection as well. This could be because horses with enterolithiasis have been shown to have a genetic predisposition toward a larger intestinal ingesta that is more alkaline than horses without enteroliths.

Diet also matters because some feeds have been shown to create an intestinal environment that is conducive to the formation of enteroliths. Alfalfa is high in protein and contains a lot of minerals, including ammonium nitrates. Feeding more than 50% alfalfa hay has been found to be statistically associated with enterolith formation (OR=4.74; 95% CI 1.44–15.63). It also causes a more alkaline pH and higher concentrations of calcium, magnesium, phosphorus, and sulfur in the colon. Wheat bran is also higher in protein, magnesium, and phosphate than most grains and is thought to be a contributing factor to the issue, though it has never been proven to be directly linked to enterolithiasis. Minerals can also be found in soil and water. Horses in California obviously have a predisposition, but enteroliths have been identified in several other states and nations.

Enterolith formation has been associated with low-fiber diets, feeding intermittent meals, and stall confinement lasting more than 50% of the day. It has been suggested that inactiveness may impact the right dorsal colon's ampulla's motility, which could lead to mineral deposition in stalling horses. Grazing can produce short chain fatty acids for fiber fermentation, which can lower colon pH and reduce the likelihood of crystallization. It can also provide exercise and dilute some of the minerals consumed.

Abdominal radiography, surgery, or necropsy can all be used to diagnose enteroliths. An enterolith in the small colon may be palpably identified, but frequently it is too far cranial to allow detection per rectum. The diagnosis may occasionally be confirmed by the owners' reports of small enteroliths passed in the feces. Horses with partial obstruction usually have a history of mild, recurrent colic; if the obstruction is complete, the clinical signs will resemble those of other large intestinal obstructions. Radiographs are less sensitive to identify enteroliths in the small colon and in horses with significant gas distention, but advanced imaging may enable diagnosis and early detection before complete obstruction occurring. The only option left is to remove it surgically using a large colon or descending colon enterotomy. If the surgery is elective and performed before obstruction and pressure necrosis develop, the prognosis could be better.

Since genetics cannot be changed, the practitioner should concentrate on prevention in high-risk horses, such as those who have a history of enteroliths, are boarded at facilities where the condition is common, or are connected to horses that have been diagnosed with enterolithiasis. Certain breeds of horses, such as ponies, Arabians, and American Miniature Horses, should also be classified as high risk if their diet, environment, or past history of chronic or recurrent colic predisposes them to further risks.

Avoiding feeds that raise the risk of enterolith formation, such as those heavy in phosphorus and magnesium, should be the main focus of prevention. Alfalfa hay ought to make up no more than half of the diet, and hay cultivated in California ought to be avoided. Because of their higher fiber content, which aids in the production of short chain fatty acids, and lower protein and magnesium levels, grasses and oat hays seem to offer some protection. Due to the ammonium produced, high protein diets should also be avoided unless absolutely necessary. A water filter can be used to extract minerals from the source of the water, but this is a costly endeavor, and in most situations, the mineral content of the feed greatly exceeds the contribution from water sources.

Acidifying the contents of the colon is one type of protective treatment. Stone formation may slow down or cease entirely if bacteria that break down starch acidify the colon, as calculi are formed in basic environments. Following that line of reasoning, studies have shown that enteroliths can be dissolved *in vivo* by lowering the pH of the colon with vinegar. Further reports suggest that administering vinegar may lower the risk of recurrence, although this has not been tested as a preventive or therapeutic measure. It is advised to consume 1–2 cups (8–16 ounces) of apple cider vinegar daily.

Controlling environmental factors is more challenging. It is not possible to completely eradicate the source of a nidus for the formation of calculi, but it is advised to lower the risk of ingesting foreign objects. In endemic areas with sandy environments, psyllium (0.25–0.5 kg, once daily for one week per month) should be administered on a regular basis. To minimize sand intake, feed should be given to horses off the ground and in tubs or on rubber matting. Finally, pasture turnout ought to be offered in light of the evidence for a protective effect (Zimmel, 2002; Shirazi - Beechey, 2008).

## 8.4.9. Ascending Colon Displacement

Given that the right dorsal and transverse colons are the only points of attachment to the body wall, the ascending colon can move freely within the abdomen. The ascending colon can become dislodged as a result of management practices such as feeding high concentrations of feed at irregular intervals, or it can develop as a primary issue secondary to other disorders like fecal impactions. Big concentrate boluses speed up the production of volatile fatty acids and fermentation by allowing fermentable carbohydrates to enter the ascending colon. Large meals may also cause disruptions to normal motility patterns, which increases the accumulation of gas and fluid due to altered fermentation. However, in certain types of displacements, the ultimate cause may be an underlying motility disorder.

Right and left dorsal (nephrosplenic) displacements, pelvic flexure retroversion, and non-strangulating volvulus at the cecocolic junction are the most common categories for displacements. The diagnosis of large colon displacement may be made despite variations in the colon's precise location within the abdomen. Entrapment of the ascending colon over the nephrosplenic ligament results in a left dorsal displacement. By tracing a band of the left ventral colon into the dorsal left quadrant between the kidney and base of the spleen during transrectal palpation, it is diagnosed. The inability to find the left kidney on an ultrasound examination through the left flank supports the diagnosis. Right dorsal displacement, which can cause the colon to move into a variety of positions inside the abdomen, is typically identified rectally by looking for a tight band that connects the right body wall to the cecum. Sometimes the colon retroflexes toward the diaphragm, causing the pelvic flexure to become imperceptible and the caudal abdomen to feel empty.

Abdominal distention and sporadic, mild to moderate abdominal pain are clinical indicators of displacement. Additionally, less feces are produced. The colon's own displacement or an impaction that may be a coexisting issue could cause obstruction of the lumen. Displacement can block lymphatic and venous outflow, which can cause tissue congestion and edema. Nevertheless, tissue damage is minimal and goes away fast if the displacement is fixed. Analgesics, hand walking to promote motility, and fluid therapy are possible medical management techniques. If the large colon is not severely distended with gas or impaction, phenylephrine and jogging may be tried to contract the spleen and replace the colon in cases with left dorsal displacement. The amount of abdominal gas distention may be lessened by trocarization. As an alternative, the colon can be attempted to be replaced ventrally by rolling the horse while sedated.

It is best to take horses to surgery right away if they have an abnormal abdominocentesis, severe distention per rectum, or excruciating pain. For most displacements, a ventral midline celiotomy will allow for correction. As an alternative, a standing flank approach might be used to relieve left dorsal displacements; however, a flank approach cannot be used for a thorough exploratory. For impaction relief, a pelvic flexure enterotomy may be chosen.

Once the displacement has resolved, horses recovering from displacements are returned to feed in a similar fashion to horses with ascending colon impactions (refer to the preceding section on ascending colon impactions). If the medical treatment was effective, transrectal palpation, sufficient fecal production, and the absence of colic symptoms for at least 12 hours can all be used to confirm that the colon has returned to its normal position. When feeding again, high-quality forage is advised, and most hays offer long fiber, which is preferred to increase the bulk of ingesta in the colon. Maintaining a sufficient amount of food in the ascending colon is thought to lower the chance that gas distention will encourage migration to an unusual location in the abdomen. It is advised to leave some food in the lumen after surgery and an enterotomy in order to give the colon some weight. To lower the chance of diarrhea, the leftover feed should also encourage regular fermentation. Concentrates lower the possibility of soluble carbohydrates entering the large intestine and causing aberrant gas production; they should be avoided for at least 10–14 days or removed from the diet.

The main strategies for preventing large colon displacements are to make sure that enough water is consumed and to provide sufficient amounts of high-quality fiber. Concentrates ought to be removed from the diet or fed in very small amounts (less than 0.5 kg/100 kg, at least once every 6 hours). Strategies for impact on prevention can also be applied to displacements, as ascending colon impactions are frequently a coexisting issue. Nephrosplenic entrapment has a low recurrence rate (between 7 and 8%), but this risk can be decreased by surgically ablation of the nephrosplenic space. According to a recent study, after right dorsal displacements are diagnosed and surgically treated, there is a higher chance of recurrent colic (recurrence rate of 41.9% compared to only

8.3% for left dorsal displacement). Recurrent colic was also more common in patients with non-strangulating volvulus and retroflexion of the pelvic flexure. Still, at roughly 7%, the recurrence rate for a second right dorsal displacement was comparable to that of nephrosplenic displacement. While increased gas production is thought to be the cause of left dorsal displacement, it is hypothesized that an underlying motility disorder may be the cause of right dorsal displacement. This could be the reason for the elevated recurrence risk, which could make other preventative measures ineffective. These studies suggest that in cases of repeated displacements, colopexy may be advised (Zimmel, 2002; Shirazi Beechey, 2008).

## 8.4.10. Ascending Colon Volvulus (Large Colon Torsion)

At the point where the right and left dorsal colons attach to the abdominal wall, the colon twists on its mesenteric axis, usually in a counterclockwise direction, resulting in ascending colon volvulus. Although the exact cause of volvulus is still unknown, recent dietary changes or the loss of abdominal fill following parturition have been suggested. The vasculature and lumen may become obstructed if the colon twists more than on its axis. This can lead to severe abdominal hypertension and blood flow strangulation because of fluid and gas distention in the lumen. Rapid therapy is necessary to increase survival because the colon sustains irreversible damage within three to four hours of the vasculature becoming occluded. Because of the severity of colic, a thorough examination of the horse is usually limited. Clinical findings include a distended large intestine on rectal exam and pain that is resistant to analgesics.

A ventral midline celiotomy is used to detorsion the large colon to treat volvulus. Once the colon has been derotated, its viability is evaluated by looking for signs of active mucosal bleeding at a pelvic flexure enterotomy, mucosal edema and bleeding, arterial pulses in the colonic arteries detected by Doppler or palpation, muscular motility, or fluorescein uptake. Histopathology is the gold standard for determining viability, but results are frequently not available until the horse has recovered from surgery or has been put to sleep. The colon can be resected up to the cecocolic fold if it is determined to be non-viable; however, damage frequently extends beyond what can be sufficiently removed. In these situations, even after a resection, dubious bowel may stay in the abdomen, raising the risk of dehiscence and lethal peritonitis.

Feeding of volvulus-affected horses after surgery is contingent upon the extent of colon damage. Early post-operative recovery (six to twelve hours after surgery) is when feeding can start for horses that are receiving prompt treatment and have little mucosal edema or mural bleeding. Typically, a pelvic flexure enterotomy is done, and some surgeons would rather not test the suture line with feed for at least 12 to 24 hours. If the colon is not severely damaged during the procedure, the horse can begin receiving high-quality feed at a rate of 0.5 kg every 4–6 hours. Hay can be substituted with full rations or alfalfa pellets fed at the same rate, or grazing for fifteen to twenty minutes. Over the following one to three days, the diet should be gradually increased until the entire ratio is reached (2.0–2.5 percent of body weight in hay). Because of the disruption of normal flora caused by both the ischemia event and lavage through the pelvic flexure enterotomy, grains should be avoided.

It is advised to delay oral nutrition until 12–48 hours after surgery in horses that have had a large colon resection or if the colon is still in situ but has suffered significant damage from ischemia to lower the risk of obstruction or anastomosis failure. This is particularly true for horses whose suture line is questionable because they are unable to fully resect the colon that is involved. Parenteral nutrition may be advised for horses with compromised intestines due to endotoxemia, anorexia, and disruption of both fermentation and absorption. Due to the loss of ascending colon surface area, a highly digestible diet is needed once the horse is reintroduced to feed based on vital signs and appetite.

Since the colon is the site of phosphorus absorption, protein digestion, and cellulose digestion, post-operative adjustments should be made to ensure that the diet includes a quality fiber source (<28%), at least 12% protein, and 0.4 percent phosphorus. Due to the shorter transit times following resections, pelleted diets—such as alfalfa pellets or complete rations—would seem to be the best option; however, it was discovered that their cellulose digestibility was decreased. In order to avoid weight loss, rations that contain pellets must frequently be fed at a rate higher than maintenance. Although grass hay is advised to lessen the possibility of post-operative diarrhea, horses with large colon resections or severe colon damage cannot maintain their body weight on grass hay alone due to its low protein content. Horses should get at least 2 percent of their body weight in alfalfa hay each day to stay in good condition. Extra phosphorus supplements are not required to maintain blood levels, even with decreased absorption. However, due to increased water loss in the feces, adequate water intake (up to 8 l more than normal horses) should be ensured after resection. Special diets are not necessary because the digestive tract appears to adapt over time through hypertrophy. In certain equines, adding oil or soybean meal supplements could enhance and sustain weight gain during the initial post-operative phase.

The mainstay of treatment for large colon volvulus is surgery. The literature reports recurrence rates ranging from 5 to 50%; therefore, to lower this risk, large colon resections or colopexy of the left ventral colon may be selected. Dietary modifications ought to be made gradually, and a diet high in forages is advised to lower the quantity of carbohydrates that enter the colon. However, broodmare occupational risk is a factor that is frequently unavoidable.

### 8.4.11. Cecal Impactions

The most common cecum-related disorder that causes colic is impaction. If impaction is the main issue, there are usually multiple contributing factors, such as low-quality roughage, compromised dentition, and decreased water intake. On the other hand, cecal impactions may occur as a side effect of treating another condition, such as an ocular or orthopedic one. Atropine administration, NSAID use (or lack thereof), painful conditions, anesthetics, and stall rest have all been linked to secondary impactions.

The physical nature of the impaction allows for additional classification of cecal impactions. Similar to ascending colon impactions, type 1 impactions are characterized by firm, dry ingesta and are frequently ascribed to mechanical obstruction. Idiopathic stasis with ingesta with a normal to fluid-like consistency is the hallmark of type 2 impactions. These horses frequently exhibit endotoxemia, peritonitis, and an edematous or thickened cecal wall. Nonetheless, there is frequently a great deal of overlap amongst classifications.

The only clinical symptoms that are frequently identified are depression, anorexia, and decreased fecal output. Since cecal impactions can burst without exhibiting any symptoms of discomfort, it is important to closely monitor the fecal production of all hospitalized horses. By using transrectal palpation to make the diagnosis, one can find either a fluid-filled cecum or an ingesta in the caudal right abdomen, along with taut cecal bands. Medical treatment for cecal impactions includes intravenous and/or oral fluids, as well as analgesics. You should not feed them. However, in cases where a type 2 stasis is suspected, the cecum is grossly distended, or medical management is not working, surgical management may be necessary. To increase survival, some surgeons advise early surgical intervention. To release the impaction, typhlotomy and lavage may be used during surgery. An ileocolostomy or jejunocolostomy, either full or incomplete, with or without occlusion of the distal small intestine, may be recommended if impaction recurs or if a type 2 impaction is diagnosed. A high rate of recurrence complicates feeding after cecal impaction. As a result, reintroducing feed to horses should be done gradually and cautiously, particularly if the triggering factor is unaddressed (i.e., stall rest, discomfort) or a possible cecal dysfunction. For the first 12 to 24 hours following the resolution of the impaction, only small amounts of water should be provided. A normal rectal examination and sufficient fecal output are prerequisites for resolution. Horses that are able to handle oral fluids may be gradually added back to the feed. Refeeding after surgery should be done gradually for patients who have had a bypass, beginning 36–48 hours after the procedure because there is a link between colic and refeeding. Since total fasting can further reduce cecal motility, hand walking and limited hand grazing (5–15 m) may be permitted early on to stimulate motility. Low residue feeds, such as whole pelleted diets and alfalfa pellets, are frequently recommended after feeding in order to minimize bulk in the large intestine and cecum. Hay should be gradually reintroduced, starting with handfuls; alfalfa is recommended because of its laxative properties. Because coastal Bermudagrass hay is linked to intestinal impactions, it is best to stay away from it.

As with ascending colon impactions, the mainstays of prevention are adequate water intake and high-quality forage provision. Good fiber should be available, such as beet pulp, fresh grass, or first cutting hays. Laxative diets, such as those based on green grass or alfalfa, might be recommended, particularly if the horse is unable to be isolated from the factors that increase the risk of cecal impactions. To improve cecal motility, if at all possible, exercise and paddock or pasture turnout should be offered. Given the possible link between tapeworms and cecal impaction, affected horses ought to undergo anti-tapeworm deworming (Zimmel, 2002; Shirazi  Beechey, 2008).

## 8.4.12. Cecocecal and Cecocolic Intussusception

The cecal apex invaginates into the cecal body, causing cecocecal intussusceptions. The cecal apex may proliferate into a cecocolic intussusception in the right ventral colon. Medication that alters motility or causes inflammation, parasites, bacterial infections, and dietary changes can all lead to cecal intussusceptions. Because of the varying degrees of intussusception and obstruction of the ingesta flow, the clinical signs are non-specific. Horses may exhibit mild to intermittent colic, fever, or weight loss during a protracted illness. The results of the physical examination are also non-specific; however, the intussusception might be palpable on the right side of the abdomen or detected by ultrasound.

Surgery is used as a form of treatment, with manual reduction of the cecum from the intussusception, typhlotomy, or amputation assisted by a colostomy, coming next. Reduction might not be feasible because of the disease's frequently protracted nature. After that, a full bypass is carried out using an ileocolostomy or jejunocolostomy, either with or without the colon's cecum being amputated.

Refeeding following intussusception can commence as soon as 24 hours following surgery, contingent upon the technique used. However, because refeeding has been linked to colic symptoms, feeding should be postponed for 36–48 hours if a bypass procedure is chosen. Water should be given first, and if the horse is able to drink, similar feed can be given to horses that have experienced a significant colon impaction. Good quality hay is given in small amounts (0.5 kg or 1 lb, 4–6 times per day) and gradually increased over 2–3 days until the horse is receiving a full ration. Alfalfa pellets or pelleted complete diets are alternatives to hay. Due to its effect on cecal motility, grazing should be encouraged, beginning at 4 hours and lasting for 15 minutes. Concentrates should be avoided for a period of 10–14 days to protect colonic microbial populations from harm.

## 8.4.13. Descending (Small) Colon Obstructions

Along with enteroliths in endemic areas, fecal impactions are the most common cause of obstruction of the small colon. The same risk factors that lead to impactions of other parts of the gastrointestinal tract also result in fecal impaction. Contributing factors have included poor quality roughage, dentition problems, parasite damage, inactivity, dehydration, and motility disorders. Although horses with small colon impactions are more likely to contract salmonellosis, it is still unclear if impaction is the cause or result of Salmonella spp. infection.

Reduced fecal output, mild to moderate abdominal pain, abdominal distention, small volume diarrhea or scant feces, fever, and straining to urinate are some of the clinical signs of descending colon obstruction. Upon rectal palpation, the presence of a foreign body or tubular impaction in the descending colon's lumen is usually indicative of an obstruction. Palpation will reveal only gas distension of the ascending colon and cecum if the obstruction is more oral or in the transverse colon. Enteroliths are not always visible, but radiographs can help visualize them.

Medical therapy for feed impactions of the small colon consists of laxatives, analgesics, and IV and oral fluids. If the horse is endotoxemic or leukopenic, antibiotics or endotoxin antiserum may be recommended. Surgery is necessary if the horse is not responding to medication or painkillers, if there is abnormal abdominal fluid, if there is an increase in abdominal distention, or if the mass seems to be an enterolith or foreign body. A high enema can be used to lavage the impaction with extraluminal massage during a ventral midline celiotomy. Another option is to perform an enterotomy to remove the impaction directly. An enema can be used to deliver a foreign body or enterolith per rectum if one is present. To remove or retropulse the obstruction into the colon if it cannot be mobilized, an enterotomy can be done in the pelvic flexure or the descending colon's taenia. To lower the chance of reimpaction, a large colon enterotomy should also be done to remove the majority of the fecal matter that is oral to the obstruction.

More cautiously, horses that underwent surgery to treat descending colon obstruction are reintroduced to feed. This part of the colon is more likely to experience stricture and dehiscence following an enterotomy because of the large lumen, high collagenase enzyme concentration, and mechanical strain from fecal balls passing through the area. It's also crucial to remember that ileus may be a factor in reimpaction because it takes a considerable amount of time (30–40 hours) after surgery to have your first bowel movement. Because of this worry, feed should be withheld from horses for 36–48 hours following surgery, and then feed should be reintroduced gradually over two to four days, starting at 0.5 kg every six hours. For the first 10–14 days following surgery, low residue feeds—like complete diets or alfalfa pellets—are recommended to prevent reimpaction. It may also be helpful to wet the feed into a slurry. If you are concerned about luminal size, you can stick to this diet indefinitely. Along with a laxative diet (fresh grass or alfalfa), laxatives and oral fluids may also help maintain soft feces in the immediate post-operative period. Furthermore, the feed can contain 0.2 grams of magnesium sulfate per kilogram per day.

A sufficient supply of fresh water and high-quality roughage may help prevent small colon impactions. In addition to dental care, older horses that no longer have enough grinding surface should be fed a complete extruded feed in place of hay. A pelleted diet ought to be fed as well to horses that have a suspected stricture. The section before this one discusses how to prevent fecoliths and enteroliths. Although the urge to eat inappropriate things is often outgrown by horses that are obstructed by a foreign body, the environment should be altered to lower the risk. A high-quality pasture will frequently decrease the amount of foreign material consumed.

## 8.4.14. Descending (Small) Colon Strangulations

Although volvulus, herniation, and intussusception have also been documented, mesenteric lipomas are the most common cause of strangulation of the descending colon. Abdominal distention and moderate to severe colic are common clinical signs, but discomfort symptoms are frequently not immediately apparent. The rectal exam and small colon obstruction are comparable in that both large colon distention and the absence of fecal balls are observed. If the blockage is far enough caudal, the clinician may occasionally be able to palpate it. Surgery will be required if there are abnormal rectal findings, progressive abdominal distention, and changes in the peritoneal fluid. To remove food particles from the ascending colon, a pelvic flexure enterotomy and colon resection are required as part of the treatment. Refeeding after surgery works similarly to surgical treatment for minor colon blockages (Zimmel, 2002; Shirazi Beechey, 2008).

# REFERENCES

1. Abutarbush, S., Carmalt, J., & Shoemaker, R. W. (2005). Causes of gastrointestinal colic in horses in western Canada: 604 cases (1992 to 2002). *Canadian Veterinary Journal*, *46*, 800–805.
2. Anthony, P. S. (2008). Nutrition screening tools for hospitalized patients. *Nutrition in Clinical Practice*, *23*, 373–382.
3. Argo, C. McG., Cox, J. E., Lockyear, C., et al. (2002). Adaptive changes in the appetite, growth, and feeding behaviour of pony mares offered ad libitum access to a complete diet in either a pelleted or chaff-based form. *Animal Science*, *74*, 517–528.
4. Asai, Y., Matsui, A., Osawa, T., et al. (1999). Digestible energy expenditure in grazing activity of growing horses. *Equine Veterinary Journal*, *30*(Suppl.), 490–492.
5. Atalay, B. G., Yagmur, C., Nursa, T., et al. (2008). Use of subjective global assessment and clinical outcomes in critically ill geriatric patients receiving nutrition support. *JPEN Journal of Parenteral and Enteral Nutrition*, *32*, 454–459.
6. Atkinson, M., & Worthley, L. I. G. (2003). Nutrition in the critically ill patient: Part I. Essential physiology and pathophysiology. *Critical Care and Resuscitation*, *5*, 109–120.
7. Barak, N., Wall-Alonso, E., & Sitrin, M. D. (2001). Validation of the stress factors used for calculating energy expenditure in hospitalized patients. *Clinical Nutrition*, *20*(Suppl.), A120.
8. Benedict, F. G. (1915). A study of prolonged fasting. *Carnegie Institute of Washington Publication*, 203, Washington, DC.
9. Bergman, E. N. (1990). Energy contributions of volatile fatty acids from the gastrointestinal tract in various species. *Physiological Reviews*, *70*, 567 590.
10. Boyd, L. E., Carbonaro, D. A., & Houpt, K. A. (1988). The 24-hour time budget of Przewalski horses. *Applied Animal Behaviour Science*, *21*, 5–17.
11. Brandi, L. S., Bertolini, R., Santini, L., et al. (1999). Effects of ventilator resetting on indirect calorimetry measurement in the critically ill surgical patient. *Critical Care Medicine*, *27*, 531–539.
12. Brown, R., Houpt, K., & Schryver, H. (1976). Stimulation of food intake in horses by diazepam and promazine. *Pharmacology Biochemistry and Behavior*, *5*, 495–497.
13. Burkholder, W. J. (2000). Use of body condition scores in clinical assessment of the provision of optimal nutrition. *Journal of the American Veterinary Medical Association*, *217*, 650–654.

14. Cargile, J. L., Burrow, J. A., Kim, I., et al. (2004). Effect of dietary corn oil supplementation on equine gastric fluid acid, sodium, and prostaglandin E2 content before and during pentagastrin infusion. *Journal of Veterinary Internal Medicine*, *18*, 545–549.
15. Carroll, C. L., & Huntington, P. J. (1988). Body condition scoring and weight estimation of horses. *Equine Veterinary Journal*, *20*, 41–45.
16. Carter, R. A., Goer, R. J., Staniar, W. B., et al. (2009). Apparent adiposity assessed by standardized scoring systems and morphometric measurements in horses and ponies. *Veterinary Journal*, *179*, 204–210.
17. Clarke, L. L., Roberts, M. C., & Argenzio, R. A. (1990). Feeding and digestive problems in horses: Physiologic responses to a concentrated meal. *Veterinary Clinics of North America: Equine Practice*, *6*, 433–450.
18. Cohen, N. D., & Peloso, J. G. (1996). Risk factors for history of previous colic and for chronic, intermittent colic in a population of horses. *Journal of the American Veterinary Medical Association*, *208*, 697–703.
19. Cohen, N. D., Gibbs, P. G., & Woods, A. M. (1999). Dietary and other management factors associated with colic in horses. *Journal of the American Veterinary Medical Association*, *215*, 53–60.
20. Cohen, N. D., Matejka, P. L., Honnas, C. M., et al. (1995). Case-control study of the association between various management factors and development of colic in horses. *Journal of the American Veterinary Medical Association*, *206*, 667–673.
21. Cohen, N. D., Vontur, C. A., & Rakestraw, P. C. (2000). Risk factors for enterolithiasis among horses in Texas. *Journal of the American Veterinary Medical Association*, *216*, 1787–1794.
22. Crowell-Davis, S. L., Houpt, K. A., & Carnevale, J. (1985). Feeding and drinking behaviour of mares and foals with free access to pasture and water. *Journal of Animal Science*, *60*, 883–889.
23. Cruz, A. M., Cote, N., & McDonell, W. N. (2006). Postoperative effects of anesthesia and surgery on resting energy expenditure in horses as measured by indirect calorimetry. *Canadian Journal of Veterinary Research*, *70*, 257–262.
24. Cuddeford, D., Pearson, R. A., Archibald, R. F., et al. (1995). Digestibility and gastrointestinal transit time of diets containing different proportions of alfalfa and oat straw given to thoroughbreds, Shetland ponies, highland ponies, and donkeys. *Animal Science*, *61*, 407–417.
25. Damask, M. C., Askanazi, J., & Weissman, C. (1983). Artifacts in measurement of resting energy expenditure. *Critical Care Medicine*, *11*, 750–752.

26. de Fombelle, A., Julliand, V., Drogoul, C., et al. (2001). Feeding and microbial disorders in horses: 1. Effects of abrupt incorporation of two levels of barley in a hay diet on microbial profile and activities. *Journal of Equine Veterinary Science*, *21*, 439–445.
27. Donaghue, S. (1992). Nutritional support of hospitalized animals. *Journal of the American Veterinary Medical Association*, *200*, 612–615.
28. Drogoul, C., de Fombelle, A., & Julliand, V. (2001). Feeding and microbial disorders in horses: 2. Effect of three hay
29. Durham, A. E. (2009). The role of nutrition in colic. *Veterinary Clinics of North America: Equine Practice*, *25*, 67–78.
30. Durham, A. E., Phillips, T. J., & Walmsley, J. P. (2004). Nutritional and clinicopathological effects of postoperative parenteral nutrition following small intestinal resection and anastomosis in the mature horse. *Equine Veterinary Journal*, *36*, 390–396.
31. Dyer, J., Fernandez-Castano Merediz, E., & Salmon, S. H. (2002). Molecular characterization of carbohydrate digestion and absorption in equine small intestine. *Equine Veterinary Journal*, *34*, 349–357.
32. Eccles, R. C., Swinamer, D. L., & Jones, R. L. (1986). Validation of a compact system for measuring gas exchange. *Critical Care Medicine*, *14*, 807–811.
33. Elia, M. (1995). Changing concepts of nutrient requirements in disease implications for artificial nutritional support. *The Lancet*, *345*, 1279–1284.
34. Ellis, J. M., & Hollands, T. (1998). Accuracy of different methods of estimating the weight of horses. *Veterinary Record*, *143*, 335–336.
35. Embertson, R. M., Colahan, P. T., & Brown, M. P. (1985). Ileal impaction in the horse. *Journal of the American Veterinary Medical Association*, *186*, 570–572.
36. Frankenfield, D. C., Coleman, A., Alam, S., et al. (2008). Analysis of estimation methods for resting metabolic rate in critically ill adults. *JPEN Journal of Parenteral and Enteral Nutrition*, *33*, 27–36.
37. Frape, D. (2010). Pests and ailments related to grazing area, diet, and housing. In D. Frape (Ed.), *Equine nutrition and feeding* (4th ed., pp. 334–342). Wiley Blackwell.
38. Gentry, L. R., Thompson, D. L., & Gentry, G. T. (2004). The relationship between body condition score and ultrasonic fat measurements in mares of high versus low body condition. *Journal of Equine Veterinary Science*, *24*, 198–203.

39. Geor, R. J. (2001). Nutritional support of the sick adult horse. In J. D. Pagan & R. J. Geor (Eds.), *Advances in equine nutrition II* (pp. 403–418). Nottingham University Press.
40. Geor, R. J. (2007). Feeding management of horses recovering from colic. *Compendium of Continuing Education for the Practicing Veterinarian, 2*, 344–355.
41. Golenz, M. R., Knight, D. A., & Yvorchuk-St. Jean, K. E. (1992). Use of a human enteral feeding preparation for treatment of hyperlipemia and nutritional support during healing of an esophageal laceration in a miniature horse. *Journal of the American Veterinary Medical Association, 200*(7), 951–953.
42. Goodson, J., Tyznik, W. J., & Cline, J. H. (1988). Effects of an abrupt diet change from hay to concentrate on microbial numbers and physical environment in the cecum of the pony. *Applied and Environmental Microbiology*, *54*, 1946–1950.
43. Green, C. J., Frazer, R. S., Underhill, S., et al. (1992). Metabolic effects of dobutamine in normal man. *Clinical Science*, *82*, 77–83.
44. Gupta, A. K., Mamta, Y. P., & Yadav, M. P. (1999). Effect of feed deprivation on biochemical indices in equids. *Journal of Equine Science*, *10*, 33–38.
45. Hanson, R. R., Wright, J. C., & Schumacher, J. (1998). Surgical reduction of ileal impaction in the horse: 28 cases. *Veterinary Surgery*, *27*, 555–560.
46. Harris, P. (2009). Feeding management of elite endurance horses. *Veterinary Clinics of North America: Equine Practice*, *25*, 137–153.
47. Hassel, D. M., Aldridge, B. M., & Drake, C. M. (2008). Evolution of dietary and management risk factors for enterolithiasis among horses in California. *Research in Veterinary Science*, *85*, 476–480.
48. Hassel, D. M., Rakestraw, P. C., & Gardner, I. A. (2004). Dietary risk factors and colonic pH and mineral concentrations in horses with enterolithiasis. *Journal of Veterinary Internal Medicine*, *18*, 346–349.
49. Hassel, D. M., Spier, S. J., & Aldridge, B. M. (2009). Influence of diet and water supply on mineral content and pH within the large intestine of horses with enterolithiasis. *Veterinary Journal*, *182*, 44–49.
50. Haugen, H. A., Can, L. N., & Li, F. (2007). Indirect calorimetry: A practical guide for clinicians. *Nutrition in Clinical Practice*, *22*, 377–388.
51. Henneberg, S., Soderberg, D., & Groth, T. (1987). Carbon dioxide production during mechanical ventilation. *Critical Care Medicine*, *15*, 8–13.

52. Henneke, D., Potter, G., & Kreider, J. (1983). Relationship between condition score, physical measurements, and body fat percentage in mares. *Equine Veterinary Journal, 15*, 371–372.
53. Hillyer, M. H., Taylor, F. G. R., & Proudman, C. J. (2002). Case control study to identify risk factors for simple colonic obstruction and distension colic in horses. *Equine Veterinary Journal, 34*, 455–463.
54. Hintz, H. F., Argenzio, R. A., & Schryver, H. F. (1971). Digestion coefficients, blood glucose levels, and molar percentage of volatile acids in intestinal fluid of ponies fed varying forage-grain ratios. *Journal of Animal Science, 33*, 992–995.
55. Hintz, H. F., Lowe, J. E., Livesay-Wilkins, P., et al. (1988). Studies on equine enterolithiasis. *Proceedings of the American Association of Equine Practitioners, 24*, 53–59.
56. Hoffman, R. M., Wilson, J. A., & Kronfeld, W. L. (2001). Hydrolyzable carbohydrates in pasture, hay, and horse feed: Direct assay and seasonal variation. *Journal of Animal Science, 79*, 500–506.
57. Hoyt, J. K., Potter, G. D., & Greene, L. W. (1995). Copper balance in miniature horses fed varying amounts of zinc. *Journal of Equine Veterinary Science, 15*(8), 357–359.
58. Hoyt, J. K., Potter, G. D., & Greene, L. W. (1995). Mineral balance in resting and exercised miniature horses. *Journal of Equine Veterinary Science, 15*(7), 310–314.
59. Hudson, J. M., Cohen, N. D., & Gibbs, P. G. (2001). Feeding practices associated with colic in horses. *Journal of the American Veterinary Medical Association, 219*, 1419–1425.
60. Hughes, K. J., Hodgson, D. R., & Dart, A. J. (2002). Hyperlipaemia in a 7-week-old miniature pony foal. *Australian Veterinary Journal, 80*(6), 350–351.
61. Hussein, H. S., Vogedes, L. A., & Fernandez, G. C. (2004). Effects of cereal grain supplementation on apparent digestibility of nutrients and concentrations of fermentation end-products in the feces and serum of horses consuming alfalfa cubes. *Journal of Animal Science, 82*, 1986–1996.
62. Julliand, V., de Fombelle, A., & Drogoul, C. (2001). Feeding and microbial disorders in horses: Effects of three hay
63. Kane, E., Baker, J. P., & Bull, L. S. (1979). Utilization of corn oil supplemented diet by the pony. *Journal of Animal Science, 48*, 1379–1384.

64. Kemper, M., & Beredijiklian, P. (1992). A comparison of measured total energy expenditure to measured resting energy expenditure. *Respiratory Care*, *37*, 1291–1295.
65. Kienzle, E., & Schramme, S. (2004). Body condition scoring and prediction of body weight in adult warm-blooded horses. *Pferdeheilkunde*, *20*, 517–524.
66. Kienzle, E., Radicke, S., & Landes, E. (1994). Activity of amylase in the gastrointestinal tract of the horse. *Journal of Animal Physiology and Animal Nutrition*, *72*, 234–241.
67. Klein, S. (1990). The myth of serum albumin as a measure of nutritional status. *Gastroenterology*, *99*, 1845–1846.
68. Kuzu, M. A., Terzioglu, H., & Genç, V. (2006). Preoperative nutritional risk assessment in predicting postoperative outcome in patients undergoing major surgery. *World Journal of Surgery*, *30*, 378–390.
69. Little, D., & Blikslager, A. T. (2002). Factors associated with development of ileal impaction in horses with surgical colic: 78 cases (1986–2000). *Equine Veterinary Journal*, *34*, 464–468.
70. Longland, A. C., & Byrd, B. M. (2006). Pasture nonstructural carbohydrates and equine laminitis. *Journal of Nutrition*, *136*, 2099S–2102S.
71. Longland, A. C., Cairns, A. J., & Humphreys, M. O. (1999). Seasonal and diurnal changes in fructan concentration in *Lolium perenne*: Implications for the grazing management of equine predisposed to laminitis. *Proceedings of the 16th Equine Nutrition and Physiology Society Symposium*, 258–259.
72. Lopes, M. A. F., & White, N. A. (2002). Parenteral nutrition for horses with gastrointestinal disease: A retrospective study of 79 cases. *Equine Veterinary Journal*, *34*, 250–257.
73. Lopes, M. A., White, N. A., & Crisman, M. V. (2004). Effects of feeding large amounts of grain on colonic contents and feces in horses. *American Journal of Veterinary Research*, *65*, 687–694.
74. Lorenzo-Figueras, M., Morisset, S. M., & Morisset, J. (2007). Digestive enzyme concentration and activities in healthy pancreatic tissue of horses. *American Journal of Veterinary Research*, *68*, 1070–1072.
75. Magdesian, K. G. (2003). Nutrition of critical gastrointestinal illness: Feeding horses with diarrhea or colic. *Veterinary Clinics of North America: Equine Practice*, *19*, 617–644.
76. Mansell, P. I., Fellows, I. W., & Birmingham, A. T. (1988). Metabolic and cardiovascular effects of infusions of low doses of isoprenaline in man. *Clinical Science*, *75*, 285–291.

77. Mazan, M. R., Deveney, E. F., & Dewitt, S. (2004). Energetic cost of breathing, body composition, and pulmonary function in horses with recurrent airway obstruction. *Journal of Applied Physiology*, *97*, 91–97.
78. Mazan, M. R., Hoffman, A. M., & Kuehn, H. (2003). Effect of aerosolized albuterol sulfate on resting energy expenditure determined by use of open-flow indirect calorimetry in horses with recurrent airway obstruction. *American Journal of Veterinary Research*, *64*, 235–242.
79. McClave, S. A., Martindale, R. G., & Vanek, V. W. (2009). Guidelines for the provision and assessment of nutrition support therapy in the adult critically ill patient: Society of Critical Care Medicine (SSCM) and American Society for Parenteral and Enteral Nutrition (A.S.P.E.N.). *JPEN Journal of Parenteral and Enteral Nutrition*, *33*, 277–316.
80. Medina, B., Girard, I. D., & Jacotot, E. (2002). Effect of a preparation of *Saccharomyces cerevisiae* on microbial profiles and fermentation patterns in the large intestine of horses fed a high fiber or a high starch diet. *Journal of Animal Science*, *80*, 2600–2609.
81. Mehdi, S., & Mohammad, V. (2006). A farm-based prospective study of equine colic incidence and associated risk factors. *Journal of Equine Veterinary Science*, *26*, 171–174.
82. Moore, B. R., Abood, S. K., & Hinchcliff, K. W. (1994). Hyperlipemia in 9 miniature horses and miniature donkeys. *Journal of Veterinary Internal Medicine*, *8*(5), 376–381.
83. Morris, D., Morris, J., & Ward, S. (1986). Comparisons of age, breed, history, and management in 229 horses with colic. *Equine Veterinary Journal*, *7(Suppl)*, 129–133.
84. Murray, R. C., Constantinescu, G. M., & Green, E. M. (1992). Equine enteroliths. *Compendium on Continuing Education for the Practicing Veterinarian*, *14*, 1104–1113.
85. National Research Council, Committee on Nutrient Requirements of Horses. (2007). *Nutrient requirements of horses* (6th ed.). National Academies Press.
86. Naylor, J. M., & Kronfeld, D. S. (1980). Hyperlipemia in horses: Effects of undernutrition and disease. *American Journal of Veterinary Research*, *41*, 899–905.
87. Pagan, J. D. (2003). Recent research developments from Kentucky Equine Research. In J. D. Pagan (Ed.), *Advances in equine nutrition III* (pp. 1–10). Nottingham University Press.

88. Pagan, J. D., & Hintz, H. F. (1986). Equine energetics. I. Relationship between body weight and energy requirements in horses. *Journal of Animal Science*, *63*, 815–821.

89. Pearson, R. A., Archibald, R. F., & Muirhead, R. H. (2001). The effect of forage quality and level of feeding on digestibility and gastrointestinal transit time of oat straw and alfalfa given to ponies and donkeys. *British Journal of Nutrition*, *85*, 599–606.

90. Potter, G. D., Arnold, F. F., & Householder, D. D. (1992). Digestion of starch in the small or large intestine of the equine. *Pferdeheilkunde*, *1*, 107–111.

91. rations on microbial profiles and activities. *Journal of Equine Veterinary Science*, *21*, 543–546.

92. ratios on digesta passage rate and digestibility in ponies. *Journal of Equine Veterinary Science*, *26*, 487–491.

93. Reeves, M. J., Salman, M. D., & Smith, G. (1996). Risk factors for equine acute abdominal disease (colic): Results from a multi-center case-control study. *Preventive Veterinary Medicine*, *26*, 285–301.

94. Reid, C. L. (2004). Nutritional requirements of surgical and critically ill patients: Do we really know what they need? *Proceedings of the Nutrition Society*, *63*, 467–472.

95. Richards, N., Choct, M., & Hinch, G. N. (2004). Examination of the use of exogenous α-amylase and amyloglucoside to enhance starch digestion in the small intestine of the horse. *Animal Feed Science and Technology*, *114*, 295–305.

96. Roberts, M. C. (1974). Amylase activity in the small intestine of the horse. *Research in Veterinary Science*, *17*, 400–401.

97. Rooney, D. K. (1998). Clinical nutrition. In S. Reed & W. M. Bayly (Eds.), *Equine internal medicine* (pp. 216–250). W. B. Saunders Co.

98. Schiesser, M., Müller, S., & Kirchhoff, P. (2008). Assessment of a novel screening score for nutritional risk in predicting complications in gastrointestinal surgery. *Clinical Nutrition*, *27*, 565–570.

99. Schmidt, O., Deegen, E., & Fuhrmann, H. (2001). Effects of fat feeding and energy level on plasma metabolites and hormones in Shetland ponies. *Journal of Veterinary Medicine*, *48A*, 39–49.

100. Shirazi-Beechey, S. P. (1995). Molecular biology of intestinal glucose transport. *Nutrition Research Reviews*, *8*, 27–41.

101. Shirazi-Beechey, S. P. (2008). Molecular insights into dietary-induced colic in the horse. *Equine Veterinary Journal*, *40*, 414–421.

102. Soeters, P. B., von Meyenfeldt, M. F., & Meijerink, W. J. H. J. (1990). Serum albumin and mortality. *The Lancet*, *335*, 348–351.
103. Sungurtekin, H., Sungurtekin, U., & Okke, D. (2008). Nutrition assessment in critically ill patients. *Nutrition in Clinical Practice*, *23*, 635–641.
104. Swinamer, D. L., Phang, P. T., & Jones, R. L. (1987). Twenty-four-hour energy expenditure in critically ill patients. *Critical Care Medicine*, *15*, 637–643.
105. Tattersfield, A. E., & Wilding, P. (1993). Agonists and ventilation. *Thorax*, *48*, 877–878.
106. Tinker, M. K., White, N. A., & Thatcher, C. D. (1997). Prospective study of equine colic risk factors. *Equine Veterinary Journal*, *29*, 454–458.
107. Treiber, K., Carter, R., & Gay, L. (2009). Inflammatory and redox status of ponies with a history of pasture-associated laminitis. *Veterinary Immunology and Immunopathology*, *129*(3–4), 216–220.
108. Vermorel, M., Vernet, J., & Martin-Rosset, W. (1997). Digestive and energy utilization of two diets by ponies and horses. *Livestock Production Science*, *51*, 13–19.
109. Vineyard, K. R., Gordon, M. E., & Williamson, K. K. (2011). Evaluation of the safety and performance of an enteral diet formulated specifically for horses. *Journal of Equine Veterinary Science*, *31*, 254–255.
110. Westenskow, D. R., Cutler, C. A., & Wallace, W. D. (1984). Instrumentation for monitoring gas exchange and metabolic rate in critically ill patients. *Critical Care Medicine*, *12*, 183–187.
111. White, N. A., & Dabareiner, R. M. (1997). Treatment of impaction colics. *Veterinary Clinics of North America: Equine Practice*, *13*, 243–259.
112. Zimmel, D. N. (2002). Post-operative parenteral nutritional support for colics. *Proceedings of the American College of Veterinary Surgeons Symposium*, 65–69.

*Chapter* 9

# NUTRITIONAL REQUIREMENTS, FEEDING LEVELS, AND DISEASES OF PIGS

## Contents

# INTRODUCTION

Understanding the nutritional requirements and appropriate feeding levels of pigs is crucial for ensuring their health, productivity, and overall well-being. Pigs require a balanced diet that includes energy, proteins, vitamins, minerals, and water to support their growth, reproduction, and daily metabolic functions. However, the nutritional needs of pigs vary based on factors such as age, weight, breed, and production stage (e.g., gestation, lactation, or finishing). Meeting these nutritional demands through carefully formulated diets is essential to maximizing growth rates, feed efficiency, and meat quality while minimizing health issues.

Feeding levels play a critical role in the management of pig health. Overfeeding can lead to obesity, which is associated with a range of metabolic disorders, including insulin resistance and fatty liver disease. Conversely, underfeeding or imbalanced diets can result in malnutrition, poor growth, reproductive failures, and increased susceptibility to infectious diseases. For instance, a deficiency in essential amino acids like lysine can stunt growth and reduce muscle development, while inadequate levels of vitamins and minerals such as zinc or selenium can compromise immune function and lead to conditions like dermatitis or reproductive disorders.

Diseases in pigs are often closely linked to their nutrition. Nutritional imbalances or deficiencies can weaken the immune system, making pigs more susceptible to infections such as swine dysentery, porcine respiratory disease complex (PRDC), and gastrointestinal diseases. Moreover, certain dietary components, like high levels of mycotoxins in feed, can be directly toxic, leading to liver damage, reduced feed intake, and impaired growth.

A comprehensive understanding of the nutritional requirements and feeding levels of pigs is essential for preventing diseases, promoting healthy growth, and ensuring efficient production. Proper nutrition not only enhances the overall health of pigs but also contributes to the sustainability and profitability of pig farming.

## 9.1. MOLECULAR NUTRITION: THE DIGESTION, ABSORPTION, AND METABOLISM OF NUTRIENTS

The field of molecular nutrition has come into being as a result of developments in molecular biology as well as the need to explain the molecular mechanisms underlying an organism's responses to nutrients. These consist of covalent protein modifications, signal transduction, and gene expression (Müller and Kersten, 2003). The first example of a nutrient regulating a gene is the lactose operon theory, which was first developed by Jacob and Monod in 1961. Shapiro et al proved the lactose operon model proposed by Jacob and Monod (1961) when they isolated pure lactose operon DNA from *Escherichia coli in 1969.* The paradigm for the interactions between the environment and the genome is gene-nutrient interactions. Each and every nutritional process depends on the interaction of several proteins expressed in a particular cell that are encoded by mRNA molecules. Changes in mRNA levels, which in turn affect the corresponding protein levels (though these changes don't always occur simultaneously), are important factors in regulating how quickly a nutrient or metabolite moves through a biochemical pathway. Therefore, molecular nutrition offered excellent mechanistic explanations of cause and effect as well as assistance in addressing basic health-related issues.

Understanding molecular nutrition has been made easier by the use of "omics" applications such as transcriptome, proteome, metabolome, and genomics (Afman and Müller, 2006). As an instance, Kitajka et al. (2004) used a high-density microarray to examine how the brain gene-expression of rats responding to various diets rich in polyunsaturated fatty acids (PUFAs) changed. It is now easier to understand the positive effects of ω-3 PUFA on the nervous system because of their discovery that PUFA-enriched diets significantly alter the expression of multiple genes in the central nervous system, and these effects seem to be mostly independent of their effects on membrane composition. Son et al. (2013) used metabonomics technology to examine the mechanism underlying pancreatic ductal adenocarcinoma (PDAC) cells' increased use of the amino acid glutamine to drive anabolic processes. They discovered that oncogenic KRAS mediates the reprogramming of glutamine metabolism through the transcriptional upregulation and repression of important metabolic enzymes in this pathway.

Initially, Yuan et al. (2015) outline a few dietary tactics, like diets with a moderately high energy content Vitamin A (Whaley et al., 1999) as well as l-arginine (Quesnel et al., 2000) supplementation in the diet lower the newborn piglets' intrauterine growth restriction (IUGR) ratio. Nie et al. (2015) provide an overview of the interactions between intestinal microbes and bile acids in host metabolism and health. We should focus on nutritional approaches that may be utilized to control the interaction between bile acids and gut microbes in order to enhance animal health.

In order to provide some insight into nutritional strategies, researchers look into the gene expression profiling of nutrient transporters, which mediate the absorption of nutrients in animal gastrointestinal tracts (Chen et al. 2015; Xie et al. 2015). Xie et al. (2015) discovered that the duodenum has the highest potential for absorbing soluble non-ammonia nitrogen (SNAN) in the form of peptide-bound amino acids (PBAA) in the non-mesenteric gastrointestinal tissues of dairy cows. It is also the primary location within the non-mesenteric digestive tract for producing milk protein precursors. Chen et al. (2015) discovered that there were positive correlations between the domestic pigeon's body weight and intestinal weight and the gene expressions of b0,+AT, EAAT3, PepT1, LAT4, NHE2, NHE3, and y+LAT2 in the small intestine. On the other hand, the mRNA expression levels of SNAT1, SNAT2, EAAT2, and CATM had the opposite effect in the small intestine.

One of the hot topics in animal nutrition research is how dietary components affect an animal's growth and health. Many researchers look into how food components affect animal gene expression profiling in this issue in an attempt to explain some of the changes in animal growth and metabolism. Wu et al. (2015) concentrate on how dietary protein content affects weaned piglets' expression of amino acid transporters. A 14% crude protein (CP) diet supplemented with crystalline amino acid (AA) may not transport enough AA into the body to maintain growth performance in piglets, as evidenced by the lowest average daily feed intake (ADFI) and average daily gain (ADG) in the 14% CP group as well as the lowest expression of ASC02, 4F2hc, and ATB0 mRNA in the jejunum. Gao et al. (2015) showed that by activating the mammalian target of rapamycin (mTOR) pathway in bovine mammary epithelial cells, l-leucine and l-histidine supplementation in medium can both upregulate milk proteins, such as α-casein, β-casein, and κ-casein synthesis. Nie C. X. et al. (2015) discovered that broiler chickens' tissue lipid metabolism and hepatic metabolomic profiling can be affected by supplementing their diet with fermented cottonseed meal (FCSM). Consuming FCSM led to a significant reduction in hepatic triglyceride and abdominal fat levels. It also downregulated the expression of lipoprotein lipase and fatty acid synthase on the mRNA in liver tissues and acetyl CoA carboxylase in abdominal fat tissues. Significant metabolic alterations of several liver pathways, including the tricarboxylic acid cycle, fatty acid synthesis, and the metabolism of glycerolipid and AAs, were also brought about by adding FCSM to the diet. Dai et al. (2015) looked into whether lowering housing density and adding taurine to the diet could improve the health of laying hens. As Wang et al. (2010) previously reported, egg production was enhanced by dietary taurine supplementation. Considering the tight connection between egg production and oviduct health, Dai et al. (2015) proposed the theory that enhanced oviduct function is the mechanism by which dietary taurine supplementation leads to higher egg production. Interestingly, taurine supplements in the diet may prevent damage to the oviducts of laying hens raised in high-density environments. Less oxidative stress, less infiltration of inflammatory cells, and reduced levels of inflammatory mediators in the oviduct of laying hens were linked to the attenuation of oviduct damage. Non-starch polysaccharide enzymes (NSPEs) have been shown in earlier research to improve immunity, nutrient absorption, and animal growth performance. These findings suggest that NSPEs are versatile metabolic pathway regulators (Ao et al., 2010; Zduńczyk et al., 2013). The way in which NSPEs control the metabolism of skeletal muscle is, however, poorly understood. Zhang et al. (2015) employed an isobaric tag for absolute and relative values.

As the primary factors influencing gene and protein activities are nutrients and other food components, molecular nutritional science will undoubtedly play a major role in the postgenomic research era that we are just now entering. Our understanding of nutrition and metabolism, as well as how these relate to animal health and disease, are the ultimate goals of applying the potent "omics" techniques.

## 9.2. NUTRITIONAL REQUIREMENTS OF PIGS

In order to meet their needs for growth, reproduction, lactation, and other functions, pigs need a variety of essential nutrients. Estimates of the amounts of these nutrients for different classes of swine under average conditions are provided by the National Research Council (NRC) in its publication, Nutrient Requirements of Swine (updated in 2012). However, a number of variables, including genetic variation, environmental conditions, the availability of nutrients in feedstuffs, disease prevalence, and other stressors, may raise the quantity of certain nutrients required for healthy growth and reproduction. While the requirements for energy, amino acids, calcium, and phosphorus are estimated by the NRC using a modeling approach, the requirements for other minerals and vitamins are estimated solely from empirical data.

When estimating nutrient requirements, the NRC takes into account a number of factors, including lean growth rate, gender, energy density of the diet, environmental temperature, crowding, parity, stage of gestation, and various measures of sow productivity. However, in order to ensure adequate intake of nutrients and for insurance purposes, nutritionists, feed manufacturers, veterinarians, or swine producers may want to include higher levels of certain nutrients than those listed by the NRC. With the exception of severe imbalances, over supplementing diets usually have very little negative effects.

Six general classes of nutrients are needed by swine: minerals, vitamins, water, carbs, fats, and protein (amino acids). Energy is a crucial nutritional component that is not specifically a nutrient but is instead obtained through the oxidation of fats and carbohydrates. Furthermore, when an animal's carbon skeleton oxidizes, excess amino acids (from protein) that are needed for tissue protein synthesis and maintenance supply energy. Although they are frequently added to swine diets to improve digestibility, speed up and optimize growth, and serve other objectives, antibiotics, chemotherapy drugs, microbial supplements (prebiotics and probiotics), enzymes, and other feed additives are not regarded as nutrients.

Dietary Nutrient Requirements of Growing Pigs Allowed Ad Lib Feed (90% dry matter) displays the NRC's estimated nutrient requirements for different body weight groups of pigs weighing between 5 and 135 kg body weight, expressed as dietary concentrations. Reproductive Measures and Dietary Nutrient Requirements of Gestating and Lactating Sows lists the requirements for gestating and lactating sows as dietary concentrations. The NRC tables' dietary concentrations are predicated on a specific feed intake; if intake falls short of the listed amount, the dietary concentration might need to be raised to guarantee a sufficient daily intake of the nutrients.

## 9.2.1. Water

Water should be readily available to pigs without charge, even before weaning. Age, feed type, ambient temperature, lactation stage, fever, high urine output (due to excessive protein or salt intake), or diarrhea all affect how much is needed. Growing pigs typically need two to three kilograms of water for every kilogram of dry feed. Because the milk that lactating sows produce has a high-water content, these sows drink more water during lactation. Water restriction lowers milk production and performance; if the restriction is severe, it may even be fatal.

**Table 9.1.** Summary of Studies Selected for Modelling Study on Local Breeds (Diet Composition and Feed Intake Expressed on an As-Fed Basis). BW: Body Weight; NE: Net Energy

| Breed [1] | Data Source | No. of | Days | Initial | Final | Diet(s) Composition | | Feed Intake, |
|---|---|---|---|---|---|---|---|---|
| | | Pigs | on Trial | BW, kg | BW, kg | Energy, MJ NE/kg | Protein, Percent (%) | kg/Pig/d |
| Alentejana_1 | Liang, Chen, et al. (2015) | 28–34 | 118 | 41.5 | 96.3 | 9.14 | 15.7 | 2.59 |
| Alentejana_2 | Liang, Chen, et al. (2015) | 13–39 | 136 | 35.3 | 95.1 | 8.51 | 15.2 | 2.45 |

| | | | | | | | | |
|---|---|---|---|---|---|---|---|---|
| Basque_conv | Müller and Kersten, (2003) | 20 | 214 | 34.8 | 139.9 | 9.59<br>8.96 | 18.0<br>14.7 | 2.39 |
| Basque_alter | Müller and Kersten, (2003) | 20 | 206 | 35.3 | 146.3 | 9.59<br>8.96 | 18.0<br>14.7 | 2.68 |
| Bísara_1 | Nie, Zhang, Wang, et al. (2015) | 24 | 209 | 29.8 | 144.3 | 9.51 | 15.0 | 2.34 |
| Bísara_2 | TREA-SURE project | 20 | 168 | 25.4 | 115.8 | 9.75<br>9.43<br>Corn: 9.90 | 16.7<br>15.3<br>8.6 | 1.5 kg + corn (0.0–0.6 kg) |
| A.Calabrese_1 | Nie, Hu, and Yan, (2015) | 36 | 142 | 40.2 | 139.1 | 9.34<br>9.69 | 15.4<br>14.2 | 3.14 |
| A.Calabrese_2 | Nie, Hu, and Yan, (2015) | 36 | 142 | 33.8 | 135.2 | 9.22<br>9.70 | 15.5<br>14.2 | 3.00 |
| Cinta Senese | TREA-SURE project | 11 | 246 | 38.0 | 143.4 | 9.64<br>9.71 | 12.4<br>10.3 | 2.48 |
| Iberian_1 | Quesnel, Quiniou, Roy, et al. (2014) | 20–29 | 273 | 16.1 | 159.5 | 10.0<br>9.5<br>10.4 | 17.8<br>15.8<br>13.5 | 2.10 |
| Iberian_2a | Shapiro, Machattie, Eron, et al. (1969) | 6–8 | 155 | 15.7 | 114 | 9.23 | 12.6 | 2.54 |
| Iberian_2b | Shapiro, Machattie, Eron, et al. (1969) | 6–8 | 174 | 16.0 | 114 | 9.08 | 16.6 | 2.55 |
| Krškopolje_con | TREA-SURE project | 12 | 115 | 36.2 | 120.4 | 8.52<br>8.82 | 15.3<br>14.0 | 3.23 |
| Krškopolje_eco | TREA-SURE project | 12 | 115 | 35.0 | 124.3 | 8.69 | 15.0 | 3.20 |
| | | | | | | 8.83 | 14.0 | |

| | | | | | | | | |
|---|---|---|---|---|---|---|---|---|
| Mangalitsa | TREA-SURE project | 12 | 198 | 24.3 | 115.5 | 9.98 | 13.7 | 2.64 |
| | | | | | | 10.00 | 14.4 | |
| Moravka | TREA-SURE project | 10–12 | 198 | 29.9 | 131.3 | 9.98 | 13.7 | 2.94 |
| | | | | | | 10.00 | 14.4 | |

Source. Brossard L, Nieto R, Charneca R, Araujo JP, Pugliese C, Radović Č, Čandek-Potokar M. (2019). Modeling Nutritional Requirements of Growing Pigs from Local Breeds Using InraPorc. Animals.; 9(4):169. https://doi.org/10.3390/ani9040169.

[1] The numbers (1, 2, 2a, 2b) or indications (conv: conventional; alter: alternative; con: conventional; eco: ecological) following the names of the breeds refer to different studies.

The quality of water matters. If there isn't much microbiological contamination in the water, chlorination might be required. Problems could arise from too many minerals in the water. Water ought to have. <1,000 ppm of total dissolved solids (TDS). Higher levels of TDS (2,000–5,000 ppm) can cause diarrhea or temporary water refusal, TDS levels of 5,000–7,000 should be avoided for breeding animals, and TDS levels > 7 ppm are not suitable for pigs. Pigs can handle modest amounts of sulfates in their water, but excessive amounts (>3,000 ppm) should be avoided (Tables 9.1 and 9.2).

**Table 9.2.** Effect of Diets Supplemented with Different Percentages of Protein on the Reproductive Performance of Gestating Sows

| **Animals** | **Experimental Design** | **Results** | **Conclusion** |
|---|---|---|---|
| 59 multiparous sows (Yorkshire × Landrace) with bodyweights (BW) around 241.67 ± 8.86 kg | (1) two levels of dietary metabolizable energy (ME) density were provided (13.40 or 13.82 MJ/kg); 2) three dietary protein levels were provided from day 35 of gestation (crude protein = CP: 10.5, 12, 13.5%). | Backfat thickness in lactating sows decreased and the percent of CP increased ($p = 0.03$). CP level in the diet had a negative effect on colostrum quality: % casein: $p = 0.03$; % protein: $p = 0.04$; % lactose: $p = 0.06$; total solids: $p = 0.03$; lean solids: $p = 0.03$, all decreased. | Backfat thickness and colostrum quality decreased as the CP level in the diet increased (10.5–13.5%). A diet for gestating sows containing 13.82 MJ/kg ME and 10.5% CP may improve reproductive and litter performance, and colostrum quality. |
| 47 Landrace × Yorkshire gilts; 190 kg at insemination | Gilts were fed one of two iso-energetic compound feeds in which dietary protein differed by 12%. | Milk yield peaked at 12.9 kg/d around day 20. Sows fed the low protein compound feed had a lower milk yield from day 20 to day 40 than controls (8.0 vs. 10.3 kg/d; $p < 0.05$). | Sows on a low-protein diet had decreased milk production at the end of lactation, so it seems problematic to reduce the protein content of the lactation diet in winter, especially in gilts with limited gastric capacity. |

| | | | |
|---|---|---|---|
| 32 Landrace × Yorkshire sows at parity two, with a similar mean bodyweight of 164.2 kg | One diet had normal crude protein (CP = 13.3%), the other had a low CP of 10.1%. | Sows receiving low levels of CP had higher serum levels of Lys and Thr and lower levels of Try, Ile, and Val ($p < 0.05$), but no effect on the serum levels of other AAs were found ($p > 0.05$). | Maternal protein deposition was decreased by a low CP. |
| 72 F1 multiparous sows (Yorkshire × Landrace) with an average BW of 218.69 kg | Experimental diets with different CP levels, as follows: (i) CP11 containing 11% CP; (ii) CP12, 12% CP; (iii) CP13, 13% CP; (iv) CP14, 14% CP; (v) CP15, 15% CP; and (vi) CP16, 16% CP. | Increasing CP levels in the gestation diet caused a significant increase in creatinine at days 35 and 110 of gestation (linear, $p = 0.01$; linear, $p = 0.01$). | Reducing dietary CP levels from 16 to 11% in a gestation diet did not have detrimental effects on the sows' body condition or piglet performance. |

*Source: Islas-Fabila P, Roldán-Santiago P, de la Cruz-Cruz LA, Limón-Morales O, Dutro-Aceves A, Orozco-Gregorio H, Bonilla-Jaime H. (2024). Importance of selected nutrients and additives in the feed of pregnant sows for the survival of newborn piglets. Animals, 14(3):418. https://doi.org/10.3390/ani14030418.*

## 9.2.2. Energy

The units of measurement for energy requirements are kilocalories (kcal) for net energy (NE), metabolizable energy (ME), or digestible energy (DE). The most widely used values are DE and ME, but there is a trend in the industry to base diet plans on NE as well. Based on NE, the NRC calculates the amount of energy needed, and from NE, estimates DE and ME. Swine weight (which affects the maintenance requirement), genetic potential for lean tissue growth or milk production, and the temperature of their housing all have an impact on how much energy they need. The energy content of the diet primarily determines how much feed is consumed by growing pigs that are permitted to eat feed ad lib.

The voluntary feed consumption falls if supplemental fat is added to the diet to increase its energy density. Pigs fed this kind of diet will usually gain more quickly and with better efficiency, but the amount of fat on their carcass may go up. Increases in fat do not have to match increases in fiber (>5–7%) in the diet to decrease the rate—and particularly the efficiency—of gain.

### 9.2.3. Protein and Amino Acids

The development of fetuses and supporting tissues in gestating sows, muscle growth, maintenance, and milk production in lactating sows all depend on amino acids, which are typically obtained from diet protein. Twelve of the animal's 22 amino acids are synthesized; the remaining 10 must be obtained through diet in order for the animal to grow normally. Arginine, histidine, isoleucine, leucine, lysine, methionine, phenylalanine, threonine, tryptophan, and valine are the ten dietary essential amino acids for swine. Part of the requirements for methionine and phenylalanine can be partially satisfied by cysteine and tyrosine, respectively. The necessary levels of lysine (the first limiting amino acid) and sufficient amounts of the other essential amino acids in diets consisting of corn and soybean meal are provided by the percentages of crude protein listed in Reproductive Measures and Dietary Nutrient Requirements of Gestating and Lactating Sowsa,b and Dietary Nutrient Requirements of Growing Pigs Allowed Ad Lib Feed (90% dry matter) a,b,c. In the middle and final starter phases, the dietary lysine requirement drops to 1.40 percent and 1.53 percent, respectively, from a relatively high 1.70 percent during the early starter phase. From 1.12 percent in the early growing phase to 0.71 percent in the late finishing phase, the requirement keeps declining during the growing-finishing stage.

Lysine, tryptophan, threonine, and methionine are the amino acids most likely to be deficient, making them the most practically important in diet formulation. Tryptophan and lysine levels in corn, the staple grain in the majority of swine diets, are noticeably low. The other three main grains fed to pigs are low in lysine and threonine: wheat, barley, and sorghum. Methionine is the first limiting amino acid in soybean meal; however, when soybean meal is combined with cereal grains, enough amounts are provided to create a complete diet that satisfies the lysine requirement. A possible exception could be young pigs fed diets high in soybean meal or diets low in sulfur-amino acid-containing dried blood products.

With the exception of very young pigs, milk protein is typically too costly to include in swine diets despite being well-balanced in essential amino acids. Although dried whey has a low total protein content, it has an excellent amino acid profile and is frequently used in beginning diets. Corn-soybean meal diets are superior to diets based on corn and animal protein byproducts (such as meat meal, meat and bone meal), but tryptophan or supplements that are good sources of tryptophan can greatly enhance these diets. Minerals and B-complex vitamins can also be found in abundance in animal proteins.

Methionine deficiency may occur in diets designed for early weaned pigs that have a high concentration of dried animal plasma or dried blood cells. Nevertheless, methionine shouldn't be added to diets arbitrarily because excessive amounts of the amino acid can inhibit growth. Although valine supplements are still too costly to be regarded as dietary supplements, they may be beneficial in diets consisting of corn-soybean meal given to nursing sows.

Since lysine is typically the first limiting amino acid in practically all practical diets, other amino acid requirements should be satisfied if diets are designed with lysine as the primary requirement. To partially satisfy the pig's lysine needs, crystalline lysine supplements should be given, but caution must be taken. As a general guideline, crude protein content can be decreased by two percentage points and lysine (0.19 percent lysine•HCl) can be added to the diet in an amount of 0.15 percent. Tryptophan, threonine, and/or methionine deficiencies, however, could arise from higher dietary protein reductions combined with extra lysine unless those nutrients are additionally supplemented. Today, it's pretty common to base swine diet formulations on the idea of "ideal" protein, that is, to express the percentage of lysine required for each essential amino acid. Furthermore, basing swine diets on standardized (or true) or apparent digestible amino acid composition is gaining traction. When the diet contains significant amounts of byproduct feeds, this approach is especially beneficial.

### 9.2.4. Minerals

These dietary components serve a variety of vital purposes in the body. Dietary Nutrient Requirements of Growing Pigs Allowed Ad Lib Feed (90% dry matter) and Reproductive Measures and Dietary Nutrient Requirements of Gestating and Lactating Sows list the dietary requirements for the essential macro- and trace minerals.

#### *9.2.4.1. Calcium and Phosphorus*

Calcium and phosphorus are vital for all phases of growth, pregnancy, and lactation, despite being largely utilized in skeletal development. They also have significant metabolic functions in the body. According to NRC estimates, growing pigs weighing between 25 and 50 kg in body weight should consume 0.66% of calcium and 0.56% of total phosphorus. While finishing pigs have lower requirements than younger ones, all weight groups have roughly the same calcium: phosphorus ratios. These levels do not permit maximal bone mineralization, but they are sufficient for maximal growth (rate and efficiency of gain). Generally speaking, increasing the amount of calcium and phosphorus in the diet by 0–1%–0–15% will optimize bone mass and strength.

The amount of calcium and phosphorus that lactating and gestating sows require depends on a number of factors, including milk production, parity, and stage of gestation (first versus last 25 days). The fetuses' rapid development is thought to be the cause of the higher requirements during late gestation. In order to ensure that sows receive an adequate amount of these minerals and to avoid posterior paralysis in heavy milking sows, pig producers may decide to feed sows at slightly higher levels. The specified daily feed intakes of 4–5–7 lb (2–1–2–6 kg) during gestation and 13–14 lb (5–9–6–6 kg) during lactation are the basis for the calcium and phosphorus requirements. These amounts include 5% wastage. The percentages of calcium and phosphorus may need to be increased if less feed is eaten daily.

For optimal use of both minerals, the ratio of total calcium to total phosphorus should be maintained between 1:25:1 and 1:1. Wide calcium:phosphorus ratios hinder the absorption of phosphorus, particularly when phosphorus intake is minimal. If there is an excess of phosphorus in the diet, the ratio becomes less important. The optimal ratio of calcium to digestible phosphorus, based on digestible phosphorus, is between 2:1 and 2:5:1.

While phosphorus in protein sources of animal origin, such as meat meal, meat and bone meal, and fish meal, is inorganic and highly available to pigs, the majority of the phosphorus in cereal grains and oilseed meals is in the form of phytic acid (organically bound phosphorus) and is poorly available to pigs. Phosphorus is not always readily available, even in cereal grains. For instance, there is only 10%–20% available phosphorus in corn, compared to 50% available phosphorus in wheat. To guarantee that the phosphorus requirement is satisfied, swine diets should be designed using "available phosphorus" as a guiding principle. The digestible phosphorus requirements are expressed as apparent total tract digestible (ATTD) and standardized total tract digestible (STTD) phosphorus in the NRC publication. The digested phosphorus is represented by ATTD phosphorus, while the digestible phosphorus that has been adjusted for endogenous phosphorus excretions is known as STTD phosphorus.

Supplements containing phosphorus, such as steamed bone meal, defluorinated phosphate, and mono- or dicalcium phosphate, are great sources of highly available phosphorus. These supplements are excellent providers of calcium as well. Ground limestone is a great calcium source as well.

Since phosphorus is thought to be a possible environmental contaminant, many swine farmers feed diets that are lower in excess phosphorus than they were in the past in order to lower phosphorus excretion. To further lower phosphorus excretion, supplemental phytase, an enzyme that partially breaks down phytic acid in feedstuffs, is frequently added to diets. It is generally advised to include at least 500 units of phytase per kg of diet in order to reduce dietary calcium and phosphorus by 0.05–0.1%.

### 9.2.4.2. Sodium and Chloride

Common salt, which is composed of 60% chloride and 40% sodium, provides these minerals. For growing and finishing diets, 0–0.75% of the salt is advised, whereas 0.5% of the salt is advised for starter diets and 0.5% for sow diets. Enough sodium and chloride should be present at these levels to satisfy the animal's needs. Byproducts from fish, milk, and animals can help meet part of the sodium and chloride needs.

### 9.2.4.3. Potassium, Magnesium, and Sulfur

There is no need for additional sources of these minerals because grains and protein sources in a practical diet provide sufficient amounts of them. Although cannibalism has been avoided by supplementing with magnesium oxide, controlled studies have not supported this approach.

### 9.2.4.4. Iron and Copper

Numerous enzyme systems involve these minerals. In order to prevent nutritional anemia, both are required for the formation of Hgb. Because milk contains very little iron, suckling pigs should be given extra iron during their first three days of life.

The best way to do this is by intramuscular injection (IM injection) of 100–200 mg of iron dextran, iron dextrin, or gleptoferron. Giving sows oral or injectable forms of iron and copper won't boost piglet stores at birth, nor will it raise iron levels in colostrum and milk to the point where it prevents anemia in young pigs. Pigs can get iron-rich sow feces from the pen due to high iron levels in lactation feed. Additional methods of providing iron include putting a solution of ferric ammonium citrate and water in a piglet waterer or regularly applying a solution of iron sulfate and a carrier—like ground corn—to the farrowing stall floor.

Growing pigs need only 3–6 ppm of copper, but sows require more. The 2012 NRC publication raised the estimated copper requirement for sows from 5 parts per million in the previous edition to 10 parts per million for gestation and 20 parts per million for lactation.

For weanling and growing pigs, copper at pharmaceutical dosages of 100–250 mg/kg in the diet works as an efficient growth promoter. High concentrations of copper seem to have an additional benefit on top of the growth-stimulating properties of antibiotics. When dietary copper sulfate levels are high, the excrement takes on a very dark hue. Additionally, the amount of copper in excreted manure increases noticeably with high copper diets.

### 9.2.4.5. Iodine

Thyroxine, which is produced by the thyroid gland from iodine, influences metabolic rate and cell activity. All classes of pigs have an iodine requirement of 0.14 mg/kg of diet. With a 0.007 percent iodine content, stabilized iodized salt can satisfy both the pigs' iodine and salt requirements when fed in sufficient amounts.

### 9.2.4.6. Manganese

Manganese is necessary for healthy growth and reproduction, but its quantitative requirement is unclear. While 2–4 mg/kg of manganese is sufficient for growth, sows require 25 mg/kg of manganese during lactation and gestation.

### 9.2.4.7. Zinc

Zinc is a vital trace mineral that serves a variety of biological purposes. Supplemental zinc is required in grain-soybean meal diets in order to prevent parakeratosis. When dietary calcium intake is excessive, higher zinc levels may be required, particularly in diets that are generally high in phytic acid, like diets containing corn and soybean meal. Pig performance during the postweaning phase has been found to bc consistently improved by pharmacologic levels of zinc (1,500–3,000 mg/kg) as zinc oxide. High zinc oxide levels have occasionally been shown to lessen postweaning diarrhea's frequency and severity. Similar to the reactions to high copper and antibiotics, responses to zinc oxide and zinc seem to be additive; however, consuming high copper and high zinc at the same time has no benefit. Like copper, excessive zinc intake raises the amount of zinc excreted in manure. The estimated zinc requirement for sows was raised to 100 ppm in the 2012 edition of the NRC publication from 50 ppm in the previous edition.

### 9.2.4.8. Selenium

The amount of selenium present in soils and, eventually, in crops varies greatly. Selenium levels in the USA tend to be higher west of the Mississippi River than east of the river, with the latter producing crops with lower selenium content. In most cases, 0.2–0.3 mg of extra selenium per kilogram of food should be sufficient to meet the requirements. The FDA controls this trace mineral, and the maximum amount of selenium that can be fed to pigs is 0.3 mg/kg.

### 9.2.4.9. Chromium

Pigs need this trace mineral, which works as a cofactor with insulin; however, the exact amount needed has not been determined. Supplemental chromium at 200 mcg/kg (ppb) has been shown in some studies to improve reproductive performance in gestating sows and carcass leanness in finishing pigs, however these results have been a little mixed.

### *9.2.4.10. Cobalt*

Cobalt is a component of vitamin B12 and is not beneficial when fed to pigs in elemental form.

## 9.2.5. Vitamins

The body uses these micronutrients for a variety of crucial functions. Dietary Nutrient Requirements of Growing Pigs Allowed Ad Lib Feed (90% dry matter) and Reproductive Measures and Dietary Nutrient Requirements of Gestating and Lactating Sows provide estimates of the necessary vitamin intake.

### *9.2.5.1. Vitamin A*

This vitamin, which is soluble in fat, is necessary for mucous secretions, growth and maintenance of epithelial tissue, vision, and reproduction. Yellow corn and green plant matter contain precursors to carotenoids, which are the source of vitamin A. Of all the carotenes, β-carotene is the most active. Regretfully, just approximately 25% of the total carotene in yellow corn is found as β-carotene. According to the NRC, pigs can receive 267 IU of vitamin A from 1 mg of chemically determined carotene found in corn or a corn-soybean mixture.

Stabilized vitamin A is frequently used in premixes or vitamin supplements, as well as in manufactured feeds. Diets may be fortified with concentrates containing natural vitamin A, most commonly found in fish oils. Additional excellent sources of β-carotene include dehydrated alfalfa meal, high-quality legume hays, and green forage. Naturally occurring β-carotene and vitamin A are readily ruined by air, light, heat, rancid fats, organic acids, and specific mineral elements. These factors suggest that natural feedstuffs shouldn't be the only source of vitamin A, especially considering how affordable synthetic vitamin A is. A retinol of 0.30 mcg or retinyl acetate of 0.344 mcg is equal to one international unit of vitamin A.

### *9.2.5.2. Vitamin D*

This fat-soluble, antirachitic vitamin is essential for healthy ossification and bone growth. Ergocalciferol (vitamin D2) and cholecalciferol (vitamin D3) are the precursor sterols that give rise to vitamin D when exposed to UV light. Pigs can utilize either vitamin D3 (irradiated animal sterol) or D2 (irradiated plant sterol), but they appear to prefer D3. Pigs can be partially vitamin D acclimated by daily exposure to direct sunlight for a brief period of time. Irradiated yeast, sun-cured hays, fish oils, activated plant or animal sterols, and vitamin premixes are among the foods that contain vitamin D. One IU of this vitamin is equal to 0.025 milligrams of cholecalciferol. In the 2012 NRC publication, the estimated 200 IU/kg vitamin D requirement for lactating and gestating sows was raised to 800 IU/kg.

### 9.2.5.3. Vitamin E

In feedstuffs, this fat-soluble vitamin functions as an organic antioxidant. Of the eight forms of vitamin E found in nature, d-α-tocopherol exhibits the highest biologic activity. Pigs of all ages need vitamin E, which is closely linked to selenium. For growing pigs, the recommended daily intake of vitamin E is 11–16 IU/kg, and for sows, it is 44 IU/kg. In the USA's eastern corn belt, where feeds are likely to be low in selenium, some nutritionists advise sows to consume higher amounts of food. Selenium deficiency can only be partially remedied by vitamin E supplements.

Vit E is present in significant amounts in green forage, legume hays and meals, cereal grains, and particularly in the germ of cereal grains. When feedstuffs are exposed to heat, high moisture, rancid fat, organic acids, and high concentrations of specific trace elements, the activity of vitamin E is decreased. One milligram of dl-α-tocopherol acetate or 0.67 mg of d-α-tocopherol is equal to one IU of vitamin E activity.

### 9.2.5.4. Vitamin K

It takes this fat-soluble vitamin to keep blood clotting at normal levels. Vitamin K is not required in large amounts (0.05 mg/kg of diet). Pigs' needs will usually be satisfied by bacterial synthesis of the vitamin and subsequent absorption, either directly or through coprophagy. Hemorrhages have been documented in both growing and newborn pigs, although they are uncommon. As a preventative measure, supplementing with vitamin K at a dose of 2 mg/kg of diet is advised. Hemorrhaging issues are typically related to feeding diets containing moldy grain or other mold-containing ingredients.

### 9.2.5.5. Riboflavin

This water-soluble vitamin is a component of two significant enzyme systems that are involved in the metabolism of fat, protein, and carbohydrates. This vitamin is typically lacking in swine diets, and premixes contain the crystalline form of the vitamin. Green forage, milk by-products, brewer's yeast, legume meals, and certain fermentation and distillery by-products are examples of natural sources.

### *9.2.5.6. Niacin (Nicotinic Acid)*

Coenzymes involved in the metabolism of protein, fats, and carbohydrates include niacin. Although the conversion is inefficient, pigs can convert excess tryptophan to niacin. Pigs are unable to utilize the niacin present in the majority of cereal grains. This vitamin is typically lacking in swine diets, and premixes contain the crystalline form of the vitamin. Brewer's yeast, distiller's solubles, fish and animal byproducts are all natural sources of niacin. The NRC raised the niacin requirement to 30 ppm during all growth phases based on recent research.

### *9.2.5.7. Pantothenic Acid*

Coenzyme A, a crucial enzyme in the metabolism of energy, contains this vitamin. This vitamin is lacking in swine diets, and vitamin premixes contain the crystalline salt d-calcium pantothenate. Green forage, legume meals, milk products, brewer's yeast, fish solubles, and a few other byproducts are natural sources of pantothenic acid.

### *9.2.5.8. Vitamin B*

This vitamin, also known as cyanocobalamin, is essential for many vital metabolic processes and contains cobalt. Plant-based feedstuffs are lacking in this vitamin, but animal products are a good source. Vitamin B12 is typically present in vitamin premixes for pigs, even though some intestinal synthesis of this vitamin takes place.

### *9.2.5.9. Thiamine*

Although this vitamin plays vital functions in the body, pigs do not actually need much of it because grains and other feed ingredients provide sufficient amounts.

### *9.2.5.10. Vitamin B*

Pyridoxines are a class of compounds that are involved in the metabolism of amino acids and exhibit vitamin B6 activity. They are widely distributed among the natural feed ingredients that pigs are typically fed. In comparison to the previous edition of the NRC publication, the requirement for vitamin B6 in young pigs (5–25 kg) was increased by a factor of three or four.

### *9.2.5.11. Choline*

All tissues require choline in order to function normally. Methionine in the diet allows pigs to produce some choline on their own. The natural food components contain enough choline to meet the needs of developing pigs. However, in certain studies, gilts and sows given choline supplements at a rate of 440–800 mg/kg of diet showed an increase in litter size. Fish solubles, fish meal, soybean meal, liver meal, brewer's yeast, and meat meal are examples of naturally occurring sources of choline.

The most common type of supplemental choline used in feeds is choline chloride, which is 75% choline. Choline chloride, which is hygroscopic and reduces the activity of vitamin A and other less stable vitamins, should not be mixed with other vitamins in a premix if choline is added as a supplement to sow diets, especially if trace minerals are present.

### *9.2.5.12. Biotin*

Pigs can readily obtain this vitamin from corn and soybean meal, but they have less access to it from sorghum, oats, barley, and wheat grains. There is proof that biotin levels in pigs, particularly breeding animals, may be low or nonexistent when these latter cereal grains are fed to them. Additions of biotin have been found to enhance sows' reproductive performance.

There is evidence that adding biotin to diets containing corn and soybean meal also improves reproductive performance, though this relationship is less clear. In certain cases, adult pigs' footpad lesions were reduced when given a biotin supplement. Supplementing with biotin is advised for insurance, particularly for diets for sows. Pigs shouldn't be given raw eggs because the whites of eggs contain a protein called avidin, which combines with biotin to make it inaccessible.

### *9.2.5.13. Folacin*

This class of substances exhibits folic acid activity. Natural feedstuffs contain enough folacin to meet growth requirements, but adding folic acid to sow diets has been shown in some studies to improve litter size.

### *9.2.5.14. Ascorbic Acid (Vitamin C)*

It is believed that under normal circumstances, pigs can synthesize this vitamin quickly enough to meet their needs. However, when this vitamin was added to the diet, a few studies have shown improvements in the performance of early-weaned pigs under stressful conditions.

### 9.2.6. Fatty Acids

Pigs need linoleic acid, arachidonic acid, and more long-chain polyunsaturated fatty acids. However, linoleic acid is regarded as the dietary essential fatty acid because it can be converted into longer chain fatty acids *in vivo*. The NRC calculates that 0.1 percent of linoleic acid is needed for swine growth and breeding. The fat found in naturally occurring food ingredients typically satisfies this requirement. Corn oil contains a lot of linoleic acid.

## 9.3. FEEDING LEVELS AND PRACTICES IN PIGS

The quality and quantity of food consumed on a daily basis affect the performance of gestating sows, lactating sows, and their nursing pigs, as well as weanling, growing, and finishing pigs. In the process of managing animal feeding, it is crucial to know how much feed the animals take in. Pigs that are weanling, growing, and finishing are typically allowed to eat feed at will. The amount that they eat depends on a number of management factors, including feeder design, crowding, temperature, gender, and feed quality (such as the absence of molds).

### 9.3.1. Growing-Finishing Pigs

Table 9.1 displays the daily feed intakes of different weight classes of growing-finishing pigs fed a diet estimated by the NRC growth model to contain 3,300 kcal of ME/kg (typical of a diet containing corn-soybean meal). These intake levels are typical for gilts and barrows. Feed intakes for gilts weighing 50–135 kg will be slightly lower and for barrows, slightly higher. Reduced feed intake can be mitigated by keeping pigs cool with automatic water sprayers during hot weather and by preventing overcrowding. In order to project total feed requirements or prescribe in-feed medication, these intake levels can be used as a guide.

### 9.3.2. Gestating Gilts and Sows

The National Research Council (NRC) estimates that feeding levels for gestating gilts and sows should be 4–4–9 lb/day (2–1–2·kg/day) during the first 90 days of gestation and 5–5·lb/day (2–5–2·kg/day) for the final 25 days before farrowing. This amount of energy should be sufficient for maintenance, some lean and fat tissue accretion (especially in gilts), and the energy requirements of developing fetuses, the placenta, and other supporting tissues. Mature sows only require the energy needed for body maintenance and a slight increase in weight. In order to meet the sow's daily energy requirement, higher feeding levels will be required if gestation diets contain oats, alfalfa meal, or other energy diluents. Ad lib access to extremely high-fiber diets has been attempted to limit voluntary feed intake during gestation, but the results have always been excess weight gain. To maintain the health of pregnant gilts and sows, producers should modify the amount of feed given to them. Overweight at the end of gestation is frequently linked to decreased feed consumption during lactation and can occasionally lead to smaller litter sizes, a higher risk of dystocia, more pig overlay, and a higher risk of postpartum dysgalactia syndrome. Poor body condition causes sows to have lower birth weights, be thinner at weaning, and return to postweaning estrus (or even anestrus) later than expected. It also increases the incidence of shoulder sores in sows. If sows are not in good health when they breed, it can also have a negative impact on the size of the litter at the next farrowing. First-litter gilts have higher requirements for amino acids than do sows. During the last 90 days of pregnancy, gilts and sows need more dietary amino acids than they did during the first 90 days.

### 9.3.3. Lactating Gilts and Sows

According to NRC estimates, 13.1–14.6 lb (6.0–6.6 kg) of feed (3,300 kcal of ME/kg) is needed daily by lactating gilts and sows nursing 11–11.5 pigs that gain 240 g/day during a 21-day lactation to meet their energy requirements. The number of pigs nursed, the weight gain (both of which affect milk production), and the sow's weight loss determine how much energy and feed is needed. During lactation, sows should be given high-energy diets ad libitum or hand-fed their entire meal three times a day. Low feed intake can be mitigated in part by using drip coolers in hot weather and properly controlling the temperature in the farrowing room.

Sows will lose excessive weight during lactation if their feed intake is too low. If this is an issue, top-dressing the lactation feed with extra fat or incorporating 3–6% fat into the lactation diet should be taken into consideration. Extra energy in the last three to six weeks of pregnancy may be beneficial if issues continue. Prolific sows raising large litters should be fed diets rich in protein and amino acids to maximize milk production and prevent the sow from losing too much weight. These sows might need diets with 16–18% crude protein or higher (with a minimum of 0–9% lysine). High-protein lactation diets will reduce or even completely stop sows from losing weight during lactation if their energy intake is adequate.

#### 9.3.3.1. Major Feed Ingredients

Feeding the most affordable cereal grains and making up for any deficiencies with high-quality protein sources, minerals, and vitamins is a cornerstone of pork production economics. Commercially available are dependable mineral and vitamin premixes or whole manufactured supplements. In pig farms, fortified corn-soybean meal diets are widely utilized; however, alternative cereals and protein sources can also be employed.

- Corn (maize) is by far the most widely used grain for feeding pigs in the USA. It is very palatable and high in energy but relatively low in crude protein. In addition, corn is deficient in lysine, tryptophan, threonine, and several other essential amino acids, as well as vitamins and minerals.
- Grain sorghum is a major energy source for pigs in the western and southwestern USA. The protein content is variable depending on factors such as variety, whether the crop was grown on irrigated or dry land, the amount of fertilizer used, and other environmental factors. In general, grain sorghum can be substituted for corn on an equal-weight basis, but because the ME value is slightly lower than that of corn, a poorer feed conversion should be expected.
- Wheat has about the same energy content as corn and contains 2%–3% more protein and 0.05%–0.1% more lysine than corn. Wheat can be substituted for corn on either an equal-weight basis or on a lysine basis, but not on a crude protein basis or it will result in a lysine deficiency. Wheat can constitute all of the grain in a swine diet. The two main types of wheat grown in the USA, hard red winter and soft red winter, have equivalent nutritional value.
- Barley has ~85%–90% of the feeding value of corn, even though it usually contains 2%–3% more protein. Scabby barley should not be fed to pigs.
- Oats have a relatively low energy content and, therefore, should not account for >20%–25% of the cereal grain in the diet. Generally, when oats are included in the diet, the rate and efficiency of gain should be expected to decline. Rolled oats groats are sometimes used in starter diets because of their excellent palatability.
- Cereal grains should be ground or rolled to maximize their feeding value. Corn and grain sorghum should be reduced to a medium-fine particle size (550–600 microns). Wheat should be ground more coarsely (650–700 microns) to prevent pasting. Fine grinding improves feed conversion, but excessive reduction in particle size may lead to an increased incidence of gastric ulcers. Pelleting of diets may result in a small improvement in gain and especially feed efficiency. In general, the benefit is greatest with pelleted diets that contain high levels of fiber, such as barley-based diets. Cereal grains should be as free as possible from mycotoxins. Aflatoxins,

vomitoxin, zearalenone, fumonisins, and other mycotoxins can reduce animal performance, depending on the level in the feed, and can especially cause reproductive problems in breeding animals.

- Soybean meal accounts for >90% of the supplemental protein fed to pigs in the USA. It is very palatable and has an excellent amino acid profile that complements the amino acid pattern in cereal grains. Ground, full-fat soybeans can also be fed to swine but only after they are heated (by extrusion or roasting) to inactivate the trypsin inhibitors and other heat-labile antinutritional factors.
- Canola meal also is an excellent protein source. Low-gossypol cottonseed meal (< 100 ppm free gossypol), peanut meal, sunflower meal, and other oilseed-based meals can be used in swine feed but generally not as the sole source of supplemental protein because of the lower lysine content of their protein. Animal protein sources such as meat meal, meat, and bone meal, or fish meal can supply a portion of the supplemental protein in swine diets.
- Distillers dried grains with solubles (DDGS) is a byproduct that has received a lot of attention in recent years because of the increased number of ethanol plants that use corn to produce ethanol for fuel. This byproduct is an excellent and generally economical feed ingredient for swine. Although DDGS has essentially no starch and considerably more fiber than corn, it is considerably higher in fat (corn oil); hence, the ME content of DDGS containing 9%–12% fat is similar to that of corn. Recently, some ethanol plants extract a portion of the oil from the solubles before adding the solubles back to the dried grains. This results in a "low-fat" DDGS, generally 5%–9% fat, which has slightly less ME than conventional DDGS. Further removal of fat, called "de-oiled DDGS" (< 5% fat) has substantially less ME than either of the other types of DDGS, so it has a lower feeding value. DDGS is also higher in protein than corn, but the quality of protein (i.e., balance of amino acids), like corn protein, is poor.

With DDGS, a significant amount of research has been conducted recently. Pigs can effectively consume diets with 20–25% DDGS; however, when finishing diets contain higher than 30% DDGS, the pigs' body fat becomes more unsaturated, as shown by higher iodine levels. As a result, the bellies become softer and more flexible, making it harder to cut them into bacon slices. In order to address this issue, producers ought to think about eliminating DDGS from the late finishing diet or lowering its content to 10% in the last three to four weeks of the finishing period.

### *9.3.3.2. Feeding Management of Sows and Litters*

Sows should be fed gestation diets sufficient in all nutrients to yield robust, healthy pigs. It is important to feed sows so they are in a healthy body condition—not too thin or too fat—when they give birth. Smaller pigs that have a lower chance of surviving than larger, more active pigs are typically born to thin sows. The sow should be put back on full feed as soon as feasible after farrowing. If a sow is eating well, constipation is usually not an issue. If constipation is a concern, add wheat bran or dried beet pulp to the diet at a rate of 5–10%. Alternatively, add chemical laxatives like potassium chloride or magnesium sulfate at a rate of 0.75%–1%.

Make sure all of the recently farrowed pigs have had a chance to nurse. If required, oxytocin can be administered to induce milk flow. Weak pigs may benefit from artificial milk if the sow is slow to produce milk; however, success depends on proper management and sanitation. An iron injection should be administered before the age of three days, along with other previously mentioned methods, to prevent nutritional anemia. After receiving colostrum, pigs from large litters can be moved to sows with smaller litters; however, the transfer must be completed within the first 24 hours of the pigs' birth. If pigs are weaned after three weeks of age, a tasty pig starter diet should be given starting at two to three weeks of age.

### *9.3.3.3. Feeding Management of Weanling Pigs*

Early weaned pigs (3–5 weeks) do best when fed a complex starter diet for 1–2 weeks following weaning. Dried blood products, dried whey, and/or lactose, along with a high lysine content, are usually found in the starter diet. To raise healthier pigs, some producers employ segregated or medicated early weaning programs. This calls for excellent nutritional management and weaning between 10 and 16 days of age. Increased amounts of lysine, lactose (as sugar or from dried whey), and 3–7% dried animal plasma should all be included in such diets. Eventually, a gradual switch to less costly starter diets and then diets containing corn and soybean meal should be made. The best way to meet the nutritional needs of growing-finishing pigs is through full-feeding programs. Limit feeding slows down the rate and effectiveness of gain but may enhance the quality of the finishing pigs' carcass. To avoid wasted feed or growth restriction, self-feeders must be designed and adjusted properly.

### *9.3.3.4. Growth Stimulants*

For a long time, swine diets have frequently included antibiotics and other chemotherapeutic agents to encourage growth and feed efficiency, lower mortality and morbidity, and enhance overall health. Pigs respond best to these growth-promoting substances when they are young, and as they get older and heavier, their responses get less. Manufacturer recommendations and regulatory limitations should be followed when determining the dosage of antibiotics to be administered and the timeframe for drug withdrawal.

The following antibiotics have been approved for use as swine feed additives: bambermycins, bacitracin zinc, bacitracin methylene disalicylate, oxytetracycline, lincomycin, neomycin, penicillin, tiamulin, tylosin, and virginiamycin. Sulfamethazine, sulfathiazole, carbadox, and roxarsone are examples of chemotherapeutic agents. Some of these can only be used in conjunction with specific other additives. The use of apramycin as a water medication is also authorized. Additionally, pharmaceutical doses of copper (100–250 ppm) as tribasic copper chloride or zinc (1,500–3,000 ppm) as zinc oxide are efficient growth promoters in young pigs.

However, the use of antibiotics has changed due to FDA action. Final Guidance 213 and the Veterinary Feed Directive (VFD) rule state that antimicrobials that were previously used at subtherapeutic levels for production purposes (better growth and efficiency) are no longer permitted. This includes all antimicrobials approved for swine, except for carbadox, bacitracin, and bambermycins. Antimicrobials that are medically important in human medicine. Rather, they can only be used to prevent disease and only under the guidance and control of veterinarians. Antibiotics utilized in feed or water are covered by this regulation. The production improvement claim on the labels of those antimicrobial products is requested to be voluntarily removed by the companies that manufacture them. Producers will only be able to obtain these products via a VFD; they will no longer be available "over-the-counter."

Direct-fed microorganisms (formerly known as probiotics), such as live cultures of *Lactobacillus acidophilus*, *Streptococcus faecium*, and *Saccharomyces cerevisiae*, have been considered as potential antibiotic replacements; however, controlled studies have not consistently demonstrated positive effects from their inclusion. For young pigs, the addition of certain sugars (mannanoligosaccharides and fructooligosaccharides, also known as prebiotics) has occasionally shown promise as an antibiotic substitute; however, the growth responses are less predictable and smaller than those resulting from the addition of antibiotics. It is believed that the GI tract's beneficial microorganisms, like bifidobacteria and lactobacilli species, are stimulated to grow by the oligosaccharides and direct-fed microbials. This partial displacement of undesirable microorganisms, such as certain pathogenic microbes, occurs.

In tests conducted on finishing pigs, it was discovered that certain "repartitioning agents" significantly increased growth rate, feed conversion, and carcass leanness. Porcine somatotropin and ractopamine are two examples of β-agonists. The only such medication that has been authorized for use in pigs in the USA as of 2015 is ractopamine. These substances have an impact on dietary requirements for nutrients, specifically raising the need for amino acids.

# 9.4. NUTRITIONAL DISEASES OF PIGS

It is challenging to diagnose nutritional deficiencies through observation. Clinical signs are frequently the outcome of a combination of malnourishment, infectious diseases (including parasitism), and poor management. The majority of nutritional deficiencies show non-specific symptoms, such as poor appetite, stunted growth, and thriftiness. A single nutrient deficiency can result in inanition, and starvation that follows can lead to several deficiencies. Furthermore, a nutritional deficit can also exist in the absence of clear symptoms. In the field, the deficiency might only be marginal or barely noticeable, which makes diagnosis challenging.

It is not always easy to diagnose a deficiency by looking at how the body reacts to nutritional therapy, especially in cases of chronic deficiencies where the damage may be irreversible. A positive diagnosis of nutritional deficiency should only be made following the manifestation of multiple clinical signs that are expected and a thorough examination of the animals' medical, dietary, and management histories.

## 9.4.1. Protein Deficiency

A lack of protein in growing and finishing pigs can be due to inadequate feed intake or deficiencies in one or more essential amino acids. This can lead to lower growth rates, poor feed conversion, and larger carcasses. Breastfeeding sows lose too much weight, produce less milk, and may not show post-weaning estrus at all or only show it later. All essential amino acids must be released during digestion at rates appropriate to needs for protein to be used as effectively as possible. Protein supplements should therefore be mixed with the grain or offered free choice with the grain at all times rather than being handfed infrequently. The idea that pigs are suffering from "protein poisoning" has not been proven. Up to 35–50% protein in a diet was found to be laxative and less effective when consumed, but no harmful effects were observed.

## 9.4.2. Fat Deficiency

Swine require specific long-chain polyunsaturated fatty acids. In addition to being necessary for proper diet, linoleic acid is converted into longer-chain fatty acids, which are most likely also necessary. In growing pigs, a lack of linoleic acid causes scaly dermatitis, hair loss, skin necrosis on the neck and shoulders, and an unthrifty appearance. Standard swine diets typically include enough fat from natural ingredients to provide sufficient amounts of important fatty acids.

### 9.4.3. Mineral Deficiency

In growing pigs, deficiencies in calcium or phosphorus cause rickets, and in mature pigs, osteomalacia. In younger pigs, signs include lameness and deformity of the long bones; in older pigs, signs include fractures and posterior paralysis (caused by fractures in the lumbar region). High milk production sows that give birth to large litters are especially vulnerable to posterior paralysis at the end of lactation or after weaning if there is a calcium or phosphorus deficiency in their diet. While phosphorus deficiency is the most frequent cause of these symptoms, vitamin D deficiency can also cause them. The main cause of the poor and inefficient growth of pigs fed diets low in salt (NaCl) is a significant decrease in feed intake. While not exclusive to salt deficiency, bad skin and hair conditions can also arise. Reports have surfaced of salt-deficient pigs trying to drink other pigs' urine. Iodine-deficient sows give birth to weak or stillborn hairless piglets. The newborn pigs may be born weak due to a borderline deficiency, but their thyroids are enlarged and have abnormalities in their histology. Certain feedstuffs, such as soybeans and soybean meal, contain goitrogens that, in the absence of iodine in the diet, may result in marginal goiter. This deficiency is avoided by using recommended amounts of iodized salt. Iron and copper deficiencies cause typical nutritional anemia and slow down the formation of Hgb. Low Hgb and RBC counts, pale mucous membranes, an enlarged heart, skin edema around the neck and shoulders, listlessness, and spastic breathing (thumps) are all indicators of nutritional anemia in suckling pigs. More often than copper deficiency, iron deficiency affects nursing pigs who do not get oral or injectable iron early in life. Growing pigs with a zinc deficiency develop parakeratosis, especially if their diets contain higher than recommended levels of calcium and phytic acid, also known as phytate, which is the main source of phosphorus in cereal grains and oilseed meals. Zinc's precise mechanism of action in preventing parakeratosis is unknown.

Selenium and/or vitamin E deficiencies can result in the unexpected death of young, fast-growing pigs. Additionally, nursing pigs are more vulnerable to iron toxicosis from iron injections due to selenium and vitamin E deficiencies.

### 9.4.4. Vitamin Deficiency

Deficiencies are less common than they used to be because the majority of commercial diets are fortified with vitamins, and vitamin premixes are easily accessible for farm-mixed feeds. Vitamin A deficiency causes problems with the epithelial tissues of the respiratory, reproductive, neurological, urinary, and digestive systems as well as with the eyes. Sows with impaired reproductive function may give birth to weak, blind, or malformed pigs. One unusual indicator of vitamin A deficiency in pregnant sows is spinal cord herniation in fetal pigs. Growing pigs lacking in vitamin A become uncoordinated, develop respiratory problems and become blind at night. Because the liver can store vitamin A, vitamin A deficiency is uncommon.

A vitamin D deficiency can cause posterior paralysis, rickets, stiffness, and weak, bent bones. These symptoms are identical to those of a phosphorus or calcium deficiency. Immune system impairment and poor reproduction can arise from vitamin E deficiency. Selenium deficiency and vitamin E deficiency share many of the same symptoms. Pigs lacking in vitamin K have slower blood clotting times and are more likely to die from bleeding. The synthesis of vitamin K may be hampered by some of the ingredients in moldy feed. These symptoms are also brought on by high dietary calcium levels interfering with vitamin K activity.

Pigs lacking in riboflavin have problems reproducing; postpubertal gilts do not cycle but exhibit no other symptoms. Anorectic deficient sows give birth to dead pigs 4–16 days early. The stillborn pigs may have enlarged forelegs, very little hair, and frequent partial resorption. Pigs raised on diets low in riboflavin gain weight slowly, exhibit a rough coat, an exudate on their skin, a poor appetite, and may develop cataracts. Pigs lacking in niacin have diarrhea, weight loss, rough skin and coats, and dermatitis on the ears in addition to inflammatory lesions of the digestive tract. Bacterial infections or niacin deficiencies can cause intestinal disorders. While adequate dietary niacin does not cure infectious enteritis, it does likely help deficient pigs maintain their resistance to bacterial invasion. Deficient pigs respond well to niacin therapy.

When fed diets low in pantothenic acid, growing pigs and pregnant sows develop a typical "goose-stepping" gait, ataxia, and a noninfectious bloody diarrhea. When the shortfall gets too great, anorexia sets in. Pigs lacking in choline display abnormal shoulder conformation and incoordination. Upon necropsy, they typically exhibit kidney damage and may have fatty livers. Pigs born to sows lacking in choline have smaller litter sizes and may have spraddle legs. Excessive hair loss, dermatitis and ulcerations of the skin, mucous membrane inflammation in the mouth, transverse cracking of the hooves, and bleeding or cracking of the footpads are all signs of a biotin deficiency. Babies pigs raised on artificial diets deficient in vitamin B exhibit hyperactivity, hoarseness, hindquarter pain and incoordination. A compromised hematopoietic system is revealed by a histologic examination of the bone marrow. At necropsy, fatty livers are also observed (Afman and Müller, 2006).

# REFERENCES


1. Afman, L., & Müller, M. (2006). Nutrigenomics: From molecular nutrition to prevention of disease. *Journal of the American Dietetic Association, 106*(4), 569–576. https://doi.org/10.1016/j.jada.2006.01.001
2. Ao, X., Meng, Q. W., Yan, L., et al. (2010). Effects of non-starch polysaccharide-degrading enzymes on nutrient digestibility, growth performance, and blood profiles of growing pigs fed a diet based on corn and soybean meal. *Asian-Australasian Journal of Animal Sciences, 23*(12), 1632–1638. https://doi.org/10.5713/ajas.2010.10123
3. Ashworth, C. J., Beattie, L., Antipatis, C., et al. (1999). Effects of pre- and post-mating feed intake on blastocyst size, secretory function, and glucose metabolism in Meishan gilts. *Reproduction, Fertility and Development, 11*(6), 323–327. https://doi.org/10.1071/RD99040
4. Chen, M. X., Li, X. G., Yang, J. X., et al. (2015). Growth of embryo and gene expression of nutrient transporters in the small intestine of the domestic pigeon (Columba livia). *Journal of Zhejiang University-Science B (Biomedicine & Biotechnology), 16*(6), 511–523. https://doi.org/10.1631/jzus.B1400340
5. Dai, B., Zhang, Y. S., Ma, Z. L., et al. (2015). Influence of dietary taurine and housing density on oviduct function in laying hens. *Journal of Zhejiang University-Science B (Biomedicine & Biotechnology), 16*(6), 456–464. https://doi.org/10.1631/jzus.B1400256
6. Gao, H. N., Hu, H., Zheng, N., et al. (2015). Leucine and histidine independently regulate milk protein synthesis in bovine mammary epithelial cells via mTOR signaling pathway. *Journal of Zhejiang University-Science B (Biomedicine & Biotechnology), 16*(6), 560–572. https://doi.org/10.1631/jzus.B1400337
7. Jacob, F., & Monod, J. (1961). Genetic regulatory mechanisms in the synthesis of proteins. *Journal of Molecular Biology, 3*(3), 318–356. https://doi.org/10.1016/S0022-2836(61)80072-7
8. Kitajka, K., Sinclair, A. J., Weisinger, R. S., et al. (2004). Effects of dietary omega-3 polyunsaturated fatty acids on brain gene expression. *Proceedings of the National Academy of Sciences of the United States of America, 101*(30), 10931–10936. https://doi.org/10.1073/pnas.0402342101
9. Lei, K., Li, Y. L., Wang, Y., et al. (2015). Effect of dietary supplementation of *Bacillus subtilis* B10 on biochemical and molecular parameters in the serum and liver of high-fat diet-induced obese mice. *Journal of Zhejiang University-Science B (Biomedicine & Biotechnology), 16*(6), 487–495. https://doi.org/10.1631/jzus.B1400342

10. Liang, X. P., Zhang, D. Q., Chen, Y. Y., et al. (2015). Effects of alfalfa saponin extract on mRNA expression of *Ldlr, LXRα*, and *FXR* in BRL cells. *Journal of Zhejiang University-Science B (Biomedicine & Biotechnology), 16*(6), 479–486. https://doi.org/10.1631/jzus.B1400343
11. Müller, M., & Kersten, S. (2003). Nutrigenomics: Goals and strategies. *Nature Reviews Genetics, 4*(4), 315–322. https://doi.org/10.1038/nrg1047
12. Nie, C. X., Zhang, W. J., Wang, Y. Q., et al. (2015). Tissue lipid metabolism and hepatic metabolomic profiling in response to supplementation of fermented cottonseed meal in the diets of broiler chickens. *Journal of Zhejiang University-Science B (Biomedicine & Biotechnology), 16*(6), 447–455. https://doi.org/10.1631/jzus.B1400255
13. Nie, Y. F., Hu, J., & Yan, X. H. (2015). Cross-talk between bile acids and intestinal microbiota in host metabolism and health. *Journal of Zhejiang University-Science B (Biomedicine & Biotechnology), 16*(6), 436–446. https://doi.org/10.1631/jzus.B1400327
14. Quesnel, H., Quiniou, N., Roy, H., et al. (2014). Supplying dextrose before insemination and l-arginine during the last third of pregnancy in sow diets: Effects on within-litter variation of piglet birth weight. *Journal of Animal Science, 92*(4), 1445–1450. https://doi.org/10.2527/jas.2013-6701
15. Shapiro, J., Machattie, L., Eron, L., et al. (1969). Isolation of pure lac operon DNA. *Nature, 224*(5221), 768–774. https://doi.org/10.1038/224768a0
16. Son, J., Lyssiotis, C. A., Ying, H., et al. (2013). Glutamine supports pancreatic cancer growth through a KRAS-regulated metabolic pathway. *Nature, 496*(7443), 101–105. https://doi.org/10.1038/nature12040
17. Wang, F. R., Dong, X. F., Zhang, X. M., et al. (2010). Effects of dietary taurine on egg production, egg quality, and cholesterol levels in Japanese quail. *Journal of the Science of Food and Agriculture, 90*(15), 2660–2663. https://doi.org/10.1002/jsfa.4136
18. Whaley, S. L., Hedgpeth, V. S., Farin, C. E., et al. (2000). Influence of vitamin A injection before mating on oocyte development, follicular hormones, and ovulation in gilts fed high-energy diets. *Journal of Animal Science, 78*(6), 1598–1607.
19. Wu, L., He, L. Q., Cui, Z. J., et al. (2015). Effects of reducing dietary protein on the expression of nutrition sensing genes (amino acid transporters) in weaned piglets. *Journal of Zhejiang University-Science B (Biomedicine & Biotechnology), 16*(6), 496–502. https://doi.org/10.1631/jzus.B1400259
20. Xie, Y. M., Xu, Q. B., Wu, Y. M., et al. (2015). Duodenum has the greatest potential to absorb soluble non-ammonia nitrogen in the nonmesenteric gastrointestinal tissues of dairy cows. *Journal of Zhejiang University-*

*Science B (Biomedicine & Biotechnology), 16*(6), 503–510. https://doi.org/10.1631/jzus.B1400299

21. Yuan, T. L., Zhu, Y. H., Shi, M., et al. (2015). Within-litter variation in birth weight: Impact of nutritional status in the sow. *Journal of Zhejiang University-Science B (Biomedicine & Biotechnology), 16*(6), 417–435. https://doi.org/10.1631/jzus.B1500010
22. Zduńczyk, Z., Jankowski, J., Juśkiewicz, J., et al. (2013). Effect of different dietary levels of low-glucosinolate rapeseed (canola) meal and non-starch polysaccharide-degrading enzymes on growth performance and gut physiology of growing turkeys. *Canadian Journal of Animal Science, 93*(3), 353–362. https://doi.org/10.4141/cjas2012-085
23. Zhang, J. Z., Gao, Y., Lu, Q. P., et al. (2015). iTRAQ-based quantitative proteomic analysis of longissimus muscle from growing pigs with dietary supplementation of non-starch polysaccharide enzymes. *Journal of Zhejiang University-Science B (Biomedicine & Biotechnology), 16*(6), 465–478. https://doi.org/10.1631/jzus.B1400266

# INDEX

## C



## D

## E



## F



## G

## H



## I



## K

## L



## M

FSC
www.fsc.org
MIX
Papier | Fördert gute Waldnutzung
FSC® C083411

Kolibri 360 GmbH
Ferdinand-Jühlke-Straße 7
99095 Erfurt, Deutschland
produktsicherheit@kolibri360.de